GROWTH MANAGEMENT IN FLORIDA

Growth Management in Florida
Planning for Paradise

Edited by

TIMOTHY S. CHAPIN,
CHARLES E. CONNERLY
and
HARRISON T. HIGGINS
Florida State University, USA

Routledge
Taylor & Francis Group

LONDON AND NEW YORK

First published 2007 by Ashgate Publishing

Reissued 2018 by Routledge
2 Park Square, Milton Park, Abingdon, Oxon OX14 4RN
605 Third Avenue, New York, NY 10017

First issued in paperback 2021

Routledge is an imprint of the Taylor & Francis Group, an informa business

A Library of Congress record exists under LC control number: 2007007954

Notice:
Product or corporate names may be trademarks or registered trademarks, and are used only for identification and explanation without intent to infringe.

Publisher's Note
The publisher has gone to great lengths to ensure the quality of this reprint but points out that some imperfections in the original copies may be apparent.

Disclaimer
The publisher has made every effort to trace copyright holders and welcomes correspondence from those they have been unable to contact.

ISBN 13: 978-0-815-38935-4 (hbk)
ISBN 13: 978-1-351-15700-1 (ebk)
ISBN 13: 978-1-138-35703-7 (pbk)

DOI: 10.4324/9781351157001

Contents

List of Figures

List of Tables

List of Contributors

Book Editors

Timothy S. Chapin, Florida State University

Chapin is an Associate Professor of Urban and Regional Planning at Florida State University. Chapin's research interests are in the areas of growth management and downtown redevelopment. Chapin has published articles on these topics in the *Journal of the American Planning Association, Urban Affairs Review, Environment and Planning B*, and the *Journal of Urban Affairs*. Recently he was a principal investigator on a three year project investigating changes in development patterns in coastal jurisdictions in Florida. Chapin's work has been supported by the Lincoln Institute for Land Policy, the Department of Housing and Urban Development, and the National Oceanic and Atmospheric Administration via their Sea Grant program.

Charles E. Connerly, Florida State University

Connerly is the William G. and Budd Bell Professor in the Department of Urban and Regional Planning at Florida State University. He has written or co-authored over twenty-five journal articles or book chapters and most recently authored a book published by the University of Virginia Press, *"The Most Segregated City in America": City Planning and Civil Rights in Birmingham, 1920–1980*. His research has been published in top journals, including the *Journal of the American Planning Association*, the *Journal of Planning Education and Research*, the *Journal of Planning Literature, Housing Studies*, and the *Journal of Urban History*. From 1991 to 1996, he served as coeditor of the *Journal of Planning Education and Research*, the refereed research publication of the Association of Collegiate Schools of Planning, and helped to make that one of the most visible and respected journals of urban planning research. Since 2001 he has been North American editor of *Housing Studies*.

Harrison T. Higgins, Florida State University

Higgins is currently Planner-in-Residence in the Department of Urban and Regional Planning at Florida State University. An urban designer and planner with more than 15 years experience, Higgins has an undergraduate degree from Princeton University and master's degrees from the London School of Economics and the Southern California Institute of Architecture. Higgins has worked with a number of Florida communities to develop and implement their comprehensive plans. Higgins planning practice has also included clients in Arizona, California, New Mexico, New York, Oregon, and Washington state. Mr. Higgins is also a former adjunct member of the faculty of the City College of New York and a former instructor at UCLA and SCI-Arc.

Contributing Authors

Earl J. Baker, Florida State University

Baker is an Associate Professor in the Department of Geography at Florida State University. His work has focused upon the effect of hurricane experience on future evacuation response, the derivation of impact fee formulas for funding storm management services, and methods for identifying public preference tradeoffs among policy options. He holds a Ph.D. in Geography from the University of Colorado.

Efraim Ben-Zadok, Florida Atlantic University

Ben-Zadok is Professor of Public Administration in the College of Architecture, Urban and Public Affairs at Florida Atlantic University. His current interests focus on evaluation of state-regional-local policies and on decision-making processes in long-range large-scale urban and regional policies. He has published papers in leading urban journals, including *Urban Studies*, *Urban Affairs Review*, and *International Planning Studies*.

Marlon G. Boarnet, University of California, Irvine

Boarnet is Professor of Urban and Regional Planning and Economics and a research associate of the Institute of Transportation Studies at University of California, Irvine. Boarnet earned his Ph.D. from Princeton University in 1992. His research interests include the economic and urban development impacts of highway infrastructure, the links between urban design and travel behavior, and the determinants of population and employment growth patterns within metropolitan areas. Boarnet serves on the editorial boards of the *Journal of Regional Science* and *Papers in Regional Science*.

Gregory S. Burge, University of Oklahoma

Burge is currently an Assistant Professor in the Department of Economics at the University of Oklahoma. He earned his Ph.D. from Florida State University. His research work has focused upon local government fiscal issues, with publications in the *Journal of Regional Science*, the *Journal of Urban Economics*, and the *National Tax Journal*.

John I. Carruthers, Department of Housing and Urban Development

Carruthers is an Economist with the US Department of Housing and Urban Development (HUD), in the Office of Policy Development and Research. Prior to joining HUD, Carruthers taught urban and regional planning at the University of Arizona. Carruthers is a prolific author on land use issues, having published papers in *Environment and Planning B*, *Growth and Change*, *Housing Policy Debate*, the *Journal of Planning Education and Research*, *Papers in Regional Science*, and *Urban Studies*.

Karen A. Danielsen, Virginia Polytechnic Institute and State University

Danielsen is a Ph.D. candidate in Urban Affairs and Planning at the Metropolitan Institute, Virginia Tech's Northern Virginia Center in Alexandria. With interests in residential development, she has seen her work published in *Housing Policy Debate* and *Opolis: An International Journal of Suburban and Metropolitan Studies*.

Casey J. Dawkins, Virginia Polytechnic Institute and State University

Dawkins is an Assistant Professor in the Department of Urban Affairs and Planning at Virginia Tech. He has a master's degree and a Ph.D. in City Planning from the Georgia Institute of Technology. His research work has investigated the impacts of urban containment on housing affordability, residential segregation, and central city revitalization. His work has been published in the *Journal of the American Planning Association, Urban Studies*, the *Journal of Planning Literature*, and *Housing Policy Debate*.

Robert E. Deyle, Florida State University

Deyle is a Professor in the Department of Urban and Regional Planning at Florida State University. He studies and teaches environmental planning and policy analysis, coastal hazards planning and management, and institutional responses to chronic environmental change including coastal erosion and global warming. He has published research in the *Journal of the American Planning Association, Natural Hazards Review, Water Resources Bulletin*, and *Environmental Management*.

Randall G. Holcombe, Florida State University

Holcombe is DeVoe Moore Professor of Economics at Florida State University. He is also the chairman of the Research Advisory Council of the James Madison Institute for Public Policy Studies, a Tallahassee-based think tank that specializes in issues facing state governments. Holcombe has authored eight books and published articles and reviews in leading academic and professional journals, including the *American Economic Review*, the *National Tax Journal, Public Finance Quarterly, Economic Inquiry*, the *Southern Economic Journal*, and the *Journal of Public Economics*.

Keith R. Ihlanfeldt, Florida State University

Ihlanfeldt is a Professor of Economics and holds the DeVoe Moore Eminent Scholar Chair at Florida State University. His research has focused on a wide range of urban and regional problems, including discrimination in the housing and labor markets, urban poverty, neighborhood decline, housing affordability, and economic development incentives. He has published widely and has received grants from numerous organizations, both public and private. Ihlanfeldt serves on the editorial boards of six academic journals, including the *Journal of Urban Economics*.

Gerrit-Jan Knaap, University of Maryland

Knaap is Professor of Urban Studies and Planning and Director of the National Center for Smart Growth Research and Education at the University of Maryland. He earned his B.S. from Willamette University, his M.S. and Ph.D. from the University of Oregon, and received post-doctoral training at the University of Wisconsin-Madison, all in economics. Knaap has published numerous articles on growth management in the *Journal of the American Planning Association, Policy Analysis and Management, State and Local Government Review, International Regional Science Review,* and the *Journal of Planning Education and Research.*

Ralph B. McLaughlin, University of California, Irvine
McLaughlin is a PhD student in the Department of Planning, Policy, and Design, University of California, Irvine. He earned a master's degree in planning from the University of Arizona.

Robert H. Mandle, Virginia Polytechnic Institute and State University
Mandle is a graduate student in the Department of Urban and Regional Planning in the College of Architecture and Urban Studies at Virginia Tech. He was awarded the National American Institute of Certified Planners Certificate award, bestowed to graduate students with exceptional academic achievement.

Arthur C. Nelson, University of Michigan
Nelson is the Director of Virginia Polytechnic Institute's Metropolitan Institute in Alexandria, Virginia. For the past twenty years, Nelson has conducted pioneering research in land use planning, growth management, public facility finance, and urban development policy. His research and practice has led to the publication of eight books and more than 150 scholarly and professional publications. Numerous organizations have sponsored Nelson's research such as the National Science Foundation; National Academy of Sciences; US Departments of Housing and Urban Development, Commerce, and Transportation; UK Department of the Environment; Lincoln Institute of Land Policy; Fannie Mae Foundation; American Planning Association; National Association of Realtors; and the Brookings Institution.

James C. Nicholas, University of Florida
Nicholas is a retired Professor of Urban and Regional Planning and Affiliate Professor of Law at the University of Florida. He also is associate director of the Environmental and Land Use Law Program in the Levin College of Law at the University of Florida. He received his undergraduate education in Business Administration at the University of Miami, Florida in 1965. He earned a doctorate in economics from the University of Illinois in 1970.

Neil B. Paradise, Florida State University
Paradise is a joint planning and law student at Florida State University. His planning and legal interests lie in the areas of land use planning and environmental land management. He holds a degree in History from the University of North Florida.

Thomas G. Pelham, Florida Department of Community Affairs
Pelham is an attorney at law and certified planner with extensive experience in administrative, environmental, and land use law and planning. He served as Secretary of the Department of Community Affairs from 1987–1991, where he played a key role in the initial implementation of the 1985 Growth Management Act. He also has served as a member of both the Tallahassee-Leon County and Capital Center Planning Commissions. In December 2006 he was again tapped by Florida's governor to lead

DCA as Department Secretary. Pelham received a B.A. in Government at Florida State University and an M.A. in Political Science at Duke University. He earned a J.D. at Florida State University and a L.L.M. in Land Use Planning Law, Local Government Law, and Property Law at Harvard Law School.

Thomas W. Sanchez, Virginia Polytechnic Institute and State University

Sanchez is Associate Professor of Urban Affairs and Planning and Fellow with the Metropolitan Institute, at Virginia Tech's Northern Virginia Center in Alexandria. His research has been in the areas of transportation, land use, residential location behavior, and questions of social equity in planning His research has been published in a variety of planning and policy journals including his article "The Connection between Public Transit and Employment" which won the 2000 award for best article in the *Journal of the American Planning Association.*

Yan Song, University of North Carolina

Song is Assistant Professor of City and Regional Planning at the University of North Carolina. She earned an M.S. in urban and regional planning at Florida State University and holds a Ph.D. from the University of Illinois at Urbana-Champaign. Her research projects have been supported by the US Department of Housing and Urban Development and the Lincoln Institute of Land Policy. Song has also had professional experiences as a planner and architect in China.

Ruth L. Steiner, University of Florida

Steiner is an associate professor in the Department of Urban and Regional Planning at the University of Florida. Her research and professional activities concentrate on the interactions between transportation, land use, and the environment, with a particular focus on alternative modes of transportation. She has a B.A. in History from Lawrence University (Wisconsin), an M.B.A. from the University of Wisconsin-Milwaukee and an M.C.P. and Ph.D. from the University of California at Berkeley.

Acknowledgements

In the last several decades, the state of Florida has undergone a remarkable transformation. Since 1970 Florida has changed from a largely rural and undeveloped southern state into an international economic powerhouse and home to many of the world's premier tourist destinations (including DisneyWorld). During this period the state's economy has flourished and the state's population has added roughly 300,000 residents every year. This remarkable transformation has made many individuals and corporations wealthy beyond their wildest dreams. It has also brought incalculable environmental damage to a state known for its pristine beaches, magnificent wetlands, and spectacular sunsets.

It was the apparent conflict between economic growth and environmental degradation that fueled early efforts at managing growth in Florida. Between 1970 and 1985 Florida made a major commitment to "Planning for Paradise," transforming itself from a state with little interest in and regard for planning into a state viewed within planning and environmental circles as a leader and innovator in methods for managing growth. Because of these efforts, Florida is known nationally and internationally for its growth management approach and commitment to planning.

It is our firsthand experiences with efforts to "plan for paradise" that spurred this book. We have seen the growth management process at its best, with energized, productive public meetings and local government commitment to development that brings the benefits of growth without compromising irreplaceable environmental assets. We have also seen growth management at its worst, with planning policies overlooked or discarded in the name of growth and the resulting permanent loss of special, natural places. We have worked with communities in Florida as they have struggled to manage other growth-related problems, including traffic congestion, depletion of water resources, and a real crisis in the provision of affordable housing.

These experiences led us to organize a symposium in January, 2005 held at Florida State University in Tallahassee, Florida. With support from the DeVoe Moore Center for the Study of Critical Issues in Economic Policy and Government, we brought together leading growth management scholars from throughout the nation to help assess the impacts of Florida's growth management approach. The symposium yielded a great deal of interesting observations about the strengths and weaknesses of Florida's system and prompted a good deal of debate about the nature of the benefits realized from the state's efforts. This book represents a collection of the papers presented at that conference, as well as other papers that were brought into the project in the subsequent year.

An edited volume like this one cannot proceed without effort from a great number of people whose names are not attached to chapters in the book. Most notably, Kathy Makinen of the DeVoe Moore Center (DMC) provided detailed and very insightful copyediting services for the book. Many of the chapters in this volume would be much less coherent and readable without her excellent suggestions for simpler and more direct language. We also thank the staff at Ashgate Publishing for their help in seeing us through the sometimes difficult, but never dull publication process.

We appreciate the encouragement and commitment for financial support from then DMC Director, and now Dean of the College of Social Sciences, David Rasmussen. We also thank current DMC Director Keith Ihlanfeldt for providing ongoing support for the project.

A nod of appreciation must also go to the students and faculty of the Department of Urban and Regional Planning at Florida State University. The relentless curiosity of our students and their quest for "answers" when it comes to growth management helped us to pursue and complete the book. We thank our colleagues in the Department for their interest in the project and their patience as we kept insisting that the book would be out "soon."

Lastly, we offer thanks to the many planners and officials in local government offices, private consulting firms, and at the Department of Community Affairs who made time to help many of the book's contributors as they completed their research. These planners are the true innovators and implementers of the Florida growth management experiment. It is these "planners on the frontline" that have helped to maintain Florida's paradise-like places in the face of substantial and sustained growth.

Chapter 1

Introduction

Timothy S. Chapin, Charles E. Connerly, and Harrison T. Higgins

The 1985 Growth Management Act[1] (GMA) fundamentally changed planning in Florida and, we would argue, more generally in the United States. While several states have growth management systems that are older than Florida's, with the most well-known being Oregon, the Florida growth management approach in many ways represents a near perfect version of the planning profession's "comprehensive planning" model. Every local government, without exception, must undertake the comprehensive planning process and prepare comprehensive plans to guide development. Further, plans are to be implemented by local governments through local development regulations that must reflect the long-term vision laid out in these plans. Comprehensive plans are also to be "living documents," with biannual revisions through the amendment process and major updates to plans are required every 5–7 years. Over the years Florida's courts have upheld the fundamental role of local comprehensive plans in shaping development patterns and these plans are now a routine part of the development review process.

While comprehensive planning was in itself not new, what made the Florida approach unique at its inception was a strong role for the state government in reviewing and commenting on the comprehensive plans developed by local governments. The Legislature provided oversight responsibility for these local comprehensive plans to the state's Department of Community Affairs (DCA). DCA also was tasked with establishing the minimum criteria for these plans. These requirements improved the quality of the plans produced by local governments and ensured that a common set of key issues were addressed by them.

Beyond placing the comprehensive plan at the center of local land use decisions and ensuring that all local governments addressed a core set of issues in their plans, the 1985 legislation also required some degree of inter-jurisdictional dialogue and cooperation. Local comprehensive plans were not to be created in a vacuum; the state's "consistency" requirement ensured that local governments and DCA had

---

1   Throughout this book, the 1985 Growth Management Act (GMA) refers to the cluster of growth management bills adopted by the Florida legislature in 1984, 1985, and 1986 that enacted the state's 20-plus year approach to growth management. Chief among this legislation was the Omnibus Growth Management Act of 1985, but the Florida legislature also adopted the 1984 Florida State and Regional Planning Act, which required the preparation of a state plan and the 1986 Glitch Bill which further clarified the 1985 bill.

opportunities to raise objections to plans that ignored or overlooked regional extra-jurisdictional planning issues.

For advocates of linking planning and budgeting, the Florida Legislature also provided specific direction that local comprehensive plans were to shape the capital budgets of jurisdictions. This tenet, that planning and budgeting were to be closely linked, was reinforced by recent changes to the state's comprehensive planning regulations (see Nicholas and Chapin, Chapter 4). In Florida, then, the local comprehensive plan was established as *the* guiding document for local land use regulations, local infrastructure planning, and government capital project expenditures.

Other major elements of the Florida growth management approach (as initially designed) include requirements for concurrency, the establishment of a state comprehensive plan, the creation and empowerment of regional planning councils, and major funding commitments by the state for infrastructure and technical support for comprehensive planning efforts. The Florida approach is generally acknowledged as the most aggressive and far-reaching growth management approach this nation has yet seen. Elements of the Florida approach have been adopted in other states, most notably Georgia and Washington, and planners throughout the nation have learned from the Florida experience.

In many ways, then, the 1985 Growth Management Act represents the high water mark for the profession of planning, as Florida's approach cemented comprehensive planning and the planning process at the core of all local, regional, and state land use decisions. While individual elements of Florida's system have been implemented elsewhere, what distinguishes the Florida approach is the broad commitment to the "good planning" model; detailed, rigorous, regionally coordinated comprehensive plans supported by state funding and linked directly to the local government capital budgeting process. Unlike other states that have undertaken much more specific and targeted approaches for managing growth (such as Maryland, Tennessee, and even Oregon), Florida's 1985 legislation represents the foremost attempt to implement the comprehensive planning model long advocated by the planning profession.

Given this context, the central purpose of this book is to document and evaluate the impacts of this innovative state-level growth management approach. While much has been written about the form and content of the Florida growth management system, few studies have attempted to assess the impacts of this legislation. In this volume we have endeavored to undertake a detailed appraisal of the Florida approach, one that evaluates the state's system, and the impacts of this system, on its merits, rather than on broad perceptions about the successes and failures of the system.

The Organization of the Book

The book's first section addresses the history and foundations of growth management in Florida. Chapter 2 features an historical overview of growth management in Florida as viewed by Thomas Pelham, who served as secretary of Florida's Department of Community Affairs in growth management's formative years, 1987–1991, and was again appointed as DCA secretary in December, 2006. The chapter highlights the promises of growth management and contains a sobering assessment of the difficulty in realizing those promises. In Chapter 3, Efraim Ben-Zadok details the changing thematic emphases of growth management in Florida from consistency, to concurrency, and most recently to compact development, while also highlighting the role that local discretion has played in growth management's implementation.

Chapter 4, by James Nicholas and Timothy Chapin, lays out the fiscal foundation for growth management in Florida and details the impacts of the state legislature's decision to defund major portions of the state's commitment to growth management soon after passage of the 1985 GMA. In Chapter 5, Timothy Chapin and Charles Connerly report on survey data they have collected on Florida public attitudes toward growth management in 1985 and 2001. They conclude that while overall public support for growth management remains high the makeup of growth management support has changed over time.

Section 2 of the book focuses on evaluating growth management outcomes in Florida. Chapters 6 and 7 analyze changes in Florida's land use patterns during the post-GMA period. In Chapter 6, Thomas Sanchez and Robert Mandle employ GIS analysis to determine whether the 1985 GMA has had an impact over time on the distribution of urban and rural densities in Florida. They find that while it appears that growth management has had an impact on curbing the low density development that is often associated with sprawl, the level of such development is still higher in Florida than elsewhere. In turn, John Carruthers, Marlon Boarnet, and Ralph McLaughlin employ a spatial adjustment model in Chapter 7 to examine the effectiveness of Florida's growth management policies. They report that the state's growth management system appears to have evened out the mix of regulatory and non-regulatory environments within the state that can produce disequilibrium in land markets.

Chapters 8 through 12 focus on outcomes originally linked to the state's growth management system when it was passed in 1985. These chapters detail the impacts of the state's system on the Florida economy, the desire for more compact urban form, minimization of housing spillovers due to regional coordination of housing production, mitigation of the exposure of people and property to natural hazards, and enhancement of overall quality of life.

In Chapter 8, Timothy Chapin examines the question of the impact of growth management on the economy of Florida and its largest cities. He finds that Florida's economic growth has continued even with aggressive, state-mandated growth management and that the economies of the state's largest cities have done comparatively well during the post-GMA period. In Chapter 9, Gerrit-Jan Knaap and Yan Song compare urban form patterns in Orlando and four other non-Florida metropolitan areas and find that, in comparison to these other areas, compact development in Orlando has "considerable room for improvement."

In Chapter 10, Yan Song reports that even though growth management is a statewide requirement in Florida, differences in regulatory stringency have resulted in housing development being shifted to jurisdictions with less stringent land use regulations. In Chapter 11, Robert Deyle, Timothy Chapin, and Earl J. Baker focus their research on the impacts of growth management on Florida's coastal areas, concluding that while growth continues in these areas, it does so at a slower rate than prior to the adoption of growth management. In Chapter 12, Arthur C. Nelson, Casey Dawkins, Thomas Sanchez, and Karen Danielsen examine the impact of compact development on perceived neighborhood quality and find that in contrast to the past, compact development in more recent years in Florida has no impact on neighborhood quality, a relationship that could conceivably turn positive in the future.

Section 3 of the book focuses on aspects of the state's growth management system introduced since the 1985 GMA. Several chapters in this section discuss planning innovations that have been developed in the state, while others discuss structural and political limitations that have restricted the implementation of growth management in Florida.

In Chapter 13, Ruth Steiner traces the development of transportation concurrency in Florida and the many varieties of concurrency exemptions that have been devised to make this tool for matching growth with infrastructure more flexible for meeting varying planning needs. In Chapter 14, Randall Holcombe examines which jurisdictions have adopted urban growth boundaries and reports that it is higher per capita income, not growth pressure, that is associated with the adoption of this form of growth management.

Harrison Higgins and Neil Paradise, in Chapter 15, explain how Florida's conservation land purchase programs have only partially complemented the state's growth management approach. Through a review of the Babcock Ranch case study they highlight why this ambitious program has sometimes served as a facilitator of both land preservation and sprawl. In Chapter 16, Charles Connerly describes the relatively limited impact that the growth management legislation has had on encouraging local jurisdictions to develop proactive affordable housing strategies like inclusionary zoning, even though this tool is recommended in the 1985 Growth Management Act. In Chapter 17, Gregory Burge and Keith Ihlanfeldt examine the emergence and use of impact fees to help growth pay for itself in Florida, finding evidence that non-water/sewer impact fees actually increase the construction of affordable housing.

Finally, in the book's conclusion the editors synthesize these chapters and outline the major conclusions from these collected works. Taken as a whole these chapters suggest that while massive population and economic growth have continued in Florida, often taking the form of sprawl, the evidence suggests that state's growth management regime appears to be moving Florida slowly in the direction of better development outcomes (what many commentators currently refer to as "smart growth") and protection of the state's natural areas. However, given the incremental, still ongoing implementation of the 1985 GMA and the lack of state funding for infrastructure, visible and measurable impacts related to this progress are likely still a decade or more away.

While the evidence concerning positive impacts of the system is far from conclusive, real progress has been made in the area of planning practice. Since 1985, Florida has emerged as a national leader in comprehensive planning, concurrency implementation, land purchases for conservation, natural hazards planning, and local approaches to paying for growth. These impacts on planning practice, largely unforeseen by the act's authors, are perhaps the GMA's greatest legacy to date.

PART I
The Foundations of Growth Management in Florida

Chapter 2

A Historical Perspective for Evaluating Florida's Evolving Growth Management Process

Thomas G. Pelham

A comprehensive evaluation of Florida's growth management process is both timely and appropriate. Florida began to enact and implement planning and growth management legislation in the early 1970s. The Legislature enacted the landmark Local Government Comprehensive Planning Act in 1975 and then substantially amended it in the even more comprehensive 1985 Growth Management Act (GMA).[1] In 2005 the Florida Legislature celebrated the twentieth anniversary of the GMA by adopting another round of major amendments to the GMA. This thirty years of statewide experience provides an adequate basis for assessing the effectiveness of Florida's growth management process.

When making an assessment of the Florida growth management experiment, it is important to keep the historical development of the Florida growth management process clearly in mind. We can learn from history. The mission of this chapter is to provide some historical perspective for evaluating Florida's growth management system, based largely upon the author's thirty years of experience with the state's system. Before discussing the evolution of the system and documenting some of the shortcomings of the Florida system, five key points must be established at the outset:

- First, we must not let our disappointments with the results of Florida's growth management process obscure how far the state has traveled. However imperfect the process, where would the state be without it?
- Second, we frequently refer to the GMA and the growth management process as if they are static or unchanging. In reality, the legislation and the process have been continuously evolving, often in ways not foreseen by the original legislative draftsmen.
- Third, we must acknowledge the difference between concept and implementation. Before concluding that the system cannot or does not work,

---

1 The GMA consists of three different legislative acts: Chapter 163, Part II (the Local Planning Act), Chapter 186 (the State and Regional Planning Act), and Chapter 187 (the State Comprehensive Plan) in the Florida Statutes.

we should consider whether and how we have implemented it. Is the concept flawed or misguided, or does the problem lie in the way we have implemented or failed to implement the concept?

- Fourth, we must distinguish between planning as process and planning as policy in Florida's growth management system. Each component is important in its own right and should be evaluated on its own merits.
- Finally, we should consider whether and the extent to which the state has filled in the gaps in the original growth management system and reinforced the key concepts of the GMA in subsequent legislation. How has the state's commitment or lack thereof to the key concepts of consistency, concurrency, coordination, and compact urban form influenced the effectiveness of Florida's growth management system?

The Evolution of Florida's Growth Management Legislation

Florida's growth management process is the culmination of more than thirty years of legislative reform. Over the past three decades, Florida has utilized a series of gubernatorial and legislative task forces and committees to evaluate growth problems and to recommend appropriate legislative responses. Periodically reviewing and evaluating the Florida's growth problems and the effectiveness of legislative responses, these broad-based citizen commissions have produced a series of reports and recommendations for establishing and fine-tuning the state, regional, and local planning systems that are known today as Florida's growth management process.

This effort commenced in 1971 with the appointment by Governor Reubin Askew of the Task Force on Resource Management (DeGrove, 1984, pp. 109–10). Based on the recommendations of this task force, the 1972 Legislature enacted a landmark package of planning legislation. Florida became the first state to follow a model code (American Law Institute, 1975) to craft legislation that provided for state regulation of large-scale developments of regional impact (DRIs) and areas of critical state concern (Fla. Laws ch. 72–317 §§ 1–13, 1972). The Legislature also adopted the State Comprehensive Planning Act (Fla. Laws ch. 72–295 §§ 1–13, 1972), which created a process for adopting a state comprehensive plan to guide growth and development. Although this process ultimately failed, it laid the groundwork for the successful adoption of the State Comprehensive Plan in 1985. Also, the 1972 Legislature also created the first Environmental Land Management Study (ELMS) Committee, which made recommendations for implementing the DRI and critical area programs.

The DRI and critical area processes were initially viewed as interim measures to give the state time to put in place a more comprehensive approach to growth management. Accordingly, the original gubernatorial task forces recommended adoption of a mandatory local comprehensive planning act. In 1975, the Florida Legislature accepted the recommendation when it adopted the Local Government Comprehensive Planning Act (LGCPA), which was at the time probably the most comprehensive piece of local planning legislation ever enacted in this country (Fla. Laws ch. 75–257 §§ 1–19, 1975; LGCPA, Fla. Stat. §§ 163,3161–3211, 1975).

It required every local government in Florida to adopt a comprehensive plan in accordance with detailed statutory requirements by 1979. However, while the LGCPA provided for review by the state land planning agency, it did not give the agency the authority to reject local plans that were inconsistent with state requirements. Moreover, the Act provided no state enforcement mechanism to ensure that local governments adopted, implemented, and enforced local plans. Not surprisingly, many did not take the LGCPA very seriously.[2]

During the 1970s, Florida's population increased from 6.7 million to almost 10 million (Florida Governor's Task Force on Urban Growth Patterns, 1989, p. 3). This growth had an enormous impact on Florida's environment and its ability to manage or pay for the growth. In response, Governor Bob Graham appointed another blue ribbon committee, the ELMS II Committee in 1982. In early 1984, the ELMS II Committee issued its final report, which contained several major recommendations. One set of recommendations called for an integrated statewide planning framework that would include a legislatively adopted state plan to be implemented through state agency functional plans, regional plans, and greatly strengthened local government comprehensive plans (Environmental Land Management Study Committee, 1984, pp. 2–4).

Several recommendations of the ELMS II Committee were subsequently adopted by the Florida Legislature. First, the Florida State and Regional Planning Act of 1984 established procedures for preparation and adoption of a state comprehensive plan (Fla. Laws ch. 84–237 §§ 1–2, 1984). In 1985, the Florida Legislature adopted by statute the State Comprehensive Plan. As defined in the legislation, the State Comprehensive Plan is a "direction-setting document" that provides "long-range policy guidance" for regional and local plans. It contains 27 goals with accompanying policies covering a wide range of social, economic, environmental, natural resources, conservation, and land planning issues. Although the state plan "does not create regulatory authority or authorize the adoption of agency rules, criteria, or standards not otherwise authorized by law," the 1985 amendments to the LGCPA required regional and local plans to be consistent with the goals and policies of the state plan (State Comprehensive Plan, Fla. Stat. § 187.101(2), 1985).

To provide a regional planning perspective, the 1985 legislation also required each of the state's eleven regional planning agencies to adopt a comprehensive regional policy plan. These regional policy plans were required to contain regional goals and policies, including growth management policies, which were consistent with the goals and policies of the State Comprehensive Plan and laid out a framework for achieving these. A regional plan was required to address and analyze the problems and needs of the region, especially with regard to land use, water resources, transportation, and infrastructure (Florida State Comprehensive Planning Act, Fla. Stat. §§ 186.507(1), (3), 1985).

The 1985 Legislature also adopted major amendments to the 1984 LGCPA (Fla. Laws ch. 85–55 §§ 1–51, 1985). These amendments required local governments to adopt revised and improved local plans that were consistent with the state and regional

---

2   For a discussion of the weaknesses of the LGCPA, see Pelham, Hyde, and Banks (1985, pp. 542–43).

plans and provided for review and approval of the local plans by the state land planning agency and for state financial sanctions against any local government that failed to adopt a local plan in compliance with state requirements. Further, the amendments required local governments to implement their plans through local land development regulations and local development orders that were consistent with the adopted local plans. Every local government was to adopt adequate public facility or concurrency requirements. To enforce the plan consistency requirements, the legislation granted broad standing to citizens to challenge land development regulations and local development orders for inconsistency with the adopted local plan.[3]

To ensure that state and regional concerns would be considered and addressed by local governments, the 1985 legislation made intergovernmental coordination "a major objective of the local comprehensive planning process" (Local Government Comprehensive Planning and Land Development Regulation Act, Fla. Stat. § 163.3177(4)(a), Supp. 1986). Local comprehensive plans must be coordinated with the plans of adjacent municipalities and counties as well as with the state and regional plans. They must also contain an intergovernmental coordination element that demonstrates consideration of their impacts on adjacent local governments and the region and how coordination with the state, regional, and other local plans will be achieved.

In 1986, the Legislature adopted further amendments to the local planning legislation, including a strengthened concurrency requirement.[4] It also approved a state minimum criteria rule for local comprehensive plans (Fla. Admin. Code Ann. ch. 9J-5, 1986). This rule established standards by which the state land planning agency reviews local plans for compliance with state law.

Subsequently, Florida's approximately 470 local governments began preparing and adopting local comprehensive plans for review and approval by the state land planning agency. The agency established a four-year schedule, commencing in 1987 and ending in 1991, for the review of these plans (Local Government Comprehensive Planning and Land Development Regulation Act, Fla. Stat. § 163.3167(2), 1991; Fla. Admin. Code Ann. ch. 9J-12, 1990). During that period, most local governments also adopted new land development regulations to implement their plans, as required by state law.

In late 1991, following the completion of the initial local plan compliance review process, Governor Lawton Chiles appointed the ELMS III Committee to undertake a year-long review and evaluation of Florida's local comprehensive planning process and the relationship between that process and the DRI and critical area processes. Although the ELMS III Committee endorsed the basic scope and thrust of Florida's state, regional, and local comprehensive planning process, it recommended significant refinements and adjustments particularly with regard to the role of the regional councils and regional plans and the DRI process (Environmental Land Management Study Committee, 1992).

---

3 For a detailed discussion of the 1985 amendments to the LGCPA, see Pelham, Hyde, and Banks (1985, pp. 544–59).

4 For a discussion of the development of the concurrency requirement, see Pelham (1992).

The 1993 Florida Legislature adopted virtually all of the ELMS III Committee's recommendations (Fla. Laws ch. 93–206 §§ 1–84, 1993). Probably the most significant changes were the restriction of the powers of the regional planning councils, reduction of the scope and significance of regional policy plans, and phased elimination of the DRI process based on the adoption of greatly strengthened intergovernmental coordination elements in local plans. The implementation of the new coordination requirements provoked considerable controversy, however, and the Legislature repealed the requirements, including the phase-out of the DRI process, in 1995.

In 2000, Governor Jeb Bush appointed a Growth Management Study Commission which issued its final report in February 2001. The Commission's recommendations were very controversial, particularly the proposal to reduce state oversight of local comprehensive plans (Florida's Growth Management Study Commission, 2001). Except for its proposals regarding greater coordination of school facilities and potable water supply with land use planning, the Commission's recommendations have not been adopted by the Legislature.

Against this historical backdrop, the 2005 Florida Legislature adopted the most significant revisions to the GMA since 1993. The new legislation deals primarily with infrastructure (Fla. Laws ch. 2005–290 §§ 1–41, 2005).[5] The need to address Florida's long neglected, overburdened, and underfunded infrastructure system resonated with some legislative leaders, who demanded that the state face up to this responsibility. However, the resulting legislation finessed the infrastructure funding issue by providing some significant short term relief, avoiding tax increases, and relaxing the underlying concurrency requirement.

Specifically, the Legislature appropriated $1.5 billion for new infrastructure funding. This appropriation—although the largest in the history of the GMA—pales in comparison to the state's estimated $30 to 50 billion infrastructure backlog (Trevarthen and Friedman, 2005, p. 39). Additional funding was not provided because of strong gubernatorial and legislative resistance to raising taxes to provide more revenue. Also, because only $750 million is recurring revenue, the appropriation will not make a dent in the state's infrastructure problem unless future legislatures step up to the plate. Further, the 2005 appropriation is primarily allocated to a portion of the state road system and leaves other local infrastructure needs largely unaddressed. Thus, from an overall infrastructure needs perspective, the 2005 legislation can be fairly characterized as "treading water."

Implicitly recognizing that it has failed to provide adequate infrastructure funding, the 2005 Legislature made a major concession to the development industry: It relaxed the concurrency requirement by creating a proportionate fair share system, referred to by its proponents as "pay as you grow." Generally, this system allows developers to obtain development permits by paying their proportionate fair share of needed transportation and school improvements. Under these circumstances, the local development permit must be issued even though there are no assurances the local government will ever raise all of the revenue needed to construct the improvements required to meet the concurrency requirement.

---

5 For a detailed discussion of this legislation, see Trevarthen and Friedman (2005).

Effective growth management requires local comprehensive plans to be financially feasible and development to be coordinated with schools and water supply facilities. To these ends, the new legislation established strict new financial feasibility requirements for local plans and mandated that local governments subject schools and water supply facilities to the concurrency requirement. Despite imposing these and other substantial new planning requirements, the legislation provided local government with no new direct funding or funding sources to meet these requirements. Indeed, the Governor and Legislature refused to support removal of referendum requirements from some existing local taxing sources based on the political pretext that mere removal of the requirement would constitute an increase in taxes. In this respect the 2005 legislation continues the Legislature's twenty-year tradition of imposing substantial planning mandates on local governments without giving them the financial resources needed for compliance.

These developments are only the major highlights in the evolution of Florida's growth management legislation. Beginning in the early 1990s, the Legislature has amended the GMA on an almost annual basis. These frequent changes have undoubtedly affected implementation of the GMA. Consequently, any serious evaluation of Florida's growth management system must acknowledge that it is and has been constantly evolving.

The Disintegration of Florida's State and Regional Comprehensive Planning Processes[6]

As envisioned by the planning legislation adopted in 1984, 1985, and 1986, the State Comprehensive Plan, state agency functional plans, and the regional plans were intended to be important components of an integrated state, regional and local comprehensive planning process. Any objective evaluation of the effectiveness of that process must take into account the subsequent evolution of its state and regional components. The State Comprehensive Plan has not been updated and is largely ignored. State agency functional plans never became a factor in the implementation of the process. Through statutory amendments and administrative neglect, the role of the regional plans has been deemphasized. State administrative compliance review of local comprehensive plan amendments has been erratic at best, particularly with regard to consistency with the state and regional plans. The result is a growth management process that does not function in the manner originally intended and designed.

The Stagnation of the State Comprehensive Plan

Adoption of the State Comprehensive Plan was a considerable achievement by the Legislature. This plan played an important role in the initial implementation stages of the 1985 legislation, forming, for example, the basis of DCA's anti-sprawl policies.[7]

---

6   This section is a revised and updated version of ideas presented in Pelham (2001).

7   See *Homebuilders and Contractors Association v. Dept. of Community Affairs*, 585 So. 2d 965 (Fla. 1st DCA 1991).

However, it was a compromised product and has in many respects failed to provide adequate and specific guidance for implementing the growth management process. With the passage of time and changing circumstances, the inefficacy of the State Plan as a planning guide has become increasingly apparent.

Despite the glaring inadequacies of the State Plan, the legislative and executive branches have failed to update or significantly improve it. In the ensuing years, other than the addition of a new planning goal for downtown revitalization in 1988,[8] only a few minor revisions have been made to the State Plan. Although the Executive Office of the Governor has had the statutory responsibility of annually evaluating and recommending changes to the State Plan, this requirement has been largely ignored.

Given the lack of attention to the content and currency of the State Plan, it is not surprising that neither the Florida Legislature nor the Office of the Governor base their actions or decisions on it even when those actions and decisions (such as those pertaining to transportation and land use) relate to growth and development issues. The state budget is not linked to nor based on the State Plan. Decisions about the location of state-owned facilities are frequently made without regard for the land use policies of the State Plan (or of local plans, for that matter). Consequently, the State Plan currently is the object of criticism and even ridicule because it is seldom used; has not been significantly updated or improved since its adoption; has little or no effect on governmental decisions; and, except for DCA's urban sprawl policies, has little impact on the review and approval of local plan amendments.

The Downsizing of Regional Policy Plans

As previously discussed, pursuant to the recommendations of the ELMS III Committee, the Florida Legislature in 1993 enacted legislation that downsized the content and role of regional plans. This legislation restricted the role of regional plans by prohibiting DCA from finding a local plan amendment not in compliance with state law based solely on inconsistency with a regional plan and by providing that regional planning councils could not adopt binding level of service standards for facilities and services provided or regulated by local governments. Further, the 1993 legislation repealed the power of regional planning councils to appeal local development of regional impact decisions.

The Minimalization of State Agency Functional Plans

The original intent of the GMA was that state government as well as local governments would act consistently with the State Comprehensive Plan. Accordingly, state executive agencies were required to adopt agency functional or strategic plans to ensure that their operations would be consistent with and implement the State Plan. In 2000, however, the Legislature substantially altered the nature and role of state agency plans and repealed the requirement that such plans be consistent with the State Plan.[9]

---

8   See State Comprehensive Plan, Fla. Stat. § 186.201(16) (2005).
9   See Florida State Comprehensive Planning Act, Fla. Stat. § 186.021 (2005).

*The Declining Effectiveness of State Review of Local Comprehensive Plan
Amendments*

With few exceptions, all local governments had adopted local plans by the end of 1991.
Since that time, the oversight role of DCA has focused on local plan amendments.
DCA has estimated that it annually reviews approximately 12,000 plan amendments.
Whether because of the sheer magnitude of the task, political pressure, inadequate
staff and financial resources, or lack of gubernatorial support or leadership, there
is a growing perception that state oversight of local plan amendments is failing to
effectively implement or enforce state growth management requirements. The fact
that the DCA reportedly approves well over 90 per cent of all plan amendments
submitted for review suggests that the state may not be effectively monitoring and
enforcing the implementation of state growth policies.

As a result of all of these factors, the state and regional planning process in effect
today is substantially different than the process envisioned by the 1985 legislation.
State government has not only failed to live by its own State Plan but has allowed
it to become outdated. Neither the Florida Legislature nor state executive agencies
base their spending and capital improvements decisions on any coherent state vision
for growth and development. The regional planning component of the process has
been weakened and does not deal effectively with multi-jurisdictional issues. State
compliance review of local plan amendments gives little weight to the state and
regional plans. Consequently, the state and regional planning overlay has little effect
on the local planning process.

The Failure to Fund Concurrency

Effective implementation of local concurrency management systems requires
adequate funding sources for infrastructure. Nevertheless, despite its imposition
of the concurrency requirement on local governments, the state has not provided
adequate state funding or local funding sources on any consistent basis (see Nicholas
and Chapin, this volume). Beginning with the Legislature's enactment and almost
immediate repeal of a new sales tax on services in 1987, the implementation of
concurrency has been plagued by controversy over the funding issue. The state's
failure to face up to this issue has undermined support for the growth management
process among the development community and local government officials and
impeded implementation of major growth management policies such as concurrency
and compact urban form. Although, as previously mentioned, the 2005 Legislature
appropriated a substantial amount of infrastructure funding, it makes only a small
dent in Florida's huge infrastructure deficit. Further, while the Legislature tightened
the financial feasibility requirements for the capital improvements requirements of
local plans, it failed to provide local governments with any additional direct funding
or funding sources.

Planning Process vs. Planning Policy

The Florida growth management system consists of both planning process and planning policy. Ultimately, the success of the system depends upon effective implementation of both process and policy. Nevertheless, it is appropriate and useful to separately evaluate both the process and policy components of the system.

The greatest success of the GMA may be the establishment of local planning processes at the local level. Prior to enactment of the 1985 GMA, many of Florida's local governments did not have or did not implement local plans, and some had no zoning or other land development regulations. Now every local government has adopted a local plan approved by the state, and all but the smallest of them have planning departments and an institutionalized ongoing local planning process to address growth and development issues. Also, as previously mentioned, proposed local comprehensive plan amendments are subject to a state administrative review process. The efficacy of the resulting local planning processes and the state administrative review process in addressing development issues merits evaluation apart from the substantive policies to be implemented through these processes.

On the other hand, Florida's growth management process must also be evaluated in terms of the substantive planning policies established by the GMA. In this regard, the disintegration of the state and regional planning overlay, which was discussed in a preceding section, is of more than academic interest. The state and regional policy plans were intended to provide the growth management policies to guide the planning process. Local comprehensive plans were to be conduits for the implementation and achievement of these substantive state and regional goals and policies. Deprived of strong state and regional planning policy guidance, the local planning process has produced very mixed results in achieving major state growth management policies. These difficulties can be demonstrated by briefly examining three important state growth management policies: transportation concurrency, compact urban form, and affordable housing.

Theoretically, transportation concurrency is to be implemented as part of an integrated comprehensive planning process that promotes compact urban form. The local comprehensive plan is required to designate areas for urban development. These urban development areas are supposed to be backed up by financially feasible capital improvements plans which will deliver the transportation and other facilities needed to serve development and satisfy adopted transportation level of service standards. In this manner, local government comprehensive plans were intended to facilitate compact urban development in designated areas. However, the state's failure to adequately fund transportation or to provide local governments with sufficient local funding sources has undermined the theory. Because local governments have frequently been unable to deliver transportation and other facilities in the designated areas, there is inadequate road capacity to accommodate development in those areas. Understandably, developers seek to develop in non-designated urban development areas where there may be adequate road capacity, thereby creating intense pressure on the local government to extend its urban growth boundaries or otherwise amend the local comprehensive plan to permit development in those areas. Consequently, by failing to adhere to its own established growth management policies, the state undermines implementation of transportation concurrency and the important growth management policy of compact urban form.

The GMA contains very strong policies designed to promote provision of affordable housing, including housing for low and moderate income groups. On paper, these policies are among the strongest in Florida's growth management legislation. However, implementation of these policies depends upon DCA's enforcement of the affordable housing requirements when it reviews local comprehensive plans. Unfortunately, the affordable housing provisions of the GMA are probably the most neglected part of the growth management system. For many years, DCA has not required local governments to adopt regulatory reform measures to remove barriers to the provision of affordable housing. In addition, the state planning agency has been willing to accept almost any affordable housing proposals put forth by local governments in their plans, regardless of whether such proposals can effectively or realistically lead to the provision of affordable housing. Partly as a result of this policy failure, affordable workforce housing is now one of Florida's most pressing problems.

Accordingly, we should carefully distinguish between planning process and planning policy when we discuss the effectiveness and successes of the GMA. The implementation of the GMA may well be deemed a success in terms of establishing the planning process mandated by the GMA but much less successful in implementing state planning policies.

The Failure to "Fill in the Gaps" in the Original Growth Management Legislation

Any objective evaluation of Florida's growth management legislation must consider not only the planning-related issues which it addresses but also those which it omits. The 1985 growth management legislation failed to address a number of major issues which significantly affect growth management, and the frequent amendments to it in the ensuing twenty years have failed to fill in these gaps. Among these missing links are the following:

1. *Annexation and Urban Service Areas.* A municipality's annexation of new land has a substantial impact on the ability of its county and neighboring municipalities to plan for growth and development. Similarly, the establishment of utility service areas by individual local governments or private utility companies also influence the timing and location of the development and the ability of individual local governments to plan for it. Nevertheless, the exercise of municipal annexation powers and the regulation of utility service areas operate outside of Florida's growth management system. The power of annexation is controlled by an independent state statute[10] which the courts have held is not subject to the GMA. The regulation of utility service areas is largely within the jurisdiction of the Florida Public Service Commission, whose decisions regarding service areas need not be consistent with local comprehensive plans.[11] Numerous proposals to bring annexation and the regulation of utility service areas within the scope of the GMA have failed.

---

10 See Municipal Annexation or Contraction Act, Fla. Stat. §§ 171.011–093 (2005).

11 See, for example, Water and Wastewater System Regulatory Law, Fla. Stat. § 367.045(5)(b) (2005).

2. *Coordination of Transportation and Land Use Planning.* Planners have long recognized that the coordination of transportation and land use planning is essential to effective growth management. Relocation of roads and other transportation facilities is an important determinant of development patterns. Nevertheless, from the inception of the GMA, the Florida Department of Transportation (FDOT) has resisted efforts to subject the planning and construction of the state transportation system to the growth management process.[12] In 2004, the FDOT successfully sought and gained passage of legislation which exempted the state highway system from all local regulations (State Highway System Act, Fla. Stat. § 335.02(4), 2005). In 2005, FDOT unsuccessfully sought control over the approval of local comprehensive plan amendments for property adjacent to or in close proximity to the state's strategic intermodal transportation system, but the 2005 legislation did give FDOT even greater control over state transportation facilities (Local Government Comprehensive Planning and Land Development Regulation Act, Fla. Stat.§§ 163.3180(5)(f), (7), and (16)(e), 2005).

Perversely, the 2005 legislation tightens the requirements for local adoption of more flexible transportation concurrency approaches in urbanized areas. The Legislature first authorized these approaches in the 1990s, in response to the recommendations of a legislatively-created study commission which concluded that strict enforcement of transportation concurrency requirements in urbanized areas was thwarting urban infill and redevelopment and promoting urban sprawl. Because some of these approaches require provision of alternative or multiple modes of transportation, including public transit, relatively few local governments have adopted them. By imposing even stricter implementation requirements and providing little new funding for public transit, the new legislation may discourage further utilization of these flexible transportation concurrency approaches. Also, these changes, in combination with the infusion of road-building funds for other areas and the proportionate fair share requirement, may promote more urban sprawl in contravention of the GMA's compact urban form strategy.

3. *Adequate Alternative Local Revenue Sources.* Historically, Florida's local governments have relied primarily on local property taxes to finance their operations. This reliance on the property tax carries with it a built-in incentive for local governments to approve development because it will increase the local tax base. Although the Legislature has provided for some other optional local taxes, such as local sales taxes, it has imposed a voter referendum requirement for exercise of these powers. Proposals to eliminate the referendum requirement were unsuccessful in the 2005 Legislature, which at the same time tightened the requirement for local government adoption of financially feasible capital improvements plans. The lack of adequate alternative revenue sources for local governments hinders their ability to effectively implement state growth management requirements.

4. *Financial Incentives and Disincentives for More Compact Urban Form.* Florida's growth management legislation requires local governments to adopt local comprehensive plans which discourage the proliferation of urban sprawl and promote

---

12 See, for example, *Department of Transportation v. Lopez-Torres*, 526 So. 2d 674 (Fla. 1988).

more compact, mixed-use development patterns. However, while maintaining these planning regulatory requirements, the state has failed to take the necessary non-regulatory measures needed to reinforce and accomplish this important state planning goal. For example, the state has not adjusted its own taxing and spending policies to discourage sprawl and promote more compact development, and it has not coordinated its capital improvements and other investment decisions to support planning mandates to discourage sprawl. Further, it has not adopted smart growth techniques, such as directing state investments and expenditures for infrastructure to those areas designated in local comprehensive plans for urban development. Consequently, Florida has a planning and regulatory system that runs counter to economic and financial forces which promote rather than discourage urban sprawl.

Conclusion

Evaluating the effectiveness of Florida's growth management system is a complex task. The system has been constantly evolving over the last thirty years, particularly since the adoption of the 1985 Growth Management Act. Any fair and objective analysis must take into consideration the historical development of Florida's growth management legislation and its implementation. If we approach the task of evaluation with the necessary historical perspectives, we can better understand the successes and failures of the system. More importantly, we can hope to avoid repetition of the mistakes of the past and make appropriate adjustments to the system in the future.

References

American Law Institute. (1975). *A model land development code*. Washington, DC: Author.

DeGrove, J.M. (1984). *Land, growth and politics*. Washington, DC: Planners Press.

Department of Transportation v. Lopez-Torres, 526 So. 2d 674 (Fla. 1988).

Environmental Land and Water Management Act, Fla. Stat. §§ 380.012–10 (1973).

Environmental Land Management Study Committee. (1984). *Final report of the Environmental Land Management Study Committee*. Tallahassee, FL: Author.

Environment Land Management Study Committee. (1992). *Building successful communities*. Tallahassee, FL: Author.

Fla. Admin. Code Ann. ch. 9J-5 (1986).

Fla. Admin. Code Ann. ch. 9J-12 (1990).

Florida Governor's Task Force on Urban Growth Patterns. (1989). *Final report of the Florida Governor's Task Force on Urban Growth Patterns*. Tallahassee, FL: Author.

Fla. Laws ch. 72–295 §§ 1–13 (1972).

Fla. Laws ch. 72–317 §§ 1–13 (1972).

Fla. Laws ch. 75–257 §§ 1–19 (1975).

Fla. Laws ch. 84–237 §§ 1–2 (1984).

Fla. Laws ch. 85–55 §§ 1–51 (1985).

Fla. Laws ch. 85–57 §§ 1–8 (1985).

Fla. Laws ch. 93–206 §§ 1–84 (1993).

Fla. Laws ch. 2005–290 §§ 1–41 (2005).

Florida State Comprehensive Planning Act, Fla. Stat. §§ 186.001–911 (1985).

Florida State Comprehensive Planning Act, Fla. Stat. §§ 186.001–901 (2005).

Florida's Growth Management Study Commission. (2001). *A livable Florida for today and tomorrow*. Tallahassee, FL: Author.

Homebuilders and Contractors Association v. Dept. of Community Affairs, 585 So. 2d 965 (Fla. 1st DCA 1991).

Local Government Comprehensive Planning Act of 1975, Fla. Stat. §§ 163.3161–3211 (1975).

Local Government Comprehensive Planning and Land Development Regulation Act, Fla. Stat. §§ 163.3167–3243 (Supp. 1986).

Local Government Comprehensive Planning and Land Development Regulation Act, Fla. Stat. §§ 163.3161–3243 (1991).

Local Government Comprehensive Planning and Land Development Regulation Act, Fla. Stat. §§ 163.3161–3247 (2005).

Municipal Annexation or Contraction Act, Fla. Stat. §§ 171.011–093 (2005).

Pelham, T.G. (1992). Adequate public facilities requirements: Reflections on Florida's concurrency system for managing growth. *Florida State University Law Review, 19*, 973–1052.

Pelham, T.G. (2001). Restructuring Florida's growth management system: Alternative approaches to plan implementation and concurrency. *University of Florida Journal of Law and Public Policy, 12*, 299–310.

Pelham, T.G., Hyde, W.L., and Banks, R.P. (1985). Managing Florida's growth: Toward an integrated state, regional, and local comprehensive planning process. *Florida State University Law Review, 13*, 515–598.

State Comprehensive Plan, Fla. Stat. §§ 187.101–201 (1985).

State Comprehensive Plan, Fla. Stat. §§ 187.101–201 (2005).

State Highway System Act, Fla. Stat. §§ 335.01–188 (2005).

Trevarthen, S.L., and Friedman, C. (2005). Senate bill 360: Growth management reform arrives and it is all about infrastructure. *Florida Bar Journal, 79*(9), 39–43.

Water and Wastewater System Regulatory Law, Fla. Stat. §§ 367.011–182 (2005).

Chapter 3

Consistency, Concurrency and Compact Development: Three Faces of Growth Management Implementation in Florida[1]

Efraim Ben-Zadok

The 1973 Oregon Land Use Act and the 1985 Florida Growth Management Act (GMA) were the first and most influential growth management initiatives in the US (Porter, 1998, p. 29). Their purpose was to protect natural resources and agricultural lands. The first for a large urbanized state (Clark, 1994), the Florida act further emphasized the balance between protection of these critical areas and development of residential and commercial land uses. Florida was the fourth-largest state in population size in 1990 and the sixth in percentage of population living in metropolitan areas (US Bureau of the Census, 1993, pp. 28–29, xiii).[2]

The focus on implementation sets apart the state growth management initiatives from past comprehensive plans: "mandated implementation of the plans to assure that the new system really makes a difference in how and where development occurs and in how and where natural areas are protected" (DeGrove, 1992, p. 5). Florida implementation doctrine is instilled in three central policies of the GMA, which together frame the Act as a multifaceted initiative. The three policies are consistency, concurrency and compact development (labeled as the 3Cs). Consistency provides the structural framework for implementing the GMA. The policy mandates co-ordination, compliance and continuity among state, regional and local plans. This tri-level review process grants the state with ultimate authority to intervene in land development decisions, power almost entirely left to localities in the past. The review

1   This article was originally published in Volume 42, No. 12 (2005) of Urban Studies, pp. 2167–2190. Reprinted with permission.

2   The 1985 Florida Growth Management Act was originally entitled the Local Government Comprehensive Planning and Land Development Regulation Act. Its formal reference in this article is Florida Statutes, chapter 163.3161–3243, 1987. Post-1987 legislative changes in title, section numbers and their contents appear in this article under the reference Florida Statutes, chapter 163.2511–3245, 1999. The original 'GMA' acronym, however, is kept throughout the article. After the GMA, growth management acts followed in New Jersey (1986), Maine (1988), Vermont (1988), Rhode Island (1988), Georgia (1989), Washington (1990) and Maryland (1992). Less comprehensive acts followed in California and Hawaii (DeGrove, 1992, pp. 1–6).

covers several plan elements such as recreation, housing and capital improvement (DeGrove, 1992, pp. 7–16).

The capital improvement program includes public facilities needed to support development. Facilities subjected to concurrency should be available 'concurrent' with the impact of development. This policy can greatly influence the volume and pace of development (implementation) in communities (DeGrove, 1992, pp. 16–20).

Compact development aims to restrain urban and suburban developments from spreading towards natural resources and agricultural lands and, instead, to direct growth to urban areas of mixed land uses and high densities. A vague requirement in the 1985 GMA, compact development later emerged as a policy shift from managing growth to managing the location of growth (DeGrove and Turner, 1998; Transportation and Land Use Study Committee, 1999).

The purpose of this study is to evaluate and compare the implementation of consistency, concurrency and compact development. The evaluation covers the GMA evolution from inception in 1985 to 2002 with a brief update to 2004. Implementation covers regulation and enforcement activities, including links to preceding legislation. The study is based on analysis of secondary sources including the GMA and its state-level regulations, government and statistical reports, and academic studies and newspaper clips.

The contributions of the study to policy theory and to the literature and practice of growth management are outlined below and elaborated in the next section. The theoretical contribution lies in the concept of the steering policy, which describes each of the 3Cs. As a leading policy in a multifaceted initiative that consists of several policies, steering policy is directed by distinct purpose and it oversees critical implementation issues. Its learning experience is adopted by other policies, thus bringing changes in the implementation course of the whole initiative. Steering policy is a useful concept to understand the rationale behind each of the 3Cs, its role in GMA evolution and how it changed the Act by transferring experience to other policies. Like a thread through the evolution of a multifaceted initiative, the concept connects different policies, delineates links between past and present implementation, and points out implications for future course.

This concept helps to show that the Florida act has experienced significant policy changes. Each policy has a distinct purpose and implementation style, different implementation issues and implications. Altogether the three policies 'write' a new 'three faces' narrative for the GMA implementation. This more diverse account contributes to growth management literature where the GMA often appears as a 'one face' story reflected through consistency. Although consistency placed Florida at the helm of the growth management literature, it only represents the early implementation phase. Concurrency and compact development have followed in the 1990s, introducing different domains and implementation styles. The 3Cs altogether shed light on the GMA as a multifaceted initiative experiencing turmoil and policy change. This more balanced evaluation of the Act corrects its image in the literature.

The study also contributes by drawing implications for growth management practice in Florida and other states that plan or implement similar policies. Implications are especially important in light of the current uncertainty around Florida growth management. In the face of the slowing national economy, GMA implementation has reached a critical phase since the late 1990s when Governor Jeb Bush attempted to reduce state control over growth management and transfer power to localities. The GMA became the subject of political debate including repeated calls for major policy revisions (see, for example, Cox, 2000; Fleshler, 2002; Kaczor, 2002; Shenot, 1999; The Sun-Sentinel, 2001a).

A framework for comparative evaluation of the 3Cs is presented in the next section. Thereafter, each policy is evaluated in one section. The article concludes with a comparative summary of the three policies and implications for growth management in Florida and other states.

Conceptual Framework for Evaluation

Florida's massive growth from 2.7 million in 1950 to 9.5 million people in 1980 has precipitated state acts in 1972 and 1975. The legislation aimed to balance growth with the protection of natural resources. It also introduced a state-centralized review process to ensure the compliance of local comprehensive plans. But implementation contents and details were controlled by growth-driven localities that often considered the review as an unnecessary intrusion. This practically bottom–up model was drastically changed in 1985 when the GMA mandated top–down consistency in review of state, regional and local plans. The review covered all implementation components with much attention to details. The Act also sought to protect Florida's natural resources and farming lands from fast population growth. Population growth posed huge problems to the state because it was followed by residential and commercial developments and increasing demands for schools and roads (DeGrove, 1992, pp. 7–11; Turner, 1990a).

Growth management studies often focus on one multifaceted state initiative such as the Florida or Oregon Act (for example, DeGrove and Turner, 1998; deHaven-Smith 1998; Epling, 1993; Gale and Hart, 1992; Howe, 1993). Other growth studies compare among specific state policies that resemble each other, including Florida's consistency (Bollens, 1992; DeGrove, 1984, 1992; Gale, 1992; Innes, 1993; May et al., 1996; Starnes, 1993). All these studies have provided extensive coverage to consistency and relatively modest coverage to concurrency and compact development.

A handful of studies that have focused on a specific policy in one state come from coast management, an area closely related to growth. Examples are evaluation of sedimentation and erosion control in North Carolina (Burby and Paterson, 1993) or evaluation of coastal natural hazard in Florida (Deyle and Smith, 1998). Other coast management studies compare one policy like response to natural hazard across several states (Berke and French, 1994) or assess nation-wide one policy like building codes (Burby et al., 1998). Coast management studies thus provide extensive evaluations on implementation of a specific policy.

This kind of treatment is extended in this study to three policies with each being evaluated along the same five characteristics of implementation. The evaluation framework begins with the theoretical concept underlying the policies and continues with a description of these characteristics.

Steering Policy

As mentioned earlier, the steering policy concept contains both leadership and learning functions. The leadership function is evident in the policy's central and specialized role in the evolution of a multifaceted initiative. It roughly resembles the idea of 'steering capacity', which a political system must have to secure its purposeful movement (Deutsch, 1963).

The leadership function of the 3Cs in the GMA is often reflected in studies that present each of them as a key policy within a multifaceted initiative (for example, Turner, 1990b, p. 83). Consistency is known as the GMA's organizing doctrine, with concurrency and compact development as the 'twin pillars of the policy base'. Studies show that implementation started with consistency, proceeded with concurrency and then evolved to compact development. Other GMA policies like affordable housing and protection of farm and forest lands are weak provisions with relatively little implementation (DeGrove, 1992, pp. 11–21; DeGrove and Turner, 1998; Pelham, 1992; Turner, 1990b). The centrality of the 3Cs goes beyond Florida. With coast management and citizen participation, all five policies comprise "the five Cs of modern growth management in the United States" (May et al., 1996, p. 30).

The learning function of the 3Cs has received little analysis in the growth management literature. This function is apparent when learning experience of one policy is transacted to other policies, thus bringing changes in the general course of implementation. That is, professional, bureaucratic and political lessons of one policy are adopted by other policies, thus steering changes in the direction of the whole initiative.

The learning function as a change mechanism is an idea built on 'policy-oriented learning', a term developed by Heclo (1974) and Sabatier (1999). The term refers to new information and/or new experience that lead to relatively enduring alterations of thought or behavioral intention aimed to achieve or revise policy objectives. That is, policy change is the result of increased knowledge of implementation problems (cognitive activity) and/or new external economic and political conditions (non-cognitive sources). Understanding a policy process that includes such changes requires a span-time of a decade or more from problem emergence until implementation experience renders a fair evaluation of cumulative impact (see Mazmanian and Kraft, 1999).

Implementation Characteristics

To generalize the evaluation effort and navigate it through numerous technical details, each steering policy is reviewed along five implementation characteristics: regulation/enforcement period, purpose, critical issue, implementation process, and implementation outcome (see Table 3.1). The first characteristic indicates the period in which policy regulation or enforcement activities took place. After a brief acknowledgement of this item, the rest of the section introduces the other four characteristics. These four guide the evaluation in succeeding sections.

Table 3.1   Steering policies of the Florida Growth Management Act by implementation characteristics, 1985–2002

Policy Implementation Characteristics	Steering Policy		
	Consistency	Concurrency	Compact Development
Regulation/ enforcement period	Regulation 1985–86 Enforcement 1986–93	Regulation 1986–89 Enforcement 1990–93	Regulation 1993–2002 Enforcement 1993– 2002
Purpose	Organization of planning	Controlled growth and economic development	Compact growth and economic development
Critical issue	Intergvtl planning Local-Regional-State plan review	Public facilities	Transport
State—local implementation process (regulation and enforcement)	Prescription	Discretion	Increased discretion
State—local implementation outcome	Success—compliance	Mixed—problems and achievements	Mixed—problems and achievements

Regulation/enforcement period   This characteristic provides a time-frame for two traditional implementation domains, the making of rules and their enforcement. Regulation refers to state and local rules of administration, land use and land development, including their links to preceding planning legislation. Enforcement refers to state and local execution of the rules. Because regulatory policies like the 3Cs impose uniform rules on all communities, making and enforcing rules raise many tensions and conflicts.[3] The contentious and competitive nature of regulation and enforcement in the 3Cs is evident in this study.

Although the three policies are simultaneously underway, they still form a time-line where each policy dominates implementation in a certain phase. That is, in a specific regulation/enforcement period of a policy, it is that policy that practically oversees most regulation or enforcement activities in the GMA. Relatively limited regulation or enforcement activities of other policies are underway at the same time (see Table 3.1).[4]

---

3   Every policy has its own problems and difficulties of implementation (Pressman and Wildavsky, 1973). But regulatory policies are more adversarial because they impose uniform rules on all groups/communities. Distributive policies are less contentious because they allocate benefits or services for a particular group/community (see, for example, Anderson, 1994, pp. 11–15).

4   The only exception in this time-line is 1990–93, a period in which both consistency and concurrency dominated the Act with enforcement activities.

Purpose and critical issue   A steering policy covers a central domain of growth management and is characterized by a distinct purpose and a number of critical issues. The purpose of consistency is the organization of planning; the critical issues are intergovernmental planning and local–regional–state plan review. To rephrase, the consistency policy organized the administrative apparatus of the GMA including regulatory mechanisms for long-range intergovernmental planning and tri-level plan review. Next, the purpose of concurrency is controlled growth and economic development; the critical issues are public facilities and funding concurrency. Concurrency aimed to restrain community growth and support the local economy by mandating construction and funding of public facilities to support development. Last, the purpose of compact development is compact growth and economic development; the critical issues are transport and urban–suburban sprawl. This policy attempted to direct growth to urban areas of mixed land uses and support local economies via strategic choices in road construction and reduction of urban and suburban sprawl (see Table 3.1).

The distinct purpose and critical issues of each policy illustrate its central and specialized role in GMA evolution. They help to understand the leadership function of the policy and how it contributed to the diverse 'three faces' story of the Act. They also highlight the learning function of the policy. Concurrency's experience in controlling local growth by mandating the construction of public facilities—for example, has increased knowledge of implementation problems and awareness of economic and political constraints. Of all public facilities, transport (roads) has emerged as forerunner of sprawl and most difficult to implement. The problem resulted in sprawling developments that threatened Florida's natural resource systems. Anti-sprawl issues then captured the GMA's public debate (DeGrove and Turner, 1998; deHaven-Smith, 1998). In the early 1990s, legislators and bureaucrats began to shift from concurrency to compact development, emphasizing transport and sprawl issues (see Table 3.1).

Implementation process and outcome   State–local relations have been debated in the environmental movement since the 1960s. The movement criticized the failure of American communities to control development and protect natural resources. The question of the appropriate implementation process for growth management began to capture public attention.[5] The environmental movement saw local governments, traditionally in control of land uses, as susceptible to development pressures, lacking expertise to deal with complex environmental issues and in need of state supervision. The movement influenced state governments to limit local power over growth policies. Alternative models of shared state–local implementation emerged in the 1970s. They varied according to specific growth policies and local conditions (Burby and Paterson, 1993; DeGrove, 1984; Weber, 1998).

The next section shows that consistency in intergovernmental planning was implemented through top–down prescription in the state's review of local

---

5   The question deserves attention because implementation process evaluations are still relatively scarce, while outcome evaluations are popular in government (deLeon, 1999, p. 323; Ingram, 1990, p. 466).

comprehensive plans (see Table 3.1). This strict 'hands on' approach aimed to deter and punish violations and to ensure compliance in local plan submissions (for example, Gormley, 1998). A host of studies describe consistency as a state dominant, highly prescriptive policy. They conclude that state control over land use planning, followed by high level of local compliance, has been much greater in Florida than in other growth states implementing their versions of consistency (for example, Gale, 1992; Innes, 1992; May et al., 1996, pp. 22–42).

With concurrency and compact development underway, attention has shifted to sprawl control and economic development issues. Florida localities faced immediate problems regarding how to build and fund roads and schools, and how to direct residential and commercial developments to urban areas. They negotiated with state and developers, a process leading to many local implementation versions. The evaluation demonstrates this flexible discretion and the increased discretion approaches that have compromised competing interests involved in both policies (see Table 3.1).

Discretion, a significant change in implementation style, helps to correct the 'one face' account of the Act. In the absence of research on other GMA policies, it was common to generalize from consistency to the whole act. By the early 1990s, the GMA already was described in the literature as a top–down coercive initiative (for example, by Bollens, 1993; Innes, 1993; Starnes, 1993: Turner, 1990a). The discretionary process of concurrency and compact development suggests again that the Act is a diverse multifaceted initiative.

Prescription in consistency was a learning experience leading to discretion in concurrency and, in turn, to further discretion in implementing compact development. With more information about problems in executing consistency and increased awareness of economic and political difficulties, state and local governments learned to negotiate and eventually have adopted discretionary style.

A state-led prescriptive process or a state–local shared discretionary process ultimately affects implementation outcome. This final evaluation characteristic shows that a high level of compliance in a state's review of local comprehensive plans was viewed as success in implementation. Next, a handful of problems and achievements in implementing both concurrency and compact development seemed roughly to balance each other. State and local governments recorded a mixed bag of outcomes for the two policies (see Table 3.1).

Each of the three succeeding sections evaluates one of the 3Cs, beginning with a sub-section on purpose and critical issues and closing with a sub-section on implementation process and outcome. Evaluation of purpose and critical issues is based on legislation (GMA) and academic studies. Evaluation of state–local implementation process is based on state-level regulations of the GMA, selected cases of state–local and local enforcement, government reports and academic studies and newspaper clips. Evaluation of state–local implementation outcome is based on statistical data drawn from government reports, census reports, survey questionnaires and academic studies; and on qualitative data from local cases, academic studies and newspaper clips.

Table 3.2   Implementation activity of the Florida Growth Management Act and its steering policies by regulation/enforcement period, 1985–2002

Regulation/Enforcement Period	Implementation Activity	
	Growth Mgt Act	**Steering Policy**

Regulation, 1985–86
Related legislature action
1985, SCP
1985, GMA
1986, GMA Amendment: "Glitch Bill"
	Continuous: 9J-5
	  Prepared by DCA
	1986, 9J-5
	  Adopted by DCA
		Consistency-driven
		  Includes concurrency
		  requirement

Regulation, 1986–89	Continuous: DCA develops
	  detailed state-wide rules
	1989, 9J-5
	  DCA amends detailed state-
	  wide rules
		Concurrency-driven
		Detailed state-wide
		  concurrency rules are
		  developed and amended by
		  DCA

Enforcement, 1986–89	Continuous: DCA's
	  LOC plan review and
	  compliance with 9J-5
		Consistency-driven

Enforcement, 1990–93	Continuous: DCA's
	  LOC plan review and
	  compliance with 9J-5
		Consistency-driven
		  Includes concurrency

Enforcement, 1990–93	Continuous: LOC develops
	  and enforces rules
	Continuous: DCA enforces
	  and monitors LOC rules
		Concurrency-driven
		  LOC CMS rules are
		  developed, enforced and
		  monitored by LOC and DCA

Regulation, 1993–2002
Related legislature action
1993, GMA Amendment: Transportation Concurrency
1996, GMA Amendment: Sustainable Communities
1999, GMA Amendment: Urban Infill and Redevelopment
2002, GMA Amendment: Local Government Comprehensive Planning Certification Program
	Continuous: 9J-5
	  Detailed rules developed
	  and adopted by DCA
		Compact-development-driven

Enforcement, 1993–2002	Continuous: DCA and
	  LOC enforce and monitor
	  detailed LOC rules
		Compact-development-driven

Notes: CMS—Concurrency management system(s); DCA—Department of Community Affairs, State of Florida; GMA—Growth Management Act; LOC—Local government(s); SCP—State Comprehensive Plan; 9J-5—Rule 9J-5.

Consistency, 1985–1993

In 1985, the Florida legislature passed the State Comprehensive Plan (SCP) and the GMA. The Florida Department of Community Affairs (DCA) started to implement the GMA by preparing minimum criteria for reviewing the consistency of local plans. In 1986, the legislature amended the GMA with the Glitch Bill, thus clarifying the consistency requirement and intergovernmental planning arrangements. The Bill required all plans to meet Rule 9J-5—the state-level new regulations for minimum criteria to review local comprehensive plans. Enforcement started in 1986 when local governments began to prepare their plans according to these regulations. The DCA reviewed the plans of all Florida localities between 1988 and 1993 and ensured their compliance with 9J-5 (see Table 3.2).

Organization of Planning: Intergovernmental Planning and Local–Regional–State Plan Review

Consistency is the policy that organized the planning system through an intergovernmental structure and local–regional–state plan review (see Table 3.1). Florida state and local governments developed both structure and minimum plan standards with much emphasis on physical planning. They tended to set aside complex issues of economic development and sprawl (Pelham, 1992; Turner, 1990a, 1993). The consistency requirement in the GMA mandates preparation and adoption of comprehensive plans by the state, regional planning councils and local governments. The plans cover capital improvement, transport, housing, recreation, open space, environmental quality and other elements. The GMA's requirements for intergovernmental co-ordination among plans include: consistency between the local plan (adopted into law and implemented by local government) and local development codes such as zoning; consistency between the local plan and its counterparts in nearby jurisdictions; consistency between local and regional plans; DCA's review and approval of local and regional plans against the SCP and other state policies; and periodic evaluation and updating of local plans including review by DCA (Florida statutes, chapter 163.3161–3243, 1987 (3184); Gale, 1992; Starnes, 1993).

These principles of consistency among different plans within a centralized intergovernmental system distinguish the GMA from typical, local-controlled, American planning (Gale, 1992). They delineate hierarchical compliance among plans from top (state) to bottom (local). The state formulates minimum criteria regulations and ensures their consistency with regional and local plans. It also is responsible for enforcing regulations through the intergovernmental review of plans. Local governments formulate regulations for comprehensive plans and land development and then supervise their routine enforcement.

Consistency shaped the GMA as a state centralized process. It granted the DCA with ultimate authority to approve local plans and impose stringent legal sanctions. When a local plan was unprepared, DCA could prepare and enforce it on the community. If a plan was not in compliance, DCA could file a 'notice of intent' through the State Division of Administrative Hearings. The Division could recommend action to the Administration Commission that might impose final sanctions. The Commission could prohibit localities from participation in the federal community development program or other state-administered programs (Florida statutes, chapter 163.3161–3243, 1987 (3184) (6–11)). With little state funding to prepare local comprehensive plans, these sanctions posed substantial threats to localities and especially limited the planning activities of small communities (Liou and Dicker, 1994).

State Prescription and Local Compliance

Prescription   The intergovernmental implementation process of consistency among plans was a top–down prescription from state to local governments (see Table 3.1). The state DCA dictated local planning agendas by promoting certain aspects of plan review. The Department frequently imposed strict sanctions on physical rather than social aspects of planning that eventually became a low priority. Local planners also perceived growth management as physical accommodation to the environment and have concentrated on infrastructure and traffic rather than affordable housing and other social issues (Turner, 1993).

Next, the DCA put much emphasis on detailed procedures in local plans. It prescribed a 'cookbook' checklist matrix to localities and requested them to adhere to the level of specificity in state standards and deadlines. The strict implementation style was followed by unequivocal procedural compliance orders and sanction threats. Localities complained that DCA had excessive control over community development and resented the rigid review process (May et al., 1996, pp. 91–145). Broward County—for example, was concerned that its state revenues would be held to attain compliance with DCA rules. After its plan was found non-compliant, the county did not even appeal and then followed most objections and recommendations raised by the Department (Turner, 1990a).

The DCA learned from the bottom–up implementation style of the 1972 and 1975 acts. By the late 1980s, it saw plan review as a state-dominated, coercive legal process. Developers were not prepared for this policy shift (Gale, 1992; Turner, 1990a). Local planners and politicians complained that the Department's staff was inflexible and inexperienced in daily growth management problems (Porter and Watson, 1993). The Department response was the 'compliance agreement' process where it filed a 'notice of intent' for a non-compliant plan and then negotiated plan revisions with local officials and stakeholders. Negotiation took place before the hearing date in the State Division of Administrative Hearings. This forum served as a settlement mechanism to bring local plans in compliance with state rules and to move the review process along (DeGrove, 1992, p. 16).

Because the GMA gave scant attention to special interest-group participation, even the compliance agreement processes often excluded environmental and farming groups. The groups that came to negotiate had difficulties reaching consensus; they complained that the DCA was repressing their voices. Local officials bitterly grumbled that the Department was pressuring them to settle by 'getting what it wants'. Not denying the allegation, the DCA staff claimed that they 'mean business' and have little interest in giving voice to negotiating groups (Innes, 1992).

Compliance   The prescriptive implementation process has resulted in a high level of compliance, an outcome viewed as 'success' by GMA observers and especially those from the DCA (see Table 3.1). Reported by several observers, the following numbers demonstrate how effective was the state driven process in achieving local compliance at specific time-intervals. By April 1990, 263 local governments out of a total of 459, constituting 75 per cent of Florida's population, had submitted their comprehensive plans for DCA review. Of these submissions, 80 per cent either complied or negotiated a compliance agreement with the DCA (Turner, 1990b, p. 90). By July 1991, all localities had submitted plans to the DCA and had met the original GMA's deadline (Nelson and Duncan, 1995, p. 24). By May 1992, 302 local plans were found in compliance and 43 were close to reaching a compliance agreement (DeGrove, 1992, p. 15). In December 1993, the implementation process of consistency was almost complete when all but 24 plans (5 per cent) were found in compliance (May et al., 1996, p. 132).

In sum, consistency is generally viewed as the successful policy that led the GMA in its initial phase and has inaugurated Florida as a growth management innovator. But complete and approved local plans did not necessarily reflect professional quality or agreement among stakeholders regarding upcoming implementation. Consistency's prescriptive and compliant style often repressed issues of plan quality and politics. In subsequent implementation phases, when these issues surfaced on state and local agendas, the GMA's Evaluation and Appraisal Report requirement became important. This mechanism for correcting consistency is revisited in the final section of the article.

Concurrency, 1985–1993

The 1986 Glitch Bill clarified state and local responsibilities for setting minimum concurrency levels. The DCA then began to develop detailed state-wide concurrency regulations. After these were adopted as amendment to Rule 9J-5 in 1989, concurrency's actual enforcement started with the bulk of this activity lasting to 1993. The DCA reviewed every local plan, including its concurrency rules, for compliance with 9J-5. After plan approval, concurrency enforcement began via the Concurrency Management System (CMS). CMS was the DCA-monitored enforcement tool that was mandated in 9J-5 for supervising the routine maintenance of concurrency in every community (see Table 3.2).

Controlled Growth and Economic Development: Public Facilities and Funding Concurrency

Concurrency aims to control growth and enhance economic development by mandating construction and the funding of public facilities (see Table 3.1). The concurrency process begins with the GMA's requirement to adopt a capital improvement program into the local comprehensive plan. The program must include public facilities needed to support development and their minimum level-of-service (LOS) standards required to accommodate projected growth. The locality must adopt rules prohibiting the issuance of a development permit that could bring reduction of LOS below the standards (Florida Statutes, chapter 163. 3161–3243, 1987 (3177)). The capital improvement program includes six public facilities (subjected to concurrency since 1989): transport (roads), sanitary sewer, solid waste, drainage, potable water, and parks and recreation (Florida Statutes, chapter 163. 2511–3245, 1999 (3180) (1) (a)). Because concurrency co-ordinates in advance the capital improvement program and the local plan, it can deal effectively with land development and building. Concurrency links a new project approval to provision of adequate public facilities. The requirement to deliver facilities is brought to the forefront of land use planning (regulation) rather than reserved for the later development stage (enforcement). This enables advanced regulation and co-ordination of time, type and location of future development (Pelham, 1992, pp. 974–975).

Concurrency can direct future development by denying permits when public facilities are inadequate. It signals to developers whether facilities are adequate and what costs are involved in capitalizing them into land values. Developers absorb the costs of facilities and pay concurrency fees as a prerequisite for permits, when they calculate that such investments will bring future profits (Nicholas, 1993, pp. 210–211). The availability of facilities (and potential market response) determines land values and, in turn, affects the volume and pace of residential and commercial building.

Concurrency played a central role in the GMA evolution due to its enormous power to affect local growth and economic development via planning and the enforcement of public facilities. But the policy could control growth more effectively if it identified the sources for funding public facilities and their shares. The absence of explicit information on funding responsibilities in the Act was a major impediment, especially in communities that either could not afford facilities or simply were reluctant to force funding on developers or residents. Although the capital improvement program included the costs of facilities and a plan to finance them, the funding sources for such budget items were not detailed (Florida Statutes, chapter 163. 3161–3243, 1987 (3177) (3)).

With no state commitment to fund concurrency, the task was automatically relegated to local governments. Most could not provide the public facilities required to support growth (Pelham, 1992, pp. 1025–1028). Funding became a critical issue that often was debated in the Florida legislature, while no clear commitment made to finance concurrency (DeGrove, 1992, pp. 23–24).[6] The toughest funding problem was the omission of school construction from the GMA's list of facilities subjected to concurrency. Expenditures for planning, construction and maintenance of new schools are among the largest in the education budgets of state and local governments,

---

6 The issue of funding concurrency was exacerbated by the state's larger fiscal environment. Florida severely restricts its localities in collecting property and sale taxes, yet continuously imposes on them 'unfunded mandates' with little financial assistance. The state was ranked as one of the lowest in per capita federal grants received (fiftieth until 1990). It is also below national average in generating state and local taxes, being one of seven states with no state personal income tax, and it relies heavily on state sales tax and corporate income tax revenues (MacManus, 1998).

especially in fast growing areas.[7] GMA stakeholders favored a short list of facilities excluding schools, public health, law enforcement and fire protection. Even strong supporters of the 1985 Act avoided confrontation around schools, a highly expensive local budget item that could have undermined the whole bill (Pelham, 1992, pp. 1012–1016).

School concurrency was then tabled to the unfinished agenda of the GMA. Despite several legislative initiatives, the list of concurrency's facilities has not been amended and schools remained a voluntary local implementation item. The 1995 legislature established school concurrency requirements, instructing localities to co-ordinate with their school districts the LOS standards, financial feasibility and location of new schools. Although principle adoption remained a local choice, the fastest-growing counties began to introduce light versions of school concurrency (DeGrove and Turner, 1998, p. 183). The 2002 legislature mandated local governments and school districts to sign, with DCA approval, an interlocal agreement to co-ordinate land use and public school facility planning (Florida Department of Community Affairs, 2002, p. 3). A few months later, a constitutional amendment on school overcrowding called for the reduction of the number of children in each classroom by 2010.

State–Local Discretion and Mixed Outcome Discretion.

Florida state and local governments followed a discretionary style in implementing concurrency (see Table 3.1). The DCA understood that, because concurrency was crucial to local economies, this policy would not stop or slow down growth. In a 1988 letter to the Chair of the Senate Select Committee on Local Government Infrastructure, the DCA Secretary said: "although the concurrency requirement is a tough one, it will not bring the state to a screeching halt if it is applied with common sense, and in a reasonable and flexible manner" (DeGrove, 1992, pp. 16–17). In a follow-up letter to the Chair, the Secretary repeated his commitment to a state–local discretionary approach by allowing flexible timing points for implementing concurrency, including their adjustment to local circumstances. The letter stated:

> The Department rejects as totally unreasonable and unworkable the position that concurrency can only mean from the moment the concurrency requirement goes into effect, all necessary facilities must actually be in place before a development permit can be issued. The legislature could not possibly have intended such an interpretation because it is totally unrealistic and unworkable. The Department rejects that approach to the statute (Pelham, 1992, pp. 1011–1012).

These words signaled to communities that the DCA was willing to negotiate when and where adequate public facilities were in place. Not detailed in the GMA, the timing of facilities was left to the Department's interpretation. DCA's emerging

---

7   Schools are the largest expenditure and the most growth-responsive public facility subjected to concurrency. During 1987–89, Florida's education accounted for the single largest cost of state government (38 per cent) and of state and local governments combined (31 per cent). The state schools were in the highest cost category (23 per cent) for expenditure by a major program (Dye, 1991, p. 286).

discretionary approach regarding the timing issue relieved many communities and real estate developers. They feared that concurrency implementation would replicate consistency's prescriptive style in local plan reviews. But the DCA seemed to learn that local compliance in consistency did not necessarily imply agreement among growth management stakeholders. State and communities still faced economic and political problems that were repressed during plan approval processes. Many communities were concerned about slow down in growth and have begged for clarification on the timing issue. They questioned whether roads, sewers, parks and other facilities should be available when a development permit is issued, when a development is opened or at some later date (Ben-Zadok and Gale, 2001).

When the DCA adopted Rule 9J-5 in 1989, it provided a two-step answer that practically invited state–local negotiation over timing points. The first step was classification of three categories for facilities complying with concurrency requirements. Category I included four facilities closely linked to public health and safety: water, sewer, waste and drainage. Category II was reserved for less critical facilities: recreation and open space. Category III contained the most costly and complex facility: roads. The second step was the provision of multiple timing-points for the availability of facilities. This arrangement rendered flexibility in implementing category I and especially in implementing category III (Pelham, 1992, p. 1019; Rule 9J-5, 001–016, 1989 (0055) (1)).[8]

With little direction from the GMA, the DCA also facilitated state–local bargaining over the minimum LOS standards that localities must meet to estimate service demands that new projects impose on public facilities. The Department did not have state-wide minimum standards and localities could comply by determining their own LOS standards for facilities. The DCA applied lenient standards even for transport, the only facility subjected to state-wide standards. When transport concurrency was mandated in 1989, the Florida Department of Transportation (DOT) required local governments to evaluate the service demand for each road-corridor according to DOT standards. When 9J-5 was amended in 1991, it already allowed LOS standards for an entire network of road-corridors rather than for each road (Boggs and Apgar, 1991, p. 12; DeGrove and Gale, 1994; Pelham, 1992, pp. 1020–1021). The DCA also approved flexible LOS standards for state and county roads in fast-growing urban areas of counties reluctant to slow down development. A negotiated agreement with Miami–Dade, for example, allowed the county lower LOS standards in the downtown area in order to "discourage urban sprawl by promoting mass transit, redevelopment of downtown areas, and efficient use of infrastructure" (Florida Department of Community Affairs, 1991).

---

8   The timing-points for availability of facilities in Category I were: when a development permit is issued; or guaranteed in the permit to be in place when the impact of development will occur; or must be under construction when the permit is issued; or must be guaranteed in a developer's agreement including one or more of the above provisions. Category II required commencement of construction within one year from the date of issuance a development permit. Category III could be satisfied by meeting the timing-points for either the first or second category, or by adopting a financially feasible five-year capital improvement program that provides for the construction of facilities by the third year (Pelham, 1992, p. 1019; Rule 9J-5. 001–016, 1989 (0055) (1) (2)).

Uneasy about communities that may suspend new project approvals and inhibit land development, the DCA also applied flexible rules for capacity deficits of existing public facilities. The issue of who pays for existing infrastructure deficits became crucial, especially when facilities were already operating at or above capacity. Many developers rushed projects through local approval processes, concerned that once concurrency rules were triggered, their impact fees would finance the earlier projects of developers who did not pay. Developers even 'pushed' infrastructure systems beyond their capacities in efforts to circumnavigate the actual enforcement date of concurrency rules. DCA recognized the problem and alleviated developers' anxieties by ruling that each project should pay in proportion to its consumption of infrastructure capacity. The courts later ruled that impact fees could not be used to construct facilities that ease infrastructure backlogs (DeGrove, 1992, p. 25; DeGrove and Gale, 1994).

Finally, a 1992 survey showed considerable local discretion in response to the question of how long after payment of impact fees a developer could reserve capacity. Of 146 Florida localities, 45 per cent had no procedure to determine whether capacity reservation was guaranteed for a permit applicant; 30 per cent did not reserve capacity until the applicant's project reached the final stage of issuing a building permit; and 59 per cent had a two-year or shorter period of reservation, after capacity was guaranteed to an approved project, subjected to a 'use it or lose it' rule (Audirac et al., 1992, pp. 18–20).

Mixed outcome   The discretionary implementation process of concurrency has resulted in mixed outcomes. The report below suggests a handful of problems in the construction and funding of public facilities, but problems affect communities on various severity levels (see Table 3.1). Two Florida studies from 1992 report difficulties in local compliance to LOS standards. Of all local governments surveyed in one study, 39 per cent indicated that several of their public facilities were not operating at the adopted LOS standards and 28 per cent identified state roads as the most inadequate facility (Environmental Land Management Study Committee, 1992, p. 61). Another study of 146 localities found that the ones wishing to decrease LOS standards were heavily populated and with a high growth rate (Audirac et al., 1992, p. 35).

The discretionary process also resulted in growing infrastructure capacity deficits, especially in transport. A 1990 survey indicated that 75 per cent of Florida local governments identified traffic congestion as a particularly severe growth issue. Another 1990 survey marked roads as the local infrastructure expected to encounter the greatest capacity deficits in the next five years (Turner, 1993, p. 192). The 1992 study mentioned above noted that, of communities with a high level of capacity deficit, 87 per cent were likely to be heavily populated and 78 per cent to have a high growth rate (Audirac et al., 1992, p. 28). A 1996 survey reported that 53 per cent of Florida's local officials ranked infrastructure and capital projects as the second most serious problem facing their jurisdiction and 30 per cent ranked growth management in general as their fifth-biggest problem. About one-third said that growth, roads and traffic and congestion were the most frequent problems mentioned by their constituents (MacManus, 1998, pp. 218–219). Florida indeed was the fourth most populated state in 1994 and ranked second with 904,543 new car registrations. It had about one-half of California's population (the most populated state), yet registered only 1655 fewer cars (Weaver, 1995).

Another indication of mixed results is in the variable levels of local compliance with CMS and how they relate to population growth.[9] Communities that did not exempt projects from CMS review were likely to be less populated (51 per cent). Communities that exempted projects from CMS review and carried serious capacity deficits were likely to be more populated (76 per cent) (Audirac et al., 1992, pp. 28 and 35).

A mixed record is also evident in the use of impact fees to generate revenues to fund public facilities. Levying impact fees is a strategy by which localities negotiate with developers on financial commitments for facilities and request them to pay the cost. Developers calculate how much of this fiscal burden would be passed on to new home-buyers and renters (Wiewel et al., 1999). Since the mid 1970s, impact fees have become a popular interim solution subject to voluntary adoption by Florida communities. Despite legal challenges, the court ruled that localities have the right to levy such fees. By 1991, 125 out of 459 local governments had adopted impact fees and many more planned to do so (DeGrove, 1992, pp. 23–25; MacManus, 1998).

Concurrency essentially required communities to collect impact fees to fund infrastructure. But the problem was in the absence of enabling state law for authorizing impact fees. After localities resorted to sales and property taxes, the legislature even restricted this option by requiring public referendum for approval of major local infrastructure taxes (Pelham, 1992, pp. 1025–1028 and 1050). Impact fees ultimately were an insufficient solution to provide facilities and ease growth pressures. The problem was exacerbated in the most populated counties like Miami-Dade and Broward. The former lacked specific directives for spending impact fee revenues collected since 1990, the latter simply lacked revenues (White, 1999). Broward's deficit may lie in its state's discretionary permit allowing the construction of new projects, served by roads already at or beyond capacity, providing that additional generated traffic did not exceed 10 per cent of the base traffic volume (Koenig, 1990, p. 8).

Mixed outcomes are also evident with respect to controlled growth and economic development, concurrency's ultimate goals. These outcomes are affected by other state–local policies, demographic and economic forces. Although the actual impact of concurrency on the following outcomes can only be guessed, they may still imply some impact of concurrency. To begin, Florida grew from 12.9 million in 1990 to 14.4 million residents in 1996, an increase of 11.4 per cent or about 650 incoming residents per day. The annual population growth rate gradually declined from 2.1 per cent in 1990 to 1.6 per cent in 1996. These 1996 numbers indicate a more moderate growth when compared with the state's 32.7 per cent increase or about 1000 incoming residents per day between 1980 and 1990. In addition, while

---

9 The CMS is a computer-based information management system that keeps track of all existing infrastructure capacities and deficits, addresses backlog problems and new development impacts, and allocates capacity priorities to all public facilities (Rule 9J-5. 001–016, 1989 (0055)).

the 1990–96 growth trend was shared by each of Florida's 67 counties, half of them were sparsely populated with less than 80 000 residents per county (University of Florida, Bureau of Economic and Business Research, 1997).

Economic development also displayed mixed records from 1987 to 1993. The number of employees in all industries increased from 4.8 to 5.5 million and the gross state product grew by 56 per cent. Although services registered the largest increase, from 1.5 to 2.1 million employees, they employed huge numbers of low-skilled/wage workers. Unemployment increased from 5.3 per cent to 7.0 per cent (University of Florida, Bureau of Economic and Business Research, 2000). The median annual income for households in Florida was lowered from $31 948 in 1987 to $29 281 in 1993, compared with $33 901 to $32 041 nation-wide. Florida employment and income levels suffered from the early 1990s' recession. Recovery seemed evident in 1995 when the state had 7.5 million full- and part-time jobs with $23 031 annual income per capita, compared with 6.8 million jobs and $19 107 income in 1990 (US Bureau of the Census, 1997).

The nation-wide recession of the early 1990s also hurt housing starts. Further doubts were cast regarding the impact of concurrency on real estate investments and on balancing local economic development and population growth. State and local governments understood the need for more aggressive policies to restrain sprawl and increase growth in urban areas. The DCA realized that communities varied in compliance with state standards and that they have become more accountable to their own concurrency rules. To keep concurrency as a viable standard state-wide policy and discourage local discretion, the DCA suggested formal revisions to Rule 9J-5. Such additions had to be supported by new legislation linking concurrency, sprawl and compact development (Ben-Zadok and Gale, 2001).

Compact Development, 1993–2002

In the 1990s, three compact development amendments were incorporated into the GMA. They were followed by new DCA rules written into 9J-5 and continuous DCA–local enforcement and monitoring of local rules. The first, 1993 Transportation Concurrency, contains several sections that were incorporated to Chapter 163.3180 of the GMA. Signed by Governor Lawton Chiles (Democrat), this amendment concurred with most recommendations of the Environmental Land Management Study Committee (1992), known as ELMS III. The second, 1996 Sustainable Communities, was incorporated to Chapter 163.3244–3245 of the GMA. Signed by Governor Chiles, this amendment was based on recommendations of the Governor's Commission for a Sustainable South Florida (1995). The third, 1999 Urban Infill and Redevelopment, was incorporated to Chapter 163.2511–2526 of the GMA. Signed by Governor Jeb Bush (Republican), the amendment concurred with several recommendations of the Transportation and Land Use Study Committee (1999) (see Table 3.2).[10]

---

10 The 2002 Local Government Comprehensive Planning Certification Program is a recent amendment incorporated to Chapter 163.3246 of the GMA. Replacing the 1996

Compact Growth and Economic Development: Transport and Urban–Suburban Sprawl

Compact development was a weak requirement in the 1985 GMA. The 1989 DCA definition for 'urban sprawl' was broad and allowed communities to spread out (DeGrove, 1992, pp. 15–21).[11] The DCA later introduced more flexible transport concurrency rules to accommodate developers and localities. Communities could expand if they only demonstrated intention to build roads, which resulted in road capacity deficits and more traffic congestion.[12] By the early 1990s, Florida's growth problem was the sprawl of urban and suburban developments towards environmental and natural resources and the decline of compact growth and economic activity in urban cores. In outlying suburbs, growth gradually invaded water dependent ecological and natural systems. In South Florida—for example, water quality was degraded after aggressive development sharply cut down green space and agricultural land (Ben-Zadok and Gale, 2001; deHaven-Smith, 2000; FAU/FIU Joint Center for Environmental and Urban Problems, 2000, pp. 87–91 and 104–105).

The 1990s GMA amendments aimed to enhance compact growth and economic development by focusing on transport and urban–suburban sprawl and directing growth and economic activity to urban areas (see Table 3.1). State and local governments had already learned that, because new roads can greatly affect sprawl or compact development, transport concurrency became the most crucial public facility. They also learned that road construction involves the largest scope of activities, longest time-periods and highest costs in comparison with other facilities subject to concurrency. Before 1993, state and local transport concurrency rules required little evidence that roads to serve new development would be constructed. This ensured developers' interest and continuous growth, but also increased sprawl and traffic congestion. The 1993 legislation aimed to rectify this situation (DeGrove, 1992, pp. 17–21; Florida Statutes, chapter 163.2511–3245, 1999 (3180) (5) (a); Pelham, 1992, pp. 1020–1023).

Lessons from concurrency implementation also showed that controlling growth via public facilities was only a partial solution. The threat of sprawl to environment and nature, a concern only implied in 1993, emerged as a central problem (deHaven-Smith, 2000, p. 14). Hence, a more comprehensive compact development policy was introduced in 1996 and 1999. The 1996 amendment addressed the problem of residential and commercial developments that stretch towards environmental and ecological systems and damage agriculture, wildlife, air and water. Participating communities had to adopt six principles of compact and sustainable development:

Sustainable Communities piece that 'sunset' in 2001, this amendment was influenced by recommendations of the Growth Management Study Commission (2001).

11 Urban sprawl was defined as one or more of the following developments: "(1) leapfrog development; (2) ribbon or strip development; and (3) large expanse of low-density, single-dimensional development" (Florida Department of Community Affairs, 1989).

12 Developers often sought sites in sparsely populated semi-rural areas with surpluses in road capacity. They avoided urban and suburban areas with inadequate infrastructure capacities, long delays before additional capacities could be provided and higher development costs due to impact fees (Altshuler and Gomez-Ibanez, 1993; Bollens, 1992, pp. 460–461).

limited urban sprawl, healthy and clean environment, restoration of ecosystems, protection of wildlife and natural areas, efficient use of land and other resources, and creation of quality communities and jobs (Florida Statues, chapter 163.2511–3245, 1999. (3244) (1) (3)).

The 1999 amendment addressed sprawl, a by-product of poor transport planning, as a problem that should be resolved via compact urban economic development. Urban and suburban sprawl has negative impact on the local economy and on community livability, especially because it causes heavy traffic congestion. Declining urban areas and suburbs can be revived via an urban infill and redevelopment strategy that stresses the 'health and vibrancy of the urban cores' and 'fiscally strong urban centers'. Sprawl should be contained through the co-ordination of compact land uses such as mixed use and multifunctional developments, with transport modes such as public transit and pedestrian ways. This strategy should improve economic, educational, cultural, recreational, social service and transport systems (Florida Statues, chapter 163.2511–3245, 1999 (2511) (2); Young and Elder, 1999).

State–Local Increased Discretion and Mixed Outcome

Increased discretion   The discretionary state–local process has intensified with the implementation of compact development (see Table 3.1). Concurrency mandated implementation tools in one physical planning area, public facilities. The 1996 and 1999 GMA amendments implied a more open strategic approach to community planning via a series of voluntary implementation tools in several areas. The 1993 amendment itself included highly flexible voluntary transport concurrency tools, suggesting an unprecedented level of discretion in concurrency.

It was a paradoxical learning experience. Advancing concurrency as a uniform statewide policy with less local discretion has resulted in unusually flexible, optional tools for local adoption. Facing considerable pressure from the legislature and Governor Chiles (see Deyle and Smith, 1998), the DCA followed the 1993 legislation by inserting various regulatory exemptions and exceptions to 9J-5 to ease state–local negotiation. Critics claimed that:

> the exceptions to transportation concurrency might be seen as so numerous and open-ended as to undermine the concept of transportation facilities keeping pace with growth (FAU/FIU Joint Center for Environmental and Urban Problems, 1993, p. 13).

The most drastic change in transport concurrency rules was the combined assessment of traffic volume for all roads in an area targeted for compact development, instead of strict assessment by single road-corridor. It was a method to assess LOS standards for a DCA-approved area designated in the local plan. An area-wide tool like the Long Term Transportation Concurrency Management System (LTTCMS), allowed a locality to delineate a network of roads within which the overall average of infrastructure backlog must be eliminated in 10–15 years according to local LOS standards. New projects could be approved in a LTTCMS with excess traffic volumes that were supposed to be accommodated within this period. The DCA did not provide details on how to link interim time-points for improvement of transport facilities and their respective LOS standards, to the LTTCMS program (Rule 9J-5.001–025, 1999 (0055) (4)).

The Transportation Concurrency Management Area (TCMA) allowed a community to designate a built-up area containing several alternative routes with access to various destinations. Instead of road-by-road assessment, new projects may be approved if altogether they do not increase traffic volume beyond the TCMA's average. No limitations were specified by DCA regarding the TCMA's size, boundaries or overall LOS standard (Rule 9J-5.001–025, 1999 (0055) (5)). The more flexible Transport Concurrency Exception Area (TCEA) even allowed a locality to exempt projects in urban areas targeted for redevelopment from full compliance to concurrency. Because no restrictions were indicated for size and boundaries of "specific geographical area, or areas", an exemption could be granted for an entire city merely by adoption of a formal plan (Rule 9J-5.001–025, 1999 (0055) (6)).[13]

The 1993 rules also allowed infrastructure to be in place no later than the time of issuing a certificate of occupancy "or its functional equivalent", rather than the time of issuing a building permit. Developers were thus given more time to provide facilities subjected to concurrency. They also had more flexibility in paying localities for these facilities. Through a 'pay-and-go' arrangement, developers could pay the prescribed fee whether or not facilities were in place. They were in compliance with the law even if facilities were not in place within the prescribed period after issuing a certificate of occupancy (Rule 9J-5.001–025, 1999 (0055) (3) (a–c)). These two discretionary rules that helped to ease state–local negotiation and enforcement were written based on previous implementation experience.

The 1996 Sustainable Communities was a pilot project open to voluntary applications from Florida communities. The DCA reviewed the applications and designated five as sustainable communities. The state gave grants and technical assistance to the communities for revamping their comprehensive plans to meet sustainability principles. It also gave communities maximum discretion for planning activities in a special urban area. All state and regional reviews in this area were eliminated, including those of local comprehensive plan amendments and development of regional impacts. The area was marked by an Urban Development Boundary that aimed to separate urban and rural land uses, and control sprawl and direct it to the area. Criteria for setting the boundary and strategic guidelines for the area were developed by each community with minimum state supervision over details (Florida Department of Community Affairs, 2000; Florida Senate, Committee on Comprehensive Planning, 2000).

---

13 These flexible area-wide tools for determining traffic's LOS standards were supplemented with generous incentives for redevelopment, reinvestment and public transit. For example, a permit could be given to an urban project if its traffic impact is no more than 110 per cent of the actual transport impact caused by existing development (Rule 9J-5.001–025, 1999 (0055) (3) (c)(5)). An exception from transport concurrency could be granted for development projects such as racetracks, stadiums, performing art centers and fairgrounds that were occupied on a part-time basis and located in areas of urban service, infill, redevelopment or downtown revitalization (Rule 9J-5.001–025, 1999. (0055) (7)).

Communities also had considerable discretionary power to designate and implement the 1999 Urban Infill and Redevelopment Area (UIRA). State guidelines to mark the urban revitalization area lacked technical prescriptions. The plan for the area was exempt from adoption as a comprehensive plan amendment and from DCA review. The DCA was criticized for limited regulatory capacity in handling UIRAs and was urged to develop more detailed rules and incentives for mixed land uses. The state was urged to provide substantial financial commitments and clear standards for funding revitalization of urban centers (American Planning Association, Florida Chapter, 2000, pp. 12–13). It was criticized for lack of priorities and funds, and for an unclear funding formula for urban revitalization. The DCA-managed Assistance Grant Program to help localities to plan and implement the UIRA also was criticized for lack of detailed grant requirements. The state was urged to allocate more funds for revitalizing UIRAs and distressed areas in communities (Growth Management Study Commission, 2001, pp. 31–36).

Mixed outcome　The Florida compact development policy shows mixed outcomes: serious problems of transport and sprawl as well as improvements in compact growth and economic development (the ultimate policy goals) (see Table 3.1). Again, these outcomes are affected by other state–local policies, demographic and economic forces. Although the actual impact of the policy on the following outcomes can only be guessed, they may still imply some policy impact. With respect to transport, a large increase in drivers and cars was followed by incremental addition in infrastructure. Between 1990 and 2001, the number of licensed drivers in Florida grew from 11 612 402 to 14 346 373 and the number of registered vehicles increased from 12 465 790 to 13 448 202. Between 1997 and 2000, the annual mileage traveled by vehicles grew by 13 per cent for all roads and highways and by 14 per cent for urban roads and highways. At the same time, the increase in road and highway mileage was miniscule, from 114 572 to 116 651 state-wide and from 48 321 to 49 227 in urban areas (University of Florida, Bureau of Economic and Business Research, 2000, p. 410; University of Florida, Bureau of Economic and Business Research, 2002, pp. 424, 427 and 439).

In Florida's most populated counties, traffic became heavier and slower, daily commuting time increased and the number of bus riders has not changed. From 1983 to 1997, the average daily miles traveled by vehicles went up by 96 per cent in Miami–Dade and by 177 per cent on Broward freeways; travel speed declined by 23 per cent and 18 per cent respectively. The average time spent 'stuck in traffic' during 2001 was 42 hours in Miami–Dade and 30 hours in Broward (FAU/FIU Joint Center for Environmental and Urban Problems, 2000, p. 114; Florida Connections, 2002). The metropolitan areas of both counties were in tie in 2000 for the third most dangerous area for pedestrians in the nation (Turnbell, 2000).

Massive urban and suburban sprawl led to this traffic congestion. From 1990 to 2000, Florida metropolitan areas grew by 23.5 per cent. In 2001, the state had 16.3 million residents with over one-half living in unincorporated areas, mainly outlying suburbs; one-half of the counties in the state still had fewer than 80 000 residents (The Sun-Sentinel, 2001b; University of Florida, Bureau of Economic and Business

Research, 2002, p. 14). Among the nation's top 30 sprawl-ridden metropolitan areas, 5 areas were registered in Florida (Sierra Club, 1998).

Despite the general sprawl trend, population growth led to some improvement in compact development. Density per square mile increased from 239 in 1990 to 267 in 1996 and 303 persons in 2001. Construction starts of multifamily housing units (mostly in urban areas) increased by 22 per cent, from 40 177 in 1997 to 48 924 units in 2000 (University of Florida, Bureau of Economic and Business Research, 2000, p. 353; University of Florida, Bureau of Economic and Business Research, 2002, pp. 14–15, 372). In the urban counties, density per square mile increased from 991 in Miami–Dade and 1039 in Broward in 1990, to 1176 and 1365 in 2001 (University of Florida, Bureau of Economic and Business Research, 2002, pp. 14–15).

Finally, positive outcomes are evident with respect to economic development.[14] All economic sectors including the labor market increased their activities (Florida Tax Watch, 1999). The gross state product grew by 65 per cent from 1992 to 2000. Unemployment was around 5 per cent in the mid 1990s and had reached a low of 3.9 per cent by 1999; although it bounced back to 5.5 per cent by December 2001. Annual income per capita reached $28 493 in 2001, an increase of 24 per cent since 1995. With a better economic climate, housing construction shot upward from 115 100 private unit starts in 1993 to 148 000 in 2000 and from 122 903 private residential permits in 1995 to 164 656 in 2001. The median sale price for a single-family home in Florida went up by 8.6 per cent in one year, from $117 600 in 2000 to $127 700 in 2001 (University of Florida, Bureau of Economic and Business Research, 2000, p. 232; University of Florida, Bureau of Economic and Business Research, 2002, pp. 89, 167, 371, 747–750).

Conclusions and Policy Implications

The story of Florida growth management implementation covers three steering policies within one multifaceted act. Consistency, concurrency and compact development—each policy led the GMA with unique purpose, dealt with critical issues and offered a learning experience that has changed the initiative. Records from 1985 to 2004 reveal 'three faces' of Florida growth management. This new narrative goes beyond the literature's 'one face' account where consistency often overshadowed the rest of the Act. After passing the 1985 legislation, consistency was launched as an organizing intergovernmental planning policy. It legitimized unprecedented state intervention in community development through the centralized local–regional–state review of comprehensive plans. Consistency transformed a decentralized bureaucracy accommodating growth-driven communities and brought recognition to Florida as a growth management leader among the states.

Implementation of consistency was still underway when concurrency emerged with a distinct purpose, emphasizing new issues. Concurrency led land development by mandating the construction, location and timing of public facilities, and their LOS

---

14 The outcomes also are related to the unprecedented national economic growth that fuelled Florida's economy in the mid 1990s when low-interest mortgages for construction and home loans became available.

standards. The agenda behind the GMA's second face was to control local 'growth machine' pressures and, simultaneously, to maintain economic development for the satisfaction of business, real estate and government actors. Following consistency's rational-physical planning approach, concurrency forced developers and localities to decide in advance where and when to build and how to fund the facilities to accommodate residential and commercial growth. Transport was revealed as the most difficult facility to implement and a major cause of urban–suburban sprawl. The professional lesson of concurrency was that transport and sprawl problems are undermining compact development and must be encountered with a more comprehensive strategy. Nonetheless, economic and political forces also contributed to this change in course and to new policy formation.

When the GMA was formulated, Florida lacked a manufacturing industry and relied on agriculture, tourism, retirement and real estate to secure its economic future. Developers built houses and shopping plazas; governments paid little attention to preservation of the environment and open space. By the mid 1990s, Florida's economy had diversified through service and information industries with emphasis on high technology and international trade. Businesses and developers realized the high cost of jobs and employment and the damage to green areas and open space, the results of residential sprawl and commuting. They expressed more sympathy towards national trends in sustainable and compact development (Leo et al., 1998). There was also more evidence that growth management and state environmental regulations enhance economic development and performance (Feiock and Stream, 2001; Nelson and Peterman, 2000).

Under the changing circumstances, concurrency lost steering power and compact development has gained leading momentum. With concurrency activities becoming local routines, the GMA's third face was revealed through transport planning, sustainable development and redevelopment. These compact development amendments aimed to reduce sprawl, direct growth to urban areas and protect environmental and natural resources.

The success of the 3Cs in leading the GMA and in learning from each other lies in a creative response to growth problems, via the promotion of new purposes and issues. However, the policy predicament of the 3Cs lies in the shift from prescriptive to discretionary state–local relations. In part, the change was brought through professional learning in the bureaucracies implementing consistency. But it was largely caused by economic and political pressures applied since the inception of concurrency and gradually increasing with the implementation of compact development. The new learning experience gained under these circumstances reinforced the adoption of flexible concurrency tools and voluntary compact development tools.

A strict centralized review of plans initially deterred special development interests and ensured adherence to uniform standards in communities. Although complaining, localities still complied with legislative intent and regulations. By the late 1980s, developers had become more frustrated with the state's strict approach and localities feared moratoria on development (Turner, 1990a). The fiscal implications of delivering concurrency's facilities and the economic impacts of the recession underway also forced the DCA to reduce oversight and increase local control.

The 1993 transport concurrency piece first formalized the transfer of power from state to local. The new arrangement facilitated case-by-case compromises through exemptions and exceptions in the GMA and 9J-5, the kind of deal the legislature and DCA were seeking with localities and developers (Ben-Zadok and Gale, 2001).

Under this discretionary implementation process, concurrency and compact development lost more steering power, displaying mixed outcomes by 2002. Funding and delivery of public facilities did not keep up with population growth; sprawl and traffic congestion posed major problems. There was some improvement in compact growth and a sound performance in economic development.

In sum, the state's prescription enhanced local compliance and consistency among plans; discretion encouraged convenient solutions in communities and raised doubts about the impact of concurrency and compact development. The biggest failure of discretionary style was in the absence of a school concurrency mandate. This resulted in overcrowded public schools and huge deficits in construction.[15]

Florida has to confront more directly the question of prescriptive or discretionary implementation in growth management (see Burby et al., 1998). DCA–local enforcement needs to be strengthened through prescriptive means—that is, by clear mandates and standards that are followed by more detailed incentives and explicit sanctions (see Burby and Paterson, 1993). Selected localities with proven track records of meeting state requirements could be entitled to more discretion, incentives and voluntary options. The vast majority of Florida communities should promote less discretionary and more prescriptive enforcement requirements (see May, 2005).

This approach was not taken in Chapter 163.3191 of the GMA, Evaluation and Appraisal Report (EAR) of Comprehensive Plan. In the most significant effort to reorganize the GMA and 9J-5, the 1998 legislation mandated EAR every seven years with little effort to reduce local discretion over comprehensive plans.[16] The legislation followed most of the recommendations of the 1997 technical report of the DCA-appointed EAR review committee. Flexible on the assessment of local plans, the committee suggested new statutory requirements practically formalizing many local exemptions and exceptions. It aimed to streamline and clarify 'cumbersome' local EAR processes and make them more 'proactive' by eliminating administrative rules and increasing flexibility in plan amendments and use of EAR funds (EAR Technical Committee and Department of Community Affairs, 1997; Florida Statutes, chapter 163. 2511–3245, 1999 (3191)).

EAR was a secondary administrative policy designed to aid consistency and improve past planning efforts. A major correction mechanism to evaluate the success or failure of every plan element, it lacked strong mandatory measures for changing local practices of concurrency and compact development and moving the GMA in a new direction.

---

15 South Floridians, for example, rated the quality of public schools as the lowest of all local services and have stressed 'good schools' as most important for quality-of-life (FAU/FIU Joint Center for Environmental and Urban Problems, 2000, p. 60).

16 This substantial revision in 1998 with first report due on December 2004 has followed the 1992–93 revision that mandated local EAR every six years. The 1975 Act first instructed Florida communities to prepare EAR every five years.

With anti-sprawl and compact development capturing the agenda, Florida must improve co-ordination among adjacent communities via effective regional planning. The Regional Planning Councils (RPCs) should be re-empowered to maintain state standards and keep implementation even-handed among communities.[17] The long-due regional policy shift via RPCs will be tough to implement. Florida is marked by distinct regions, diverse communities, local 'growth machines', strong interest-groups and a competitive two-party system with periodical shifts between Democratic and Republican administrations (Huckshorn, 1998; Parker, 1998). Coordination among communities also is difficult because developers and businesses exert enormous influence over local–regional growth issues and often co-opt citizens during planning processes (see Turner and Murray, 2001).

At the time of closing this evaluation, the state continues the transfer of oversight responsibilities to localities. Governor Bush's administration began to overhaul the GMA in the late 1990s by giving communities more discretionary power. The DCA attempted to gain professional feedback and public support on the strengths and weaknesses of the Act. The strategy provoked anxiety among growth management supporters, although the idea of GMA revision seemed to achieve consensus. Meanwhile, developers and environmentalists continue to argue over annual growth management bills. Still open are the most acute issues of state–local power distribution and the whole direction of compact development and regional growth (Bousquet, 2001; Cox, 2000; Fleshler, 2002; Kaczor, 2002; The Palm Beach Post, 1999; Shenot, 1999; The Sun-Sentinel, 2001a, 2001b).

After two decades of GMA implementation, the future of Florida continues to depend on maintaining the balance between economic development and environmental protection. To keep and attract more private industries, the state should provide their investors, managers and employees with economic incentives. To secure good quality-of life for all residents, the state should improve air, water, open space, roads, traffic, public transit and schools. To steer the coming growth agenda, the state has to support jobs and citizens in viable communities that work together to sustain their region.

References

Altshuler, A. and Gomez-Ibanez, J. A. (1993) Regulation for Revenue: The Political Economy of Land Use Exactions. Washington, DC: The Brookings Institute.

American Planning Association, Florida Chapter. (2000) Report to governor's growth management study commission. October. Tallahassee, FL: American Planning Association, Florida Chapter.

Anderson, J. E. (1994) Public Policymaking. Boston, MA: Houghton Mifflin.

---

17 Although regional–local consistency of plans was a major function of RPCs in the 1985 GMA, developers did not expect these bodies to become influential and have pressured the state to ignore decisions at this level. While the 1996 legislature reaffirmed this consistency requirement, it also met developers' demands by eliminating the requirement in cases of determining the impact of development proposals on other communities or on regional and state facilities (deHaven-Smith, 1998; Starnes, 1993).

Audirac, I., O'Dell, W. and Shermyen, A. (1992) Concurrency management systems in Florida. Gainesville, FL: University of Florida, Bureau of Economic and Business Research.

Ben-Zadok, E. and Gale, D. E. (2001) Innovation and reform, intentional inaction, and tactical breakdown: the implementation record of the Florida concurrency policy, Urban Affairs Review, 36(6), pp. 836–871.

Berke, P. R. and French, S. P. (1994) The influence of state planning mandates on local plan quality, Journal of Planning Education and Research, 13, pp. 237–250.

Boggs, G. H. and Apgar, R. C. (1991) Concurrency and growth management: a lawyer's primer, Journal of Land Use and Environmental Law, 7(1), pp. 1–27.

Bollens, S. A. (1992) State growth management: intergovernmental frameworks and policy objectives, Journal of the American Planning Association, 58(4), pp. 54–66.

Bollens, S. A. (1993) Integrating environmental and economic policies at the state level, in: J. M. Stein (Ed.) Growth Management: The Planning Challenge of the 1990s, pp. 143–161. Newbury Park, CA: Sage.

Bousquet, S. (2001) Independent senate rankles gov. Bush yet again, The Miami Herald, 25 March.

Burby, R. J. and Paterson, R. G. (1993) Improving compliance with state environmental regulations, Journal of Policy Analysis and Management, 12(4), pp. 753–772.

Burby, R. J., Mau, P. J. and Paterson, R. G. (1998) Improving compliance with regulations: choices and outcomes for local governments, Journal of the American Planning Association, 64(3), pp. 324–334.

Clark, T. A. (1994) The state-local regulatory nexus in US growth management: claims of property and participation in the localist resistance, Environment and Planning C, 12(4) pp. 425–447.

Cox, D. (2000) Battle building on state growth, The Sun-Sentinel, 2 March. DEGROVE, J. M. (1984) Land, Growth and Politics. Washington, DC: American Planning Association/Planners Press.

DeGrove, J.M. (1992) The New Frontier for Land Policy Planning and Growth Management in the States. Cambridge, MA: Lincoln Institute of Land Policy.

DeGrove, J. M. and Gale, D. E. (1994) Linking infrastructure to development approvals: Florida's concurrency policy under statewide growth management. Paper presented at the Annual Conference of the Association of Collegiate Schools of Planning, Phoenix, November.

DeGrove, J. M. and Turner, R. (1998) Local government: coping with massive and sustained growth, in: R. J. Huckshorn (Ed.) Government and Politics in Florida, pp. 169–192. Gainesville, FL: University Press of Florida.

Dehaven-Smith, L. (1998) The unfinished agenda in growth management and environmental protection, in: R. J. Huckshorn (Ed.) Government and Politics in Florida, pp. 233–265. Gainesville, FL: University Press of Florida.

Dehaven-Smith, L. (2000) Facing up to the political realities of growth management, Florida Planning, 7(5), pp. 1 and 14–17.

Deleon, P. (1999) The missing link revisited: contemporary implementation research, Policy Studies Review, 16(3/4), pp. 311–338.

Deutsch, K. W. (1963) The Nerves of Government: Models of Political Communication and Control. New York: Free Press.

Deyle, R. and Smith, R. A. (1998) Local government compliance with state planning mandates: the effects of state implementation in Florida, Journal of the American Planning Association, 64(4), pp. 457–469.

Dye, T. R. (1991) Public policy in Florida: education and welfare, in: R. J. HUCKSHORN (Ed.) Government and Politics in Florida, pp. 284–315. Gainesville, FL: University Press of Florida.

Ear Technical Committee and Department of Community Affairs (1997) The evaluation and appraisal report (EAR) process. December. Tallahassee, FL: Florida Department of Community Affairs.

Environmental Land Management Study Committee (1992) Final report: building successful communities. December. Tallahassee, FL: Environmental Land Management Study Committee.

Epling, J. W. (1993) New Jersey state planning process: an experiment in intergovernmental negotiations, in: J. M. STEIN (Ed.) Growth Management: The Planning Challenge of the 1990s, pp. 96–112. Newbury Park, CA: age.

FAU/FIU Joint Center for Environmental and Urban Problems (1993) Staff analysis of key provisions of the 1993 Planning and Growth Management Act. Technical report, Fort Lauderdale, FL: FAU/FIU Joint Center for Environmental and Urban Problems.

FAU/FIU Joint Center for Environmental and Urban Problems (2000) Imaging the Region: South Florida via Indicators and Public Opinions. Fort Lauderdale, FL: FAU/FIU Joint Center for Environmental and Urban Problems.

Feiock, R. C. and Stream, C. (2001) Environmental protection versus economic development: a false trade-off?, Public Administration Review, 61(3), pp. 313–319.

Fleshler, D. (2002) Voters express alarm at pace of growth, The Sun-Sentinel, 30 October. Florida Connections (2002) Editorial, 2(1), pp. 1–5.

Florida Department of Community Affairs (1989) Technical memorandum, 5(1). Tallahassee, FL: Florida Department of Community Affairs.

Florida Department of Community Affairs (1991) The evolution and requirements of the CMS rule. Technical Memorandum 6(3), pp. 1–10. Tallahassee, FL: Florida Department of Community Affairs.

Florida Department of Community Affairs (2000) Florida Sustainable Communities Demonstration Project: Annual Report. Tallahassee, FL: Florida Department of Community Affairs.

Florida Department of Community Affairs (2002) Summary of growth management legislation, Community Planning, 11(1), pp. 3–5.

Florida Senate, Committee on Comprehensive Planning (2000) Sustainable Communities Demonstration Project. Tallahassee, FL: Florida Department of Community Affairs.

Florida Statutes Chapter 163.2511–3245 (1999) Growth policy; County and municipal planning; Land development regulation. Tallahassee, FL: State of Florida.

Florida Statutes Chapter 163.3161–3243 (1987) County and municipal planning and land development regulation. Tallahassee, FL: State of Florida.

Florida Tax Watch (1999) Budget watch: state economy sets stage for 1999 legislative session. March. Tallahassee, FL: Florida Tax Watch.

Gale, D. E. (1992) Eight state-sponsored growth management programs: a comparative analysis. Journal of the American Planning Association, 58(4), pp. 425–439.

Gale, D. E., and Hart, S. (1992) Public support for local comprehensive planning under statewide growth management: insights from Maine, Journal of Planning Education and Research, 11, pp. 192–205.

Gormley, W. T. (1998) Regulatory enforcement styles, Political Research Quarterly, 51(2), pp. 363–383.

Governer's Commission for a Sustainable South Florida (1995) Initial report. October. Tallahassee, FL: Governor's Commission for a Sustainable South Florida.

Growth Management Study Commission (2001) A liveable Florida for today and tomorrow. February. Tallahassee, FL: Growth Management Study Commission.

Heclo, H. (1974) Social Policy in Britain and Sweden. New Haven, CT: Yale University Press. Howe, D. A. (1993) Growth management in Oregon, in: J. M. Stein (Ed.) Growth Management: The Planning Challenge of the 1990s, pp. 61–75. Newbury Park, CA: Sage.

Huckshorn, R. J. (1998) Political parties and campaign finance, in: R. J. Huckshorn (Ed.) Government and Politics in Florida, pp. 10–29. Gainesville, FL: University Press of Florida.

Ingram, H. (1990) Implementation: a review and suggested framework, in: N. B. Lynn and A. Wildavsky (Eds) Public Administration: The State of the Art, pp. 462–480. Chatham, NJ: Chatham House.

Innes, J. E. (1992) Group processes and the social construction of growth management: Florida, Vermont and New Jersey, Journal of the America Planning Association, 58(4), pp. 440–453.

Innes, J. E. (1993) Implementing state growth management in the United States: strategies for coordination, in: J. M. Stein (Ed.) Growth Management: The Planning Challenge of the 1990s, pp. 18–43. Newbury Park, CA: Sage.

Kaczor, B. (2002) Growth issues no state priority, The Miami Herald, 20 January.

Koenig, J. (1990) Down to the wire in Florida, Planning, 56 (October), pp. 4–11.

Leo, C., Beavis, M. A., Carver, A. and Turner, R. (1998) Is urban sprawl back on the political agenda? Local growth control, regional growth management, and politics, Urban Affairs Review, 34(2), pp. 179–212.

Liou, K. T. and Dicker, T. J. (1994) The effect of the Growth Management Act on local comprehensive planning expenditures: the South Florida experience, Public Administration Review, 54(3), pp. 239–244.

Macmanus, S. A. (1998) Financing government, in: R. J. Huckshorn (Ed.) Government and Politics in Florida, pp. 193–232. Gainesville, FL: University Press of Florida.

May, P. J. (2005) Regulation and compliance motivations: examining different approaches, Public Administration Review, 65(1), pp. 31–44.

May, P. J., Burby, R. J., Ericksen, N. J. et al. (1996) Environmental Management and Governance: Intergovernmental Approaches to Hazards and Sustainability. London: Routledge.

Mazmanian, D. A. and Kraft, M. E. (1999) The three epochs of the environmental movement, in: D. A. Mazmanian and M. E. Kraft (Eds) Toward Sustainable Communities: Transition and Transformations in Environmental Policy, pp. 3–41. Cambridge, MA: The MIT Press.

Nelson, A. C. and Duncan, J. B. (1995) Growth Management Principles and Practices. Chicago, IL: American Planning Association/Planners Press.

Nelson, A. C. and Peterman, D. R. (2000) Does growth management matter? The effect of growth management on economic performance, Journal of Planning Education and Research, 19(3), pp. 277–285.

Nicholas, J. C. (1993) Paying for growth: creative and innovative solutions, in: J. M. Stein (Ed.) Growth Management: The Planning Challenge of the 1990s, pp. 200–214. Newbury Park, CA: Sage.

The Palm Beach Post (1999) Editorial: Bush plan won't stop sprawl and strip malls, 31 October.

Parker, S. L. (1998) Interest groups and public opinion, in: R. J. Huckshorn (Ed.) Government and Politics in Florida, pp. 83–104. Gainesville, FL: University Press of Florida.

Pelham, T. G. (1992) Adequate public facilities requirements: reflections on Florida's concurrency system for managing growth, Florida State University Law Review, 19(4), pp. 974–1053.

Porter, D. R. (1998) The states: growing smarter, in: Urban Land Institute (Ed.) Smart Growth: Economy, Community, Environment, pp. 28–35. Washington, DC: Urban Land Institute.

Porter, D. R. and Watson, B. (1993) Rethinking Florida's growth management system, Urban Land, February, pp. 21–25.

Pressman, J. L. and Wildavsky, A. (1973) Implementation. Berkeley, CA: University of California Press.

Rule 9J-5.001–016 (1989) Amendment to chapter 9J-5: Florida administrative code. Tallahassee, FL: Florida Department of Community Affairs.

Rule 9J-5.001–025 (1999) Minimum criteria for review of local government comprehensive plans and plan amendments, evaluation and appraisal reports, land development regulations and determinations of compliance. Tallahassee, FL: State of Florida.

Sabatier, P.A. (1999) The need for better theories, in: P. A. SABATIER (Ed.) Theories of the Policy Process, pp. 3–17. Boulder, CO: Westview Press.

Shenot, C. (1999) Creating a design to guide Florida growth, The Orlando Sentinel, 14 March.

Sierra Club (1998) The Dark Side of the American Dream: The Costs and Consequences of Suburban Sprawl. Washington, DC: Sierra Club.

Starnes, E. M. (1993) Substate frameworks for growth management: Florida and Georgia, in: J. M. Stein (Ed.) Growth Management: The Planning Challenge of the 1990s, pp. 76–95. Newbury Park, CA: Sage.

The Sun-Sentinel (2001a) Editorial: Support bill to control growth, 9 April.

The Sun-Sentinel (2001b) Editorial: Growth brings pluses, minuses, 31 December.

Transportation and Land Use Study Committee (1999) Final report of the transportation and land use study committee. January. Tallahassee, FL: Transportation and Land Use Study Committee.

Turnbell, M. (2000) Study: In S. Florida, you walk at own peril, The Sun-Sentinel, 16 June. Turner, R. S. (1990a) New rules for the growth game: the use of rational state standards in land use policy, Journal of Urban Affairs, 12(1), pp. 35–47.

Turner, R. S. (1990b) Intergovernmental growth management: a partnership framework for state–local relations, Publius: The Journal of Federalism, 20(Summer), pp. 79–95.

Turner, R. S. (1993) Growth management decision criteria: do technical standards determine priorities?, State and Local Government Review, 25(3), pp. 186–196.

Turner, R. S. and Murray, M. (2001) Managing growth in a climate of urban diversity: South Florida's Eastward Ho! Initiative, Journal of Planning Education and Research, 20, pp. 308–328.

University of Florida, Bureau of Economic and Business Research (1997) Estimates of Population, April 1, 1996. Gainesville, FL: University of Florida, Bureau of Economic and Business Research.

University of Florida, Bureau of Economic and Business Research (2000) Florida Statistical Abstracts 2000. Gainesville, FL: University of Florida, Bureau of Economic and Business Research.

University of Florida, Bureau of Economic and Business Research (2002) Florida Statistical Abstracts 2002. Gainesville, FL: University of Florida, Bureau of Economic and Business Research.

US Bureau of the Census (1993) Statistical Abstracts of the United States. Washington, DC: US Government Printing Office.

US Bureau of the Census (1997) Statistical Abstracts of the United States. Washington, DC: US Government Printing Office.

Weaver, J. (1995) A driving force, The Sun-Sentinel, 1 June.

Weber, E. P. (1998) A wish list for 21st century environmental policy: decentralization, integration, cooperation, flexibility, and enhanced participation by citizens and local governments, Policy Studies Journal, 26(1), pp. 185–195.

White, D. G. (1999) High fees, low impact, The Miami Herald, 7 February.

Wiewel, W., Persky, J. and Sendzik, M. (1999) Private benefits and public costs: policies to address suburban sprawl, Policy Studies Journal, 27(1), pp. 96–114.

Young, V. and Elder, M. (1999) 1999 Florida legislative session, Environmental and Urban Issues, 25(2), pp. 2–3.

Chapter 4

The Fiscal Theory and Reality of Growth Management in Florida

James C. Nicholas and Timothy S. Chapin

The 1985 Growth Management Act was based upon certain expectations about the availability of funding for infrastructure and land acquisition (DeGrove, 2005; Nicholas and Steiner, 2000). The legislation was drafted on the assumption that these funds would be available and that concurrency would then be a matter of timing. New development would be timed to occur as needed infrastructure was provided and infrastructure provision was in turn timed to be in accord with the availability of funds (Ben-Zadok, 2005; Chapin, in press). At the time the Act was passed, anticipated funding included a "services" tax and a ten cent per gallon increase in motor fuels taxes. However, the failure to implement these two sources of new revenues has fundamentally undercut the basic approach of the state's growth management legislation.

Despite a lack of funding, the legislation and its implementing rules proceeded as if these funding sources remained in place. It is the disconnect tension between the "fiscal theory" and the "fiscal reality" of Florida's growth management that is the focus of this chapter. The requirement that the timing of development be linked with infrastructure provision has triggered two major reactions. The first effect is best understood as one of denial, as the state has pursued various exceptions to the concurrency mandate, exceptions that are becoming so common that, in many areas, they have become the rule (see Chapter 13 of this book by Ruth Steiner for more detail on concurrency exceptions).

The second reaction centers upon a scramble for funding by local governments to implement the state's concurrency mandate. Local option sales taxes, local option motor fuel taxes, impact fees, and special districts have been the primary tools utilized by local governments to enhance their infrastructure funding, with impact fees being the most noticed. This scramble for funding has shaped and will continue to shape the implementation of growth management in Florida. This chapter argues that because the state de-funded key revenue streams during the formative years of the state's system, it is unfair and unreasonable to lay the blame for continued environmental degradation, infrastructure shortages, and fiscal strain at the feet of the 1985 Act. The "fiscal reality" of the Florida growth management approach falls far short of the "fiscal theory" outlined by the program's framers twenty years ago.

Table 4.1   Florida's population growth, 1830–2030

	Population	Annual Rate of Growth	Persons per Year
1830	34,730		
1840	54,477	4.60%	1,975
1850	87,445	4.85%	3,297
1860	140,424	4.85%	5,298
1870	187,424	2.93%	4,700
1880	269,493	3.70%	8,207
1890	391,422	3.80%	12,193
1900	528,542	3.05%	13,712
1910	752,619	3.60%	22,408
1920	968,470	2.55%	21,585
1930	1,468,211	4.25%	49,974
1940	1,897,414	2.60%	42,920
1950	2,771,305	3.86%	87,389
1960	4,951,560	5.98%	218,026
1970	6,791,418	3.21%	183,986
1980	9,746,324	3.68%	295,491
1990	12,937,926	2.87%	319,160
2000	15,982,378	2.14%	304,445
2010	19,397,600	1.96%	341,522
2020	22,588,500	1.53%	319,090
2030	25,494,400	1.22%	290,590

Source: US Bureau of the Census

The Growth of Florida

Perhaps no other occurrence captures the essence of Florida more than the simple, inexorable population growth of the last several decades. When Florida's population was first determined in 1830, 34,730 people were counted under the restrictive rules of the time. Even as late as 1950, Florida remained a largely rural state of less than 3 million residents (see Table 4.1). However, between 1960 and 2000 the state has grown by roughly 3 million residents per decade, with year 2000 population of 15,982,378 citizens.

The lowest rate of growth yet observed in the state—in terms of percentage increases—was in the decade of the 1990s. The 2.14 percent annual rate of growth during the 1990s resulted in an additional 300,000 new persons to the population each year. Since 1970, annual population increase has run at a consistent 300,000 per year, a pace of growth expected to continue for the next 30 years. While the percentage rates of increase have abated, the numbers of people being added to Florida's population have not and are not expected to decline in the future. The 300,000 persons per year has been Florida's recent past and is continuing to be Florida's future.

The early development of Florida was encouraged by grants of free land to railroads and canal interests. Free or cheap land inducements continued well into the 20th century.

Whether growth and economic development occurred because of such inducements cannot be known. Regardless, growth and development did come to Florida, so that it now has the fourth largest population of the 50 states. Florida is projected to surpass New York to become the third largest state between 2015 and 2020.

In addition to the dramatic demographic growth of the state, Florida has arisen as an integral component of the national and international economies. At the outset of the 20th century Floridians were among the nation's poorest, with state per capita income at 55 percent of the national norm. By 2000 Florida per capita incomes had grown to 100 percent of the national. Clearly, development of the state has been economically beneficial to Florida's citizens. Today Florida is an evolving economic giant, with a gross state product approximating that of the country of Belgium.

The Florida of the early 21st century is still in a process of transition, and perhaps also a state of denial. What was an undeveloped haven for alligators, mosquitoes, and an occasional snow-bird has evolved into a mega-state, one that is a national and international center for trade and industry. It is the world's largest tourist destination, but it is also a major financial and transportation hub. The opportunities made available to Florida and Floridians resulted from technological developments that overcame the limitations of distance and climate. Rather than being a remote outpost of the nation, Florida is now in the midst of international trade. Air conditioning overcame the climatic disadvantages of the subtropical summers and now the state is hospitable both for residents and guests year-round.

It is important to note that this evolution occurred largely by fortuitous happenstance. The state had nothing to do with the development of the long-range jet aircraft or economical air conditioning, both so important to present prosperity. The major investments in transportation infrastructure—most notably the interstate highway system—were made by the federal government, largely for purposes of national defense. However, all indications are that the federal largess that was so important to Florida's rise has ended. The state cannot expect Washington or simply fate to provide further economic development. Now more than ever before, what Florida will become depends upon what Floridians do.

The Nature of Florida's Growth

Much of Florida's early development involved consumption of land and natural resources. However, development of Florida has become so extensive that water shortages are becoming annual events. Extreme measures have also been necessary to save native fauna. Coastal dredging and filling has progressed to a point where the coastal ecosystem of much of the state has been fundamentally altered. The interior lands have been subdivided and sold around the world, frequently by "swamp peddlers," and the term "Florida water front property" has taken on a whole new meaning (see Bernard, 1983).

The late 1960s saw the beginning of a new era for Florida. The state began to move away from the consumption of natural resources to the use (and reuse) of those resources. During this period the Legislature passed and the Governor signed to law a number of planning and environmental initiatives, including:

- the Florida Air and Water Pollution Control Act (1967);
- the County and Municipal Planning for Future Development Act (1969);
- the Coastal Control Act (1971);
- the Environmental Land and Water Management Act (1972);
- the Water Resources Act (1972);
- the State Comprehensive Planning Act (1972); and
- the Land Conservation Act (1972).

Collectively this legislation began to reverse more than one hundred years of neglect and frequent scorn of the natural environment. It represented the initial steps by the state to manage growth by directing where development could and could not occur and by requiring that the services that will be needed by these developments must be provided by someone.

The natural environment of the state had been stressed to a point that further development was impossible without significant alterations in the mode of land development. In 1970 development in many parts of Dade County was stopped because Biscayne Bay simply could not accept any more effluent (Healy and Rosenberg, 1979; Reilly, 1972). Dade County responded by establishing a Water and Sewer Authority that developed sewage collection, treatment, and disposal systems. This use of technology reduced the flow of effluent into Biscayne Bay, allowing some environmental regeneration, and allowed development to proceed—but at a higher cost than before. Other areas of the state had similar experiences.

The above statutes, together with the Local Government Comprehensive Planning Act (1975), sought to establish an integrated system of local, regional, and state planning that would guide the future development of the state. This system, to become known as "growth management," involved identifying those areas of the state where development should be limited or not allowed at all, and requiring the installation of the capital facilities to accommodate additional development.

Florida's Fiscal Reality

Florida is one of seven states that does not have a personal income tax (Florida TaxWatch, 2006). The result is that Florida is much more dependent on sales taxes than most states. When considered together, states receive roughly 14 percent of recurring receipts from general sales taxes. However, Florida receives 47.6 percent of its self generated revenues and 33 percent of all revenues from sales and use taxes (Florida Department of Revenue, 2005). The problem with Florida's dependency on sales taxes is that retail sales are a volatile source of revenue, with surpluses during boom times and deficits during economic respites (see Table 4.2).

Florida exempts services from sales taxation, thus excluding the most rapidly growing sector of the economy from its revenue base. This means that Florida's revenue system, which is based upon proceeds from taxes on retail sales, is doomed to lag behind the overall growth of the state. Florida's sales tax collections have been growing at 6.3 percent since 2000, roughly equal to the rate of growth of the Florida economy, which has been 6.6 percent during the same period. Further, the

Table 4.2   Annual rate of growth in sales tax collections, state of Florida

Year	% Change From Previous
1999–2000	8.26%
2000–2001	5.77%
2001–2002	1.58%
2002–2003	2.03%
2003–2004	8.81%
2004–2005	9.07%
2005–2006	5.53%

Source: Florida Department of Revenue

state exempts a large portion of retail sales from taxation, as only roughly 40 percent of retail sales in Florida are subject to taxation. The result is that Florida will tend to have periods of fiscal surplus followed by deficits, as sales tax collections go through their cyclical gyrations. This situation is due to the basic structure of the Florida revenue system.

Floridians have amended their constitution to further restrict taxation. In 1992 voters passed a "save our homes" amendment to the State Constitution that capped annual increases in property tax assessments for homesteaded (owner occupied) properties at three percent or the rate of increase in the Consumer Price Index, whichever is lower (Gatzlaff and Smith, 2003). In 1994 the voters approved an amendment that limited total state revenue increases to that of the increase in personal income. The state also caps millage rates for local governments at 10 mills.

These realities of Florida fiscal affairs have led many to conclude that the state needs additional sources of revenue in order to cope with the demands of development. This was the premise behind the funding recommendations that followed the passage of the 1985 Growth Management Act. Yet the facts show that the productivity of Florida's revenue system has not been the basis for funding the needs of growth. The Legislature has a history of cutting state taxes even in times of great need, especially in the areas of transportation and public education. The only possible interpretation of these actions is that the Legislature holds the opinion that transportation and public education funding are largely local issues. But local jurisdictions find themselves in the same basic situation as the state, with constitutional and statutory limitations on raising revenues.

In sum, then, the Legislature has never offered an explanation for its decisions on funding growth needs nor has a clear policy been established. However, with the passage of 20 years, one thing has become obvious: Local jurisdictions will continue to finance the substantial portion of local growth. Perhaps this is just the Legislature avoiding unpopular tax increases or seizing opportunities to enact popular tax cuts.

The Growth Management Act of 1985

In 1985 the Legislature determined that growth in Florida was not being appropriately managed by the legislation of the previous two decades. Like Oregon, Florida had enacted what was basically a voluntary system of local government planning. There was a clear desire in Florida to have planning and development regulation be primarily a local function. The state's 1975 Local Government Comprehensive Planning Act required all local governments to develop, adopt, and implement "comprehensive plans," but there were no quality standards for these plans and the state did not review them. Additionally, "implementation" simply meant following the plan; and if the plan had no substance, as was characteristic of many local plans, then implementation was effectively meaningless.

Ten years previously Oregon had gone through exactly the same experience. In both states, those local governments that wanted to engage in the regulation of development did so, while those that did not want to regulate development did not. The premise of the initial legislation in each state was that all the legislation had to do was identify the problem—a lack of planning—and then require local governments to solve that problem by mandating planning (Callies, Freilich, and Roberts, 2004). Results have revealed the naiveté of that view. In response, Oregon passed its Land Use Law and its State Comprehensive Plan to rectify the situation. This new legislation continued the requirement for local planning but added the requirement that those plans be consistent with and implement goals laid out in a state plan. Moreover, the development regulatory authority of local governments was suspended until their plans were "acknowledged" as being in compliance with state planning requirements.

Florida undertook a similar approach in establishing its growth management approach in the mid-1980s. The Legislature adopted a State Comprehensive Plan as Chapter 187, Florida Statutes. This document was intended to provide the same direction as Oregon's State Plan. The State Plan, together with the provisions in other statutes provided by the Omnibus Growth Management Act, continued the practice of identifying those areas where development was to be avoided or restricted and imposing requirements for adequate provision of capital facilities. Rule 9J-5 was developed to establish the criteria by which a local plan could be evaluated for compliance with the provisions of the State Comprehensive Plan.

Under the Florida model, local plans were required to be "consistent" with the state plan. Additionally, the concurrency mandate was established. Together, "consistency" and "concurrency" were to be the means of implementing these statutory mandates (Ben-Zadok, 2005). While consistency was expected to be controversial, concurrency was expected to be expensive. The State Comprehensive Plan Committee (1987) estimated the ten-year cost of implementing the provisions of the State Plan at $52.9 billion, with the largest portion of this cost for roads. Note might be taken of the fact that schools were not considered by the Committee and, if they had been included, this number would have been substantially higher.

While the final cost of implementation was not available until a year after enactment, it was well known that providing adequate capital facilities was going to be very expensive for the state, which had been assigned primary responsibility for providing or funding needed improvements. In fact, the authors of the legislation argued that if funding was not to be provided, that the legislation should not be passed. The Legislature itself conditioned the state comprehensive plan on funding:

Table 4.3   Growth management costs and recommended revenues (in thousands)

	Annual Total	Ten-Year Totals
State Cost		$35,000.0
Local Cost		$17,900.0
Total Cost		$52,900.0
State Revenue		
"Services" Tax	$889.2	$15,750.0
Increased Motor Fuel Tax	$480.0	$4,800.0
Other	$1,445.0	$14,450.0
Local Revenue		
Local Share of Services Tax	$175.0	$1,750.0
Additional State Motor Fuel	$120.0	$1,200.0
Other	$1,495.0	$14,950.0
Total Revenues		$52,900.0
Identified		$23,500.0
Unidentified		$29,400.0

Source: State Comprehensive Plan Committee, 1987, pp. 2, 27 and 33

The State Comprehensive Plan is intended to be a direction-setting document. *Its policies may be implemented only to the extent that financial resources are provided pursuant to legislative appropriation or grants or appropriation of any other public or private entities* [italics added] (State Comprehensive Plan, Fla. Stat. § 187.101(2), 1985).

There was an acceptance of the proposition that the reason infrastructure investments were lagging was that the revenue system was inelastic with respect to growth (and inflation). This acceptance led to the inexorable conclusion that additional revenues were going to be needed in order to achieve that level of funding deemed adequate.

The State Comprehensive Plan Committee

The 1985 Growth Management Act called for a committee to study the issue of growth and its provision with adequate infrastructure. This was the State Comprehensive Plan Committee, commonly known as the "Zwick Committee" after Chairman Paul Zwick. The far ranging inquiry of this Committee addressed not only the revenue needs of funding Florida's dynamic growth, but also the reason why it was necessary to address revenue insufficiencies. The Committee argued that it was good business to provide adequate streets and roads, potable water, parks, and police and fire protection. Committee members continuously referred to sensible business practices in their study of Florida's future needs, laying out economic reasons as a primary justification for heavy state involvement in the funding of infrastructure.

The Committee also estimated the additional costs of providing adequate infrastructure, shown in Table 4.3. The total bill for existing infrastructure and near-term infrastructure needs came in at almost $53 billion, although these figures did not include any costs for public education. The table also illustrates that the original "fiscal theory" envisioned a roughly two-thirds state government to one-third local government split of revenues to fund these improvements. This Committee therefore saw the primary responsibility for infrastructure funding to reside with the state.

To pay the state's share of these costs, the Committee made several recommendations:

1. Extend the sales tax to services (the infamous "Services Tax"), with 90 percent of the proceeds directed to a state infrastructure trust fund and 10 percent allocated to local governments for infrastructure costs;
2. Increase motor fuels taxes by 10 cents per gallon, with 2 of the additional cents directed to local governments; and
3. Remove a number of restrictions on local governments' ability to raise revenues.

The revenue recommendations were expected to raise about one-half of the anticipated ten-year cost. The remainder of state funding would come from "growth," understood to be increases in revenues from a larger base, inflation, or other cause not resulting from a legislative enactment, or simply from some undetermined other source of funding. Again, in the fiscal theory as originally espoused by the state, funding was viewed primarily as a state responsibility.

In 1987 the Legislature passed and the Governor signed into law the extension of the general sales tax to selected services. This enactment was seen as a major fulfillment of the commitments made by the Growth Management Act. No action was taken at that time on the increase in motor fuel taxes. The Services Tax applied the general sales tax to a number of transactions previously exempt from this tax, including legal and accounting services, dry cleaning, and, perhaps most significantly, newspaper advertising. This last item was seen by many in the press as a tax on the First Amendment and, consequently, the state's leading newspapers led a stringent assault on the tax, calling for its total repeal.

Given the hammering in the press over the tax, as well as a more general public backlash, Governor Martinez called a special session in 1987 and asked for the repeal of the tax on services. The Legislature responded and repealed the tax. Of the two principal state funding mechanisms to support growth management's implementation, the Services Tax and the increase in motor fuel taxes, the first lasted less than six months and the second was not (and has never been) fully implemented by the state.

Florida was the first state to adopt an extensive services tax, which was being considered by a number of other states at the time. Florida's action attracted unwanted national attention and even fueled a backlash that contributed to the tax's demise. For example, the Association of National Advertisers canceled its 1989 annual meeting in the state when Florida enacted the service tax. It was estimated that Florida lost some $35 million in convention business during the short life of this tax (Szabo, 1988). Since the repeal of Florida's Services Tax, no state has attempted its enactment. The Florida experience has been cited as evidence of the impracticality of this means of revenue raising (Levine, 2003).

The repeal of the Services Tax brought an end to broad based and state funding of infrastructure. Since then, the burden has been shifted largely to local and increasingly narrow sources of revenue. In contrast to increasing tax revenues to the state, the Legislature has acted on several occasions to remove various limitations on the ability of local government to raise revenues. The state has authorized local governments to impose additional optional motor fuels taxes. Additionally, legislation has authorized counties to increase sales taxes by an additional one percent for infrastructure, known as the "infrastructure surtax." The state has also enabled and gone so far as to encourage local governments to impose impact fees and special assessments to generate revenue to support growth.

Florida's "New Fiscal Reality"

At the time of passage of the 1985 legislation, the state promised a "new fiscal reality," one in which the state was to be the primary agency for raising revenues to fund needed public capital improvements. This was going to be done by extending the sales tax to the highest growth sector of Florida's economy—services. These revenues would be growth elastic, that is, keep up with the growth of the state and its industries. In addition, increased state motor fuels taxes and revenues from other sources would help to pay for the state's two-thirds share of this estimated $53 billion bill. Had this fiscal theory been fulfilled, there would indeed have been a new fiscal reality in Florida.

However, as discussed earlier the new fiscal reality initially outlined has never come to pass. The funding role for the state remains largely as it was before the landmark 1985 legislation. While enabling and encouraging a variety of new revenue streams for local governments, the Legislature has remained committed to a low impact system of taxation. This system ranks Florida among the bottom third of the fifty states (35th in overall tax burden and 44th in taxes as a percent of personal income, according to Florida TaxWatch, 2006), despite population levels and growth rates that place Florida among the nation's leaders. As a consequence, local governments were and remain the primarily agent for infrastructure funding.

What does the actual new fiscal reality look like? The reality for local governments in Florida is that they have needed to become entrepreneurial agents, constantly on the lookout to generate new revenues. Among the entrepreneurial activities, each sanctioned and encouraged by the state, have been local option taxes, special districts, and impact fees (summarized in Table 4.4).

Local government infrastructure financing has been obtained largely through:

- Extensive use of optional motor fuels taxes
  - All 67 counties impose the an optional motor fuels tax
  - 17 counties impose the second local option motor fuels tax;
- Enactments of local option sales taxes
  - 38 counties (57 percent) have enacted the Infrastructure Surtax, although 16 have since repealed the tax or let it expire
  - 23 small counties are using the Small County Surtax as general revenue

Table 4.4   Summary of county taxing systems, 2005

County	Pop2000	Sales Surtax 2005 (1)	Local Gvt Infra Surtax (2)	Small County Tax (2)	Local Option Gas Tax (3)	Add'l Loc Opt Gas Tax (3)	9th Cent Gas Tax (3)	Impact Fees (4)	Impact Fee Adoption Year (4)	Number of CDDs 2005
Alachua	217,955	$0.003	$0.000	N/A	$0.029	$0.00	$0.01	Y	1992	0
Baker	22,259	$0.010	$0.000	$0.01	$0.029	$0.00	$0.01	N	NA	0
Bay	148,217	$0.005	$0.000	N/A	$0.019	$0.00	$0.00	N	NA	2
Bradford	26,088	$0.010	$0.000	$0.01	$0.019	$0.00	$0.00	N	NA	0
Brevard	476,230	$0.000	$0.000	N/A	$0.019	$0.00	$0.00	Y	1989	4
Broward	1,623,018	$0.000	$0.000	N/A	$0.079	$0.05	$0.01	Y	1977	15
Calhoun	13,017	$0.010	$0.000	$0.01	$0.019	$0.00	$0.00	N	NA	0
Charlotte	141,627	$0.010	$0.010	N/A	$0.069	$0.05	$0.00	Y	1986	4
Citrus	118,085	$0.000	$0.000	N/A	$0.019	$0.00	$0.00	Y	1987	1
Clay	140,814	$0.010	$0.010	N/A	$0.029	$0.00	$0.01	N	NA	6
Collier	251,377	$0.000	$0.000	N/A	$0.079	$0.05	$0.01	Y	1985	16
Columbia	56,513	$0.010	$0.000	$0.01	$0.079	$0.05	$0.01	N	NA	0
DeSoto	32,209	$0.010	$0.000	$0.01	$0.079	$0.05	$0.01	N	NA	0
Dixie	13,827	$0.010	$0.000	$0.01	$0.019	$0.00	$0.00	Y	1986	0
Duval	778,879	$0.010	$0.005	N/A	$0.019	$0.00	$0.00	N	NA	6
Escambia	294,410	$0.015	$0.010	N/A	$0.029	$0.00	$0.01	N	NA	0
Flagler	49,832	$0.010	$0.005	$0.00	$0.029	$0.00	$0.01	Y	1990	4
Franklin	9,829	$0.000	$0.000	$0.00	$0.000	$0.00	$0.00	N	NA	0
Gadsden	45,087	$0.010	$0.000	$0.01	$0.019	$0.00	$0.00	N	NA	0
Gilchrist	14,437	$0.010	$0.000	$0.01	$0.029	$0.00	$0.01	Y	1999	0
Glades	10,576	$0.010	$0.010	$0.00	$0.029	$0.00	$0.01	N	NA	0
Gulf	14,560	$0.005	$0.000	$0.00	$0.019	$0.00	$0.00	N	NA	0
Hamilton	13,327	$0.010	$0.010	$0.00	$0.019	$0.00	$0.00	N	NA	0
Hardee	26,938	$0.010	$0.000	$0.01	$0.029	$0.00	$0.01	N	NA	0
Hendry	36,210	$0.010	$0.000	$0.01	$0.049	$0.02	$0.01	N	NA	0

County	Pop2000	Sales Surtax 2005 (1)	Local Gvt Infra Surtax (2)	Small County Tax (2)	Local Option Gas Tax (3)	Add'l Loc Opt Gas Tax (3)	9th Cent Gas Tax (3)	Impact Fees (4)	Impact Fee Adoption Year (4)	Number of CDDs 2005
Hernando	130,802	$0.005	$0.000	N/A	$0.049	$0.02	$0.01	Y	1987	5
Highlands	87,366	$0.010	$0.010	N/A	$0.079	$0.05	$0.01	N	NA	0
Hillsborough	998,948	$0.010	$0.005	N/A	$0.029	$0.00	$0.01	Y	1985	35
Holmes	18,564	$0.010	$0.000	$0.01	$0.019	$0.00	$0.00	Y	1985	0
Indian River	112,947	$0.010	$0.010	N/A	$0.019	$0.00	$0.00	Y	1985	0
Jackson	46,755	$0.015	$0.000	$0.01	$0.029	$0.00	$0.01	N	NA	0
Jefferson	12,902	$0.010	$0.000	$0.01	$0.029	$0.00	$0.01	N	NA	0
Lafayette	7,022	$0.010	$0.010	$0.00	$0.019	$0.00	$0.00	Y	1986	0
Lake	210,528	$0.010	$0.010	N/A	$0.029	$0.00	$0.01	Y	1985	7
Lee	440,888	$0.000	$0.000	N/A	$0.079	$0.05	$0.01	Y	1985	27
Leon	239,452	$0.015	$0.010	N/A	$0.029	$0.00	$0.01	Y	1989	3
Levy	34,450	$0.010	$0.010	$0.01	$0.019	$0.00	$0.00	N	NA	0
Liberty	7,021	$0.010	$0.010	$0.01	$0.029	$0.00	$0.01	N	NA	0
Madison	18,733	$0.010	$0.010	$0.01	$0.019	$0.00	$0.00	N	NA	0
Manatee	264,002	$0.005	$0.000	N/A	$0.029	$0.00	$0.01	Y	1986	18
Marion	258,916	$0.005	$0.000	N/A	$0.029	$0.00	$0.01	Y	1990	5
Martin	126,731	$0.000	$0.000	N/A	$0.079	$0.05	$0.01	Y	1987	0
Miami-Dade	2,253,362	$0.010	$0.000	N/A	$0.059	$0.03	$0.01	Y	1989	31
Monroe	79,589	$0.015	$0.000	N/A	$0.019	$0.00	$0.00	Y	1986	0
Nassau	57,663	$0.010	$0.010	$0.01	$0.029	$0.00	$0.01	Y	1987	2
Okaloosa	170,498	$0.000	$0.000	N/A	$0.029	$0.00	$0.01	N	NA	0
Okeechobee	35,910	$0.010	$0.000	$0.01	$0.029	$0.00	$0.01	N	NA	0
Orange	896,344	$0.005	$0.000	N/A	$0.019	$0.00	$0.00	Y	1983	10
Osceola	172,493	$0.010	$0.010	N/A	$0.029	$0.00	$0.01	Y	1989	16
Palm Beach	1,131,184	$0.005	$0.000	N/A	$0.079	$0.05	$0.01	Y	1979	19
Pasco	344,765	$0.010	$0.010	N/A	$0.029	$0.00	$0.01	Y	1986	29
Pinellas	921,482	$0.010	$0.010	N/A	$0.019	$0.00	$0.00	Y	1987	1

County	Pop2000	Sales Surtax 2005 (1)	Local Gvt Infra Surtax (2)	Small County Tax (2)	Local Option Gas Tax (3)	Add'l Loc Opt Gas Tax (3)	9th Cent Gas Tax (3)	Impact Fees (4)	Impact Fee Adoption Year (4)	Number of CDDs 2005
Polk	483,924	$0.010	$0.000	N/A	$0.079	$0.05	$0.01	Y	1990	8
Putnam	70,423	$0.010	$0.010	N/A	$0.019	$0.00	$0.00	N	NA	0
Santa Rosa	117,743	$0.005	$0.000	N/A	$0.019	$0.00	$0.00	N	NA	0
Sarasota	325,957	$0.010	$0.010	N/A	$0.079	$0.05	$0.01	Y	1989	4
Seminole	365,196	$0.010	$0.010	N/A	$0.029	$0.00	$0.00	Y	1987	1
St. Johns	123,135	$0.000	$0.000	N/A	$0.019	$0.00	$0.01	Y	1988	14
St. Lucie	192,695	$0.005	$0.000	N/A	$0.079	$0.05	$0.01	Y	1986	14
Sumter	53,345	$0.010	$0.000	$0.01	$0.029	$0.00	$0.01	N	NA	10
Suwannee	34,844	$0.010	$0.000	$0.01	$0.079	$0.05	$0.01	N	NA	0
Taylor	19,256	$0.010	$0.010	$0.00	$0.019	$0.00	$0.00	N	NA	0
Union	13,442	$0.010	$0.000	$0.01	$0.019	$0.00	$0.01	N	NA	0
Volusia	443,343	$0.005	$0.000	N/A	$0.079	$0.05	$0.01	Y	1986	1
Wakulla	22,863	$0.010	$0.010	$0.00	$0.029	$0.00	$0.01	Y	1989	0
Walton	40,601	$0.010	$0.000	$0.01	$0.029	$0.00	$0.01	Y	1993	4
Washington	20,973	$0.010	$0.000	$0.01	$0.029	$0.00	$0.01	N	NA	0

(1) Source: Florida Department of Revenue web page (http://www.myflorida.com/dor/forms/2005/dr15dss.pdf)

(2)Source: Florida Legislative Committee on Intergovernmental Relations, 2005

(3) Source: Florida Department of Revenue web page (http://www.myflorida.com/dor/pdf/04b05-06.pdf)

(4) Source: Jeong (2006)

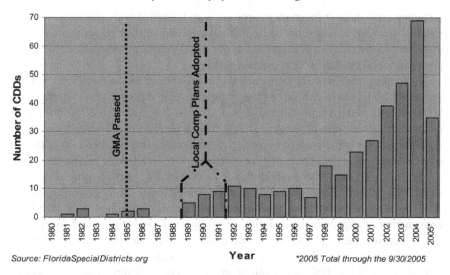

Source: FloridaSpecialDistricts.org   **Year**   *2005 Total through the 9/30/2005

Figure 4.1   Number of Florida community development districts by year, 1980–2005

- ° 16 counties use the School Capita Surtax
- ° 54 counties (81 percent) have enacted Tourist Development Taxes, continuing a long-time practice of exporting the tax burden to the maximum extent possible;
- • Extensive use of impact fees, hookup fees, and connection charges (see Chapter 17 by Burge and Ihlanfeldt for more detail on the use of impact fees); and
- • An increasing reliance upon special taxing districts, especially Community Development Districts, with broad powers to finance and construct a wide array of infrastructure improvements. These districts have become especially popular in the state in the last ten years (see Figure 4.1).

SB360 and the "New" New Fiscal Reality

In the 2005 session, the Florida Legislature passed Senate Bill 360 (SB360), by far the most important and far-reaching legislation relating to growth management in Florida in a decade. Among many changes, SB360 mandates school concurrency and greater regional planning for potable water. The legislation also provides regulatory incentives for good planning practices (including waivers of state review of plan amendments when certain anti-sprawl measures are in place), which serve to reinforce the comprehensive planning model.

Of direct consequence for this chapter are new requirements regarding the fiscal dimensions of local comprehensive plans. In an effort to better link comprehensive plans to the local capital budgeting process, the state's Department of Community Affairs is now tasked with reviewing local comprehensive plans, comprehensive plan amendments, *and* annual updates to the capital improvement elements (CIEs) of these plans. Further, each jurisdiction's CIE must be "financially feasible," including only those infrastructure projects with identified and committed funding sources. If a CIE is found not to be financially feasible, then comprehensive plan map amendments cannot proceed and major development proposals are effectively placed in limbo until the CIE is deemed feasible.

To support these new requirements, the state did offer two carrots. First, SB360 provides $1.5 billion in new infrastructure funding, aimed primarily at transportation infrastructure, water supply planning, and schools, although only $750 million of this total takes the form of recurring funds. This represents the first significant, new state funding for infrastructure in many years—funding that resulted in part from a robust state economy and larger than expected tourism-related tax revenues.

Second, SB360 also establishes "proportionate share" provisions for transportation and school facilities. Under "prop-share," as it is known in the state, developers can mitigate their infrastructure impacts through land donations, construction or expansion of facilities, or payment of funds to be used for road or school improvements. Prop-share is intended to facilitate development by allowing developers to pay for "their" road and school infrastructure costs. If a developer has met these obligations (paid their share of the costs for new roads or schools), and needed capital improvements are scheduled in the CIE, then a project is to be allowed to proceed even if these improvements are years from being completed. SB360 required all local governments to establish proportionate share procedures.

When SB360 is viewed in its entirety, this legislation only continues the trend of passing the costs of growth on to local governments and developers. Although the state has committed to some new infrastructure funding, this amount must be balanced by the new requirement for school concurrency. The $750 million a year in recurring infrastructure funding almost certainly does not cover the increased costs to local governments once school concurrency is implemented. Also, the proportionate share language represents an explicit commitment by the state to a policy of placing the costs of growth on new development. While this may be a sound policy and reflect a basic tenet of good planning ("growth pays for itself"), the proportionate share language reflects the existing state approach of state-mandated, but locally-funded growth management.

Conclusion

While it is necessary to evaluate the influence and effectiveness of the state of Florida's landmark 1985 Growth Management Act, any such evaluation must be grounded in the fiscal context of the implementation of this legislation. As detailed in this chapter, the state of Florida promised a "new fiscal reality," one in which the state would take the lead in raising the funds necessary to provide infrastructure to support ongoing growth in the state. The fiscal theory underpinning the 1985 legislation rested largely upon new and increased state taxes. However, the fiscal reality of the Florida growth management model has proven to be far different than the one originally envisioned.

Table 4.5 compares the fiscal theory and the fiscal reality of the Florida growth management model. While the original intent was that the state would fund 66

Table 4.5   **Comparing the fiscal theory and the fiscal reality of Florida's growth management system (in thousands)**

	Theory	Reality
State Expenditures	$35,000	$24,000
Percent	~66%	~45%
Local Gvt Expenditures	$17,900	$28,900
Percent	~34%	~55%
Total Cost	$52,900	$52,900
State Revenue Sources		
"Services" Tax	$15,750	$0
Increased Motor Fuel Tax	$4,800	$2,400
Other	$14,450	$0
Local Revenue Sources		
Local Share of Services Tax	$1,750	$0
Additional State Motor Fuel	$1,200	$0
Local Motor Fuel Tax		$3,500
Infrastructure Surtax		$10,000
Impact Fees		$7,000
Unidentified Revenues	$14,950	$30,000
Total Revenues	$52,900	$52,900

percent of all infrastructure costs (excluding public education), the reality has been more like 45 percent. The result of the repeal of the Services Tax has been to effectively shift these costs to local government. Local governments, in turn, have shifted about one-half of that burden to private development, in the form of impact fees and new special districts that employ special assessments and service charges as their revenue sources. The proportionate share provisions of the 2005 legislation will likely continue this trend.

Table 4.5 also illustrates that the funding gap that existed in 1985, denoted by the unidentified sources category, has only widened since. Over time, then, the state and local governments have been looking to fill an ever-increasing hole in their infrastructure budgets. This gap has been filled to some degree by special districts, which produce several billion dollars a year in revenues, but these revenues have proven to be insufficient to address shortages in key infrastructure areas, such as transportation, potable water, and wastewater treatment. As noted earlier, in 2005 the state also committed to $750 million per year in new infrastructure funding. While these new funds have been generally well-received by local governments, they are insufficient to address the deficit in state infrastructure funding.

The purpose of this chapter has been to illustrate that the "fiscal theory" behind Florida's growth management approach has not been reflected by "fiscal reality." Florida's approach was designed to be a state-mandated, state-funded model, with infrastructure concurrency sitting at the heart of this model. Under the original model, concurrency was to be matched by substantial new state infrastructure funding, and comprehensive plans were to be implemented through a state-local government funding partnership. The fiscal reality, however, has been state-mandated concurrency, but with significantly diminished state funding to implement this policy mandate (Chapin, in press). The Florida growth management model has been saddled since the very beginning with a fiscal model that has hampered implementation, promoted urban sprawl, and created an entrepreneurial environment in which local governments pursue revenue streams that are politically feasible and acceptable to existing residents.

References

Ben-Zadok, E. (2005). Consistency, concurrency, and compact development: Three faces of growth management implementation in Florida. *Urban Studies, 42*(12), 2167–2190.

Bernard, E. (1983). *Lies that came true: Tall tales and hard sales in Cape Coral, Florida.* Ocoee, FL: Anna Publishers.

Callies, D. L., Freilich, R. H., and Roberts, T. E. (2004). *Cases and materials on land use.* St. Paul, MN: West Group.

Chapin, T. (in press). Local government as policy entrepreneurs: Evaluating Florida's "concurrency experiment". *Urban Affairs Review.*

DeGrove, J. (2005). *Planning and politics: Smart growth and the states.* Cambridge, MA: Lincoln Institute of Land Policy.

Florida Department of Revenue. (2005). *Florida tax handbook.* Tallahassee, FL: Author.

Florida TaxWatch. (2006). *How Florida compares: State and local taxes in Florida and the nation.* Tallahassee, FL: Author.

Gatzlaff, D., and Smith, M. (2003). *Florida's "Save Our Homes" Amendment and property tax incidence.* Unpublished manuscript.

Healy, R., and Rosenberg, J. (1979). *Land use and the states.* Baltimore: Johns Hopkins University Press.

Jeong, M. (2006). Local choices for development impact fees. *Urban Affairs Review, 41*(3), 338–357.

Levine, R. A. (2003, January 30). Commentary: Sales tax on services? Not so fast. *Orange County Register.* Retrieved from http://www.rand.org/commentary/013003OCR.html.

Local Government Comprehensive Planning Act of 1975, Fla. Stat. §§ 163.3161–3211 (1975).

Nicholas, J., and Steiner, R. (2000). Growth management and smart growth in Florida. *Wake Forest Law Review 35*, 645–670.

Reilly, W. (1972). *The use of land,* New York: Crowell.

State Comprehensive Plan, Fla. Stat. §§ 187.101–201 (1985).

State Comprehensive Plan Committee. (1987). *The keys to Florida's future: Winning in a competitive world. Final report of the State Comprehensive Plan Committee.* Tallahassee, FL: Author.

Szabo, J. C. (1988). Setback for service tax—Florida repeals service tax. *Nation's Business, 76*(2), 12.

Chapter 5

Attitudes Towards Growth Management in Florida: Comparing Resident Support in 1985 and 2001[1]

Timothy S. Chapin and Charles E. Connerly

Researchers have long been interested in resident attitudes towards growth management (see, for example, Bollens, 1990; Connerly, 1986; Connerly and Frank, 1986; Dowall, 1980; Knaap, 1987). While this research has yielded sometimes conflicting results (Deakin, 1989; Gale and Hart, 1992), there is evidence that certain factors (e.g., higher socioeconomic status, interest in environmental issues, and perceived need for growth controls) help to predict support for growth management initiatives. This research is important in that identifying the factors that predict support for growth management can help planners galvanize support for these initiatives amongst pro-growth management groups.

While analyses of resident attitudes towards growth management are prevalent, to date no study has investigated whether resident attitudes have changed over time and, if so, in what ways. This study begins to fill this gap by investigating Floridians' attitudes towards growth management at two key points in time: (1) in 1985, as Florida's groundbreaking growth management legislation was being passed by the legislature, and (2) in 2001, when substantial revisions to the state's growth management approach were being considered. This article therefore provides some insight into the ebb and flow of resident support for growth controls over time. Florida is also a useful state to study because its long-established, state-level approach to growth management has served as a model to other states pursing similar efforts.

To preview the findings of this study, analysis indicates that in 2001 Florida residents continued to support growth management, albeit at a slightly lower level than in 1985. Further, the broad-based support for growth management that existed in 1985 had diminished and, by 2001, support varied much more markedly across the population. While overall support for controlling growth has remained high, support for government intervention in growth management has diminished across almost all population subgroups. Florida's experience provides insights into how resident support for growth management can change as systems age, flaws are recognized, revisions are made, and places continue to grow despite state-mandated

1   This chapter was originally published in Volume 70, No. 4 (2004) of the *Journal of the American Planning Association*, pp. 443–452. Reprinted with permission.

growth controls. The experience of Florida points to the need for planners to monitor support for these complex and long-term-oriented policy interventions.

The article opens with a brief discussion of Florida's growth management context. The next section summarizes the data and methodology used for this study. The results of our analyses are then presented. In closing, we discuss potential reasons for the decline in support for growth management in the state between 1985 and 2001, placing them in the context of ongoing efforts to remake the state's growth management system.

The Florida Growth Management Context

In 1985 the state of Florida passed one of the most innovative growth management programs this country has ever seen (Ben-Zadok and Gale, 2001; Porter, 1998). The Florida 1985 Growth Management Act (GMA) called for state oversight of local planning efforts, required consistency between formerly disconnected local plans, and established infrastructure concurrency, a mandate that specified that certain urban services must be in place prior to development. The legislation also outlined a very detailed process for resident input into local planning decisions.

Although in many ways this was a landmark piece of legislation, Florida's growth management system has not been without its problems. Initially heralded around the country as one of the most comprehensive and potentially most effective attempts at managing growth, this system has experienced mixed success at guiding growth in Florida (Porter, 1998; Weitz, 1999). Despite the requirement of infrastructure concurrency, the state chose not to provide sufficient resources to address infrastructure backlogs or the need for new infrastructure (Ben-Zadok and Gale, 2001; Koenig, 1990; Pelham, 1992). The system has also been criticized for placing the state in the role of a command and control entity, rather than the role of a facilitator and advisor to localities on planning issues (Florida Growth Management Study Commission, 2001).

Over time these criticisms have prompted reforms to the state's growth management system, including relaxation of concurrency standards to promote infill development and more attention to issues of affordable housing and rural economic development (Porter, 1998). More recently Florida's governor has spearheaded an effort to reshape the state's growth management system, with specific attention paid to modeling the true fiscal impact of new development and to making the system more accessible and responsive to public input. As a consequence, growth management in Florida is currently at a crossroads; changes to the system are clearly in the works, but the changes have not yet been finalized and the ultimate form of growth management in the state remains unclear.

Data and Methodology

The data employed in this study came from the *Florida Annual Policy Survey (FAPS)*, a survey conducted by the Florida State University Survey Research Lab annually since 1979 (Florida State University Survey Research Lab, 2004).

This survey is intended to monitor the policy interests and attitudes of Floridians on issues of importance to state and local governments. In addition to questions included annually in the *FAPS*, in 1985 and 2001 the survey incorporated a panel of questions on growth management (Florida State University Survey Research Lab, 1985; Florida State University Survey Research Lab, 2001). These questions queried respondents as to the need for growth controls (the need to stop or limit growth in their community) and whether or not they feel that government should play a role in managing growth. Responses to the survey were generated from a random-digit-dial telephone survey of Floridians 18 years old or older. Survey results were obtained from a representative sample of 983 and 1,085 Florida residents in 1985 and 2001, respectively.

Two statistical measures were employed in the analysis. For each year's data, simple chi-square tests were performed to identify any variance in resident attitudes across demographic, economic, and political variables. Similar to the methodology of Gale and Hart (1992), this analysis provides insight into how attitudes towards growth management vary across the population at a given point in time. In order to ascertain changes in attitude over time, difference of proportion tests were employed. This test calculates whether any changes between 1985 and 2001 in the proportion of respondents indicating a given response are statistically significant.

Results

The Perceived Need for Growth Management

Using explanatory variables routinely found in the literature on resident attitudes towards growth management, we explored the perceived need for growth controls in both 1985 and 2001. Table 5.1 summarizes results from these analyses. This table shows the percentage of respondents who perceived a need to stop or control growth in their community in each survey year. For example, in 1985, 71.8 percent of all respondents perceived a need to stop or control growth in their community, compared to 67.7 percent in 2001. Table 5.1 also illustrates the change in perceived need for all residents over time, with a decline of 4.1 percent between 1985 and 2001.

The table also summarizes the results of the chi-square tests for variables at a given point in time (along rows below each variable heading) and the difference of proportions tests to determine a statistically significant change over time (in the DoP Test columns). For example, in 1985, the chi-square analysis found no statistically significant difference in perceived need for growth management across race in 1985 (71.8 percent for Whites, 71.9 percent for Blacks), but a significant difference was found in 2001 (74.3 percent for Whites, 57.5 percent for Blacks). For Whites, there was no statistically significant change in perception of need for growth controls from 1985 to 2001 (the decline of 2.5 percent was not significant), but for Blacks there was a statistically significant decrease (-14.4 percent was found to be statistically significant at the .95 level).

In reviewing the findings for 1985 (the 1985 columns), Table 5.1 illustrates the broad-based perceived need for growth controls in that year. Across the entire sample,

Table 5.1   Perceived need for growth controls in Florida, 1985 and 2001

	1985	2001	Change 85–01	DoP Test[a]		1985	2001	Change 85–01	DoP Test[a]
All respondents	71.8%	67.7%	-4.1%	**					
DEMOGRAPHIC VARIABLES									
Race					*Gender*				
Whites	71.8%	74.3%	2.5%		Males	72.2%	63.9%	-8.3%	***
Blacks	71.9%	57.5%	-14.4%	**	Females	71.6%	70.3%	-1.3%	
Chi-Sq Test[b]		***			*Chi-Sq Test[b]*		**		
Ethnicity					*Age*				
Hispanics	67.1%	43.5%	-23.6%	***	18–29	69.0%	55.7%	-13.3%	***
Chi-Sq Test[b]		***			30–45	71.7%	66.8%	-4.9%	*
Florida native					46–59	70.9%	71.9%	1.0%	
Yes	65.7%	71.1%	5.4%		60+	73.9%	72.3%	-1.6%	
No	73.1%	66.6%	-6.5%	***	*Chi-Sq Test[b]*		***		
Chi-Sq Test[b]	*								
SOCIOECONOMIC VARIABLES									
Educational attainment					*Household income*				
No college	69.6%	65.5%	-4.1%	*	Low	69.6%	59.4%	-10.2%	***
Some college	74.0%	67.5%	-6.5%	*	Medium	77.2%	70.5%	-6.7%	**
College grad	75.4%	71.1%	-4.3%		High	69.5%	76.4%	6.9%	**
Chi-Sq Test[b]					*Chi-Sq Test[b]*	**	***		
Housing tenure									
Own	74.0%	71.7%	-2.3%						
Rent	66.7%	59.0%	-7.7%	***					
Chi-Sq Test[b]	**	***							
POLITICAL VARIABLE					**CONTEXTUAL VARIABLE**				
Registered political party					*Region of the state*				
Republican	71.5%	71.1%	-0.4%		North	59.7%	65.0%	5.3%	
Democrat	71.5%	67.0%	-4.5%	*	Central	77.8%	73.2%	-4.6%	*
Independent	76.1%	70.4%	-5.7%		South	72.5%	62.5%	-10.0%	***
Chi-Sq Test[b]					*Chi-Sq Test[b]*	***	***		

Note: "Do you feel growth in your community should be stopped, limited, or not controlled at all?"

Results show the percentage of those who believed growth should be stopped or limited.

[a] This illustrates the level of statistical significance from a difference of proportions test, 1985 versus 2001.

[b] This represents the level of statistical significance from a chi-square test for a given variable in a given year.

For both Chi-Square and Difference of Proportions Tests:

     *p < .10 **p < .05 ***p < .01

almost 72 percent of respondents indicated that growth in their community needed to be stopped or controlled. This high perception of need was mirrored across almost every population subgroup. In only one case across all of the variables did less than 65 percent of respondents indicate that growth needed to be controlled (respondents in largely rural north Florida). While certain variables show statistically significant intravariable variation (Florida native, Income, Housing tenure, and Region), perceived need for growth controls is very high across all variables. These survey results indicate that the perceived need for growth controls was very high in 1985 and that this perceived need existed in all segments of the population.

Table 5.1 indicates that in 2001, Floridians continued to support growth management at nearly the same level as they did in 1985. While the drop from 71.8 to 67.7 percent is statistically significant, over two thirds of the state's residents still believe that growth in their communities should be stopped or controlled. However, the broad-based coalition supportive of growth management in 1985 had fragmented by 2001 (see the 2001 columns). Whereas there was no variation across basic demographic variables in 1985, by 2001 Blacks, Hispanics, males, and younger respondents perceived significantly less need for growth controls than Whites, females, and older respondents. Similarly, lower-income households and renters also were less likely to support growth management than higher-income households or homeowners. At a more basic level, the threshold of 65 percent that only one variable fell below in 1985 was now broken seven times, with an eighth coming in at that threshold. Whereas the 1985 survey found similar levels of perceived need for growth controls across the population, by 2001 this perception varied markedly across fundamental demographic and socioeconomic lines.

Underscoring this finding that the coalition supporting growth management had fragmented by 2001, Table 5.1 also reveals that between 1985 and 2001 the perceived need for growth management fell significantly across 14 of the 25 variables investigated (see the DoP Tests columns). Particularly noteworthy was the decline in perceived need for growth controls amongst Blacks, Hispanics, males, renters, lower- and medium-income households, and south Florida residents. While the need for growth controls had traction amongst all groups in 1985, by 2001 perceived need for growth controls varied markedly across the population. In this later time period, minorities and respondents of lower socioeconomic status perceived less of a need for growth controls, whereas high-income households and Whites perceived a steady or increasing need for these controls. This provides evidence that attitudes towards growth in Florida had bifurcated between 1985 and 2001.

Support for a Government Role in Managing Growth

A second question asked in both 1985 and 2001 provides a more direct assessment of resident support for government-led growth management. Respondents who indicated that they thought growth in their communities needed to be stopped or controlled were then asked whether or not government should play a role in managing growth in their communities. The idea behind this question was that although people might wish to limit growth in their communities, they may also be loathe to see government play a strong role in limiting private development. In a state such as Florida, where Republicans have replaced conservative Democrats as the dominant political faction, it was reasoned that residents' fear of government intrusion might limit overall support for growth management.

Table 5.2   **Support for a government role in managing growth in Florida, 1985 and 2001**

	1985	2001	Change 85–01	DoP Test[a]		1985	2001	Change 85–01	DoP Test[a]
All respondents	75.5%	63.7%	-11.8%	***					
DEMOGRAPHIC VARIABLES									
Race					*Gender*				
Whites	75.2%	62.1%	-13.1%	***	Males	77.6%	65.2%	-12.4%	***
Blacks	72.5%	69.8%	-2.7%		Females	73.7%	62.8%	-10.9%	***
Chi-Sq Test[b]		**			*Chi-Sq Test*[b]				
Ethnicity					*Age*				
Hispanics	84.9%	76.1%	-8.8%	**	18–29	78.1%	69.6%	-8.5%	***
Chi-Sq Test[b]		**			30–45	74.9%	68.6%	-6.3%	***
Florida native					46–59	71.6%	64.2%	-7.4%	***
Yes	78.8%	66.1%	-12.7%	***	60+	76.4%	55.6%	-20.8%	***
No	75.0%	62.9%	-12.1%	***	*Chi-Sq Test*[b]		**		
Chi-Sq Test[b]									
SOCIOECONOMIC VARIABLES									
Educational attainment					*Household income*				
No college	74.6%	67.2%	-7.4%	***	Low	74.6%	63.6%	-11.0%	***
Some college	72.0%	59.7%	-12.3%	***	Medium	76.4%	65.6%	-10.8%	***
College grad	81.3%	64.0%	-17.3%	***	High	76.9%	64.7%	-12.2%	***
Chi-Sq Test[b]					*Chi-Sq Test*[b]				
Housing tenure									
Own	75.6%	61.3%	-14.3%	***					
Rent	75.8%	71.3%	-4.5%	***					
Chi-Sq Test[b]		**							
POLITICAL VARIABLE					**CONTEXTUAL VARIABLE**				
Registered political party					*Region of the state*				
Republican	73.5%	60.2%	-13.3%	***	North	71.0%	54.9%	-16.1%	***
Democrat	77.9%	66.4%	-11.5%	***	Central	73.7%	64.9%	-8.8%	***
Independent	76.2%	62.2%	-14.0%	***	South	80.7%	69.9%	-10.8%	***
Chi-Sq Test[b]					*Chi-Sq Test*[b]	*	***		

Note: "Should the government do something about growth in your community?" Results indicate the percent of respondents who answered "yes". Asked of those who believe that growth should be controlled or stopped.

[a] This illustrates the level of statistical significance from a difference of proportions test, 1985 versus 2001.

[b] This represents the level of statistical significance from a chi-square test for a given variable in a given year.

For both Chi-Square and Difference of Proportions Tests:
       $*p < .10 **p < .05 ***p < .01$

Table 5.2 summarizes the findings from this line of inquiry, indicating the level of support for a government role in managing growth in both 1985 and 2001. Similar to Table 5.1, this table also shows the results of both the chi-square analysis for each variable in a given year and the difference of proportions test to determine statistically significant changes between 1985 and 2001 (DoP Tests columns).

In 1985, three quarters of all respondents who perceived a need for growth controls were supportive of a government role in managing growth. Underscoring the broad-based support for growth controls found when investigating perceived needs, Table 5.2 shows that the level of support for a government role in managing growth was high across all groups, with no individual variable falling below 70 percent. Supporting this broad-based view, the statistical analysis also found that support for growth management only varied across one variable—region of the state—as respondents in north Florida were significantly less supportive of a government role in growth management efforts than residents in other areas of the state.

A comparison of these results with those from 2001 reveals a decline in support for a government role in growth management. Whereas in 1985 no variable fell below 70 percent support, by 2001 only two variables broke the 70 percent support level (Hispanics and renters). Table 5.2 reinforces the conclusion that the broad-based support for growth management that existed in 1985 had splintered by 2001. While in 1985 there was little variation across population subgroups, by 2001 there were statistically significant variations in support for a government role across race, ethnicity, age, housing tenure, and region.

As for the change in attitudes over time, overall support for a government role in managing growth fell significantly between 1985 and 2001, down almost 12 percent. Further, across 24 of the 25 variables there were statistically significant declines in support for a government role in managing growth. Only among Blacks did support for a government role in managing growth not decline significantly between 1985 and 2001. The decline in support is particularly striking for college graduates, homeowners, and Whites, groups that would be thought to support government-led growth management initiatives. Among groups not expected to back government intervention, such as Republicans and more rural north Floridians, support also declined substantially and significantly.

Overall, this finding is remarkably robust, as support for a government role in managing growth declined across almost every subset of the population, including demographic, socioeconomic, and contextual variables. The results from this question provide further evidence of bifurcation of support for growth management in the state, albeit in a different direction than that shown in Table 5.1. By 2001, among those respondents who felt growth was a problem, minorities, younger residents, renters, and central and south Floridians showed greater support for a government role in growth management.

This second finding may at first seem at odds with the finding on perceived need for growth controls. Recall that minorities, younger voters, and respondents in lower income households perceived less of a need for growth controls in 2001 (see Table 5.1). Intuitively it makes sense that these groups do not perceive a need for growth controls, as they are in many cases the "have nots" in Florida. For these groups, managing growth is not a problem—attracting growth is. However, once individuals in these "have not" groups do perceive growth as a problem, then it is little surprise that they show greater support for a government role in managing growth; these groups are accustomed to government intervention to address problems.

Explaining Declining Support for Growth Management in Florida

Taken as a whole, these findings indicate that while Floridians continue to perceive a need for growth controls, they are less supportive of a government role in achieving this outcome. It seems reasonable, then, to inquire why resident support for government-led growth management fell between 1985 and 2001. To address this question, we first discuss the historical and political contexts of growth management in 1985 and 2001, as we believe these contexts partly explain the decline in support. We also briefly characterize the state economies in 1985 and 2001, finding that the economy was not likely a factor behind changes in resident attitudes. Following this discussion, we provide results from other *FAPS* survey questions that augment the political contextual argument.

The Policy and Political Context: 1985 and 2001

In 1985, Florida was still in labor with growth management, as the state legislature and governor were finalizing legislation for one of the most comprehensive approaches to managing growth the country had yet seen. At the time of the first survey, the long, difficult local comprehensive planning process and the contentious state review of these local planning documents still lay in the future. The implications of the decision by the state legislature to disassociate funding from the final legislation and the actual impact of the concurrency provision on development patterns remained unknown. In short, to survey respondents in 1985, growth management in Florida remained unblemished, a vision of good planning yet to be spoiled by the realities of day-to-day implementation.

The survey results from 1985 almost certainly reflect heightened expectations amongst Florida's residents as to the effectiveness of this approach to managing growth. Support for growth management would be expected to be high in these heady early days; otherwise this comprehensive approach to managing growth probably would not have made it through the legislature. The high level of support for managing growth that existed across all subsets of the population in 1985 therefore reflects the success of the sponsors of growth management in marketing the concept to Floridians.

Moreover, in 1985 Florida's drive for growth management was led in part by the state's governor, Bob Graham. Graham, who was elected governor in 1978 and reelected in 1982, was a champion of growth management and environmental protection in Florida. Graham had led the Save Our Coasts, Save Our Rivers, and Save Our Everglades programs in the early 1980s, and the legislature had adopted bills in 1983 and 1984 to protect the state's waters and wetlands. In his speech opening the 1985 legislative session, Governor Graham made passing a growth management bill a priority (Hauserman, 2001; Woodburn, 1998).

Contrast the situation in 1985 with that in 2001, by which time the state's system of growth management had been in play for over a decade. The first set of local comprehensive plans had been prepared; the state's community planning agency, the Department of Community Affairs (DCA), had reviewed, commented on, and in some cases rejected some of these local plans; and required updates to these plans had been prepared as dictated by the legislation. During this period, development was directed to areas where infrastructure was available (sometimes promoting sprawl rather than containing it), loopholes in the concurrency requirement had been identified resulting in new regulations to close these, and shortages in funding for new and existing infrastructure needs were experienced throughout the state. In short, the system designed in the mid 1980s had been put into practice, in the process exposing shortcomings and problems of the approach as initially designed (Porter, 1998).

While this visionary but imperfect system was attempting to manage growth across the state, much of central and southern Florida continued to boom as retirees, workers in the tourist industry, and wealthy households came to take advantage of warm weather, Disney World, and the lack of a state income tax. This growth had also begun to spread into northern Florida, generating growth issues in rural counties formerly passed over by the state's phenomenal growth.

A final, equally important change occurred in the political orientation of the state. In the 1990s, Republicans came into control of both houses of the legislature and the governor's mansion with Jeb Bush's election in 1998. Governor Bush proposed revamping Florida's growth management laws, predicated on the idea that the current system was ineffective and served only as a deterrent to growth in the state. With the appointment of a new Secretary of Department of Community Affairs in 1999, the governor's office began a campaign to revamp and rewrite the state's growth management system, convening a Growth Management Study Commission (GMSC) in 2001 to investigate growth issues and to make recommendations for improving this system for managing growth. One of the major recommendations to emerge from this commission was that the state should have much less oversight responsibility and serve primarily as a technical resource in the preparation of local plans (Florida Growth Management Study Commission, 2001).

The 2001 survey results therefore provide insights into resident attitudes towards growth management after years of struggling to implement and debug a very complex approach to managing growth across the state. Given the significant problems with the approach as initially designed, the continued growth in the state, and the significant shift in political leadership, *a priori* it would be expected that support for government intervention in growth management in the state would wane.

The Economic Context: 1985 and 2001

Previous research has suggested that the state of the economy is a factor in resident attitudes towards growth controls (Connerly and Frank, 1986; Logan, 1978; Protash and Baldassare, 1983). It is argued that in tough economic times, residents are less supportive of growth management, as individuals tend to be more interested in policies that promote growth rather than those that attempt to control it. A review of Florida's economic context does not provide much evidence for this hypothesis. A review of state economic indicators in both 1985 and 2001 suggests that at the time of each of these surveys, the state economy was generally healthy.[1] For example, the unemployment rate was 6.0 percent in 1985 and 4.8 percent in 2001, both at or below national rates (Bureau of Economic and Business Research, 1986, 2002). The state economy was growing in both of these periods, although at a faster rate in the mid 1980s. Perhaps the only evidence supportive of this economic context argument lies in the relative trajectories of the state economy in 1985 compared to 2001. In 1985, the state (and national economy) had emerged from the economic recession of the early 1980s and the overall economic outlook was improving. In contrast, in 2001 the Florida economy was in many ways better than that of 1985, but the economic boom of the 1990s had begun to fade away and the economic forecast for the state painted a picture of an economy that was slowing down.

Despite this somewhat dreary economic outlook in 2001, survey results suggest that residents were more concerned about the economy at the time of the 1985 survey. In both surveys respondents were asked to identify the most important problem facing the state. In 1985, 13.9 percent of respondents identified the economy as the most important problem, compared to only 7.8 percent in 2001. Reflecting the stated priorities of the governors in each time period, in 1985 the modal response for the most important problem was community development, and in 2001 education was deemed the most important issue facing the state. Taken as a whole, these data seem to suggest that the state of the economy was not a major factor in explaining the different results generated by the 1985 and 2001 surveys.

Insights from the 2001 FAPS Survey

While no questions in the surveys were directed specifically towards why residents did or did not support a government role in managing growth, other survey questions provide insights as to why there was decline in support between 1985 and 2001. These questions focused upon three broad issues: (1) resident familiarity with growth management (asked in 2001 only), (2) the impact of the state's system of growth management on problems related to growth, such as traffic, water supply, and the condition of the environment (asked in 2001 only), and (3) the appropriate government level for managing growth (asked in both 1985 and 2001).

Resident familiarity with and understanding of the state's system of growth management has long been an issue in Florida. It is widely held that the comprehensive plan amendment process, reviews of developments of regional impacts (DRI), and the many government layers involved in the process have created barriers to resident input. One of the primary policy recommendations of the GMSC appointed by Governor Bush revolved around this issue of resident familiarity with and participation in the state's growth management process. The GMSC's final report states that Florida needs to "empower residents to better understand and participate in the growth management process" (Florida Growth Management Study Commission, 2001, p. 2).

The 2001 survey data support the contention that the state's growth management system is not well understood by residents. Survey respondents were asked about their level of familiarity with the state's system. Less than one third of respondents stated that they were familiar (at any level) with the state's growth management system (see Figure 5.1). What is remarkable about this finding is that growth issues have been at the forefront of state and local politics in Florida for the better part of 30 years, and that the state's growth management system has routinely been discussed and debated in the state legislature since its inception in 1985.

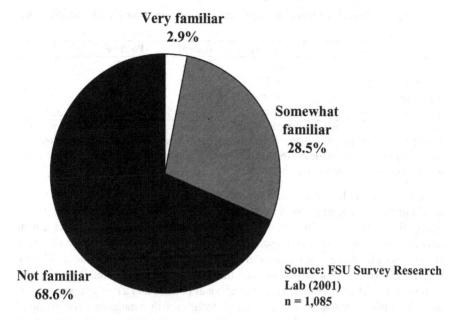

Figure 5.1   **Resident familiarity with Florida's growth management system, 2001**

This finding is ironic because the system of growth management established in 1985 centered on a comprehensive planning process that required resident involvement in the initial preparation of local comprehensive plans and continued resident involvement in refining and updating these plans on an annual basis. While groups such as 1000 Friends of Florida and the Sierra Club have played important roles at the state and local levels in implementing the state's growth management laws, it appears that the vast majority of the state's residents remain unfamiliar with state-led growth management.

Resident support for government-led growth management may also be affected by the perceived ineffectiveness of the growth management system. To acquire some sense of the perceived impacts of the state's system of growth management, a question on the 2001 *FAPS* asked respondents about their perceptions of the impacts of growth management on traffic, local water supply, the environment, and disaster preparedness. This question was asked only of those respondents who stated that they were very familiar or somewhat familiar with the state's growth management system. Table 5.3 illustrates the responses to this question, generally indicating that respondents believed that growth management has not been effective in addressing these issues. Particularly striking are the results for traffic, as over three fifths of respondents indicated that growth management has negatively impacted traffic conditions. Also of interest is the finding that of those familiar with the state's growth management system, a greater percentage responded that it has had a *negative* impact on the environment. In only one of the four issues—disaster preparedness—was the perceived positive impact greater than the perceived negative impact.

Table 5.3   **Perceived impacts of growth management on growth issues, 2001**

Issue	Negative	Neutral	Positive	N
Traffic	60.7%	23.5%	15.8%	336
Local water supply	30.7%	49.7%	19.6%	332
Environment	40.9%	26.4%	32.6%	337
Disaster preparedness	24.4%	32.6%	43.0%	328

Note: "Please indicate whether you think that Florida's system of growth management has had a positive effect on these items, a negative effect, or a neutral effect." This question was asked only of those respondents who indicated they were somewhat familiar or very familiar with Florida's growth management system.

Taken as a whole, these results strongly indicate Florida's residents perceive the state's growth management system to have been ineffective at addressing key issues facing the state. What should be particularly disturbing to growth management advocates is that this group of respondents is not only more knowledgeable about the system, but also more supportive of growth controls. An analysis of the familiarity variable determined that familiarity with the state system is predicted in part by two factors: (1) level of interest in state and national politics and 2) general support for growth controls (Chapin, 2004). Respondents to the growth management effectiveness question are in many ways the marketing demographic for this legislation—active individuals supportive of growth management. Yet Table 5.3 underscores that even within this core constituency there is a strong sentiment that the state's system has not positively impacted growth issues.

One other survey question provides insights into the erosion of support for growth management in Florida. In both 1985 and 2001, those respondents who stated that they were supportive of a government role in managing growth were then asked what level of government should take *primary* responsibility for controlling growth (federal, state, county, city/town). Between 1985 and 2001 there was an observed shift in resident preference from the *state level* in 1985 to the *county level* in 2001 (results not shown here but available from the authors). Respondents in 1985 reflected the nascent system, with the greatest number of respondents assigning the state primary responsibility for managing growth. By 2001, however, the county level emerged as the modal response, up over 12 percent, while the state level lost an almost equal share. By 2001, almost two thirds (66.4 percent) of respondents supportive of a government role in managing growth felt that the local or county level should have primary responsibility for growth management in the state, up from 51.6 percent in 1985.

This growing dissatisfaction with state intervention in growth management may have been a key factor in Governor Bush's ongoing efforts to devolve responsibility for controlling local growth to local governments (Kennedy and Newman, 2002). While changes in resident support for growth management between 1985 and 2001 suggest the impact that a change in gubernatorial leadership has had on public support for growth management, the public's growing preference for local control of growth also makes it easier for Governor Bush to push for the devolution of growth management responsibilities to the local level.

When taken together, the survey data provide some insights as to why support for growth management in Florida fell significantly between 1985 and 2001. Any system that is poorly grasped by residents, deemed ineffective at addressing key problems, and viewed as a poor organizational response to these problems is almost certain to lose support, even within its core constituency. In Florida, all of these shortcomings are perceived to exist. Residents are largely unfamiliar with the system, they don't think it works, and there is a growing belief that the existing structure of the system is unsuitable to managing growth.

Despite these problems, however, it is clear that growth management is here to stay in Florida. Florida continues to be among the fastest growing states in the nation and growth issues remain at the forefront of the public debate at the local and state levels; two thirds of Florida's residents continue to perceive growth to be a problem in their communities. Even as Governor Bush attempts to remake the DCA into a user-friendly organization oriented towards providing technical support for local comprehensive planning efforts, the debate has centered not on whether or not growth management is needed, but what form growth management should take in Florida and what the appropriate role of the state government is in these efforts.

Conclusion

Taken as a whole, these results indicate that there is still substantial support for the general idea of growth management in Florida. Overall, the state's residents continue to support growth management, with two thirds supporting efforts to limit or control growth in both 1985 and 2001. While debate continues over the form and content of growth management policy in Florida, these survey results make it clear that growth is an issue still of great importance to the state's citizens.

At the same time, these survey results also point to two potential concerns. First, the across-the-board support found in 1985 has now eroded. Support for limiting or controlling growth is now more likely to be found among affluent White, homeowners. This means that growth management now appears to be more of a class issue than it was in 1985. This may enable opponents of growth management to use economic class issues, such as housing affordability, to attack growth management. Planners must be prepared to identify ways to bring these disaffected groups back into the growth management fold—it does growth management no good to be viewed as a "boutique" cause.

Second, support for government intervention in growth management declined among nearly all groups between 1985 and 2001. This decline appears to reflect the difficulty associated with the implementation of growth management that usually results when government attempts to carry out a complicated piece of legislation. At the same time, it also likely reflects the very different tone set by Governor Bush in 2001 than the one set by former Governor Graham in 1985. A critical question is whether a strong gubernatorial advocate for growth management like former Governor Graham could boost public support for government control of growth. Given that most Floridians remain unfamiliar with the state's growth management act, it appears there is room for a pro-growth management governor to step forward and influence the public's familiarity with and support for this legislation.

In addition to interpreting these results for Florida, this study offers some insights to other states that have chosen a growth management course. Survey results show that citizen attitudes towards growth management do indeed change over time. These results illustrate that support for growth management may differ across subgroups of the population and these differences may increase over time. Taken together, these findings make it clear that citizen attitudes should be monitored so that policymakers can track and respond properly to these attitudinal shifts.

Florida's experience also indicates that popular support for growth management is likely to persist in mature growth management states, even in the face of frustration with government implementation of the legislation and the emergence of political regimes that are less supportive of such efforts. While changing conditions have contributed to a decline in support for government intervention and a narrowing of the constituency for growth management to one that is less representative of the state's residents, growth issues remain of great importance to citizens in the state. The planning community in Florida and other mature growth management states must recognize these changing conditions and their related impacts and respond in ways that bring together this still broad constituency supportive of managing growth.

Acknowledgements

Special thanks to colleague Gayla Smutny, David Rasmussen of the Devoe Moore Center, and Mary Stutzman of the FSU Survey Research Lab for their help on this project. This research was sponsored in part by a grant from the Devoe Moore Center for the Study of Critical Issues in Economic Policy and Government in the Florida State University College of Social Sciences.

References

Bollens, S. (1990). Constituencies for limitation and regionalism: Approaches to growth management. *Urban Affairs Quarterly, 26*(1), 46–67.

Ben-Zadok, E., and Gale, D. (2001). Innovation and reform, intentional inaction, and tactical breakdown: The implementation record of the Florida concurrency policy. *Urban Affairs Review, 36*(6), 836–871.

Bureau of Economic and Business Research. (1986). *Florida statistical abstract.* Gainesville: University of Florida.

Bureau of Economic and Business Research. (2002). *Florida statistical abstract.* Gainesville: University of Florida.

Chapin, T. (2004). Variations in citizen familiarity with growth management processes, Evidence from Florida. *Urban Affairs Review, 39*(4), 441–460.

Connerly, C. (1986). Growth management concern: The impact of its definitions on support for local growth control. *Environment and Behavior, 18*(6), 707–732.

Connerly, C., and Frank, J. (1986). Predicting support for local growth controls. *Social Sciences Quarterly, 67*(3), 572–586.

Deakin, E. (1989). Growth control: A summary and review of empirical research. *Urban Land, 48*(7), 16–22.

Dowall, D. (1980). An examination of population-growth-managing communities. *Policy Studies Journal, 9*(3), 414–427.

Florida Growth Management Study Commission. (2001, February). *A liveable Florida for today and tomorrow.* Retrieved October 2001, from http://www. floridagrowth.org/pdf/gmsc.pdf.

Florida State University Survey Research Lab. (1985). *Florida annual policy survey, 1985.* Retrieved March 2004, from http://www.fsu.edu/~survey/FAPS/ Codebooks/faps1985.htm.

Florida State University Survey Research Lab. (2001). *Florida annual policy survey, 2001.* Retrieved March 2004, from http://www.fsu.edu/~survey/FAPS/faps2001. htm.

Florida State University Survey Research Lab. (2004). *Florida annual policy survey.* Retrieved March 2004, from http://www.fsu.edu/~survey/FAPS/index.htm.

Gale, D., and Hart, S. (1992). Public support for local comprehensive planning under statewide growth management: Insights from Maine. *Journal of Planning Education and Research, 11*(3), 192–205.

Hauserman, J. (2001, December 28). Bush to try again with schools, growth bill. *St. Petersburg Times*, p. B1.

Kennedy, J. and Newman, J. (2002, December 14). Governor Bush wants secretary of state to tackle growth. *Orlando Sentinel*, p. A1.

Knaap, G. J. (1987). Self-interest and voter support for Oregon's land use controls. *Journal of the American Planning Association, 53*(1), 92–97.

Koenig, J. (1990). Down to the wire in Florida. *Planning, 56*(10), 4–11.

Local Government Comprehensive Planning and Land Development Regulation Act, Fla. Stat. §§ 163–3161–3242 (1985).

Logan , J. (1978). Growth, politics, and the stratification of places. *American Journal of Sociology, 84*(2), 404–416.

Pelham, T. (1992). Adequate public facilities requirements: Reflections on Florida's concurrency system for managing growth. *Florida State University Law Review, 19*, 973–1052.

Porter, D. (1998). The states: Growing smarter? In Urban Land Institute (Ed.), *Smart growth: Economy, community, environment* (pp. 28–35). Washington, DC: Urban Land Institute.

Protash, W., and Baldassare, M. (1983). Growth policies and community status: A test and modification of Logan's theory. *Urban Affairs Quarterly, 18*(3), 397–412.

Weitz, J. (1999). *Sprawl busting: State programs to guide growth.* Chicago: American Planning Association.

Woodburn, K. (1998). Were the '80s our environmental peak? *Florida Sustainable Communities Center State/Regional News.* Retrieved January 25, 2003, from http://sustainable.state.fl.us/fdi/fscc/news/state/ken3.htm.

PART II
Evaluating Growth Management's Outcomes

Chapter 6

Growth and Change Florida Style: 1970 to 2000

Thomas W. Sanchez and Robert H. Mandle

In response to development patterns leading to what may be termed "urban sprawl," several local, regional, and state governments in the United States have embarked on growth management or urban containment strategies. These strategies typically aim to synchronize key public facilities with urban development pressures, preserve open spaces, and facilitate development in ways that preserve public goods, minimize public costs, and account for development impacts by those who cause them (Nelson and Dawkins, 2002; Nelson and Duncan, 1995). We refer the reader to Nelson and Dawkins (2002) for a review of how growth management and urban containment work and how they vary in application across the United States.

One of the cornerstones of urban containment is limiting development beyond an urban containment boundary such as an urban growth boundary, urban service limit, or (in the UK) urban growth stopline (see Easley, 1992). Development is restricted in one of two principal ways. First and foremost in all containment schemes is preventing the extension of urban facilities into the rural countryside, especially wastewater treatment provided via sanitary sewers. This restriction is sometimes but not always extended to public water systems.

The second and more difficult way is restricting actual density. In the Twin Cities (Minneapolis-St. Paul, MN), minimum lot size restrictions do not discourage low density urban development since lot sizes can range from one to five acres on septic systems with or without public water. Such small acreage development is perhaps the most pernicious of all forms of urban sprawl because it consumes land at a very rapid pace, removes land from a variety of open space uses, signals to farmers impending conversion to development, and exacerbates inefficient provision of services (Nelson, 1999). These are generally considered "weak" containment programs. At the other extreme is metropolitan Portland, Oregon, where development outside urban growth boundaries occurs only in "exception" areas (areas excepted from strict application of farm and forest use policies because they are already built or committed to low density uses) or in farms and forests where needed to manage a commercial-scale operation (which can range from about 20 acres for high-intensity nurseries to 160 acres for timber production). Such efforts have been considered "strong" containment programs.

Urban containment can also occur because of natural conditions. Honolulu, HI comes to mind because the city has virtually nowhere to expand. On the mainland, perhaps Los Angeles is the best example of natural containment since an ocean, mountain ranges, and federally-owned desert hem in development. Phoenix can also be considered naturally contained because individual water wells are not financially feasible and government agencies own a majority of the land around that metropolitan area.

The Florida Context

As one of the pioneering states in passing growth management legislation, Florida (along with Oregon) has influenced several other states in the design and implementation of growth management strategies (DeGrove, 1992). Managing growth effectively involves balancing competing values for resource protection and economic development with the overall aim of furthering the public's welfare. To do this, Florida's 1985 Growth Management Act (GMA) depends on three complementary policies: consistency, concurrency, and compact development (Ben-Zadok, 2005). While these three policies collectively aim to affect orderly urban development, the GMA's objective of compact development lends itself to evaluation more easily than either consistency or concurrency. The coordination of city and county land use plans (consistency) and timely provision of public services for new development (concurrency) do not directly manifest themselves through urban form as does compact development.

This evaluation will focus on the Florida GMA's compact development policy, using historical trends in residential development densities as an indicator of whether population growth has been absorbed as dense, contiguous, urban, or potentially in-fill type land uses. Compact development should be detectable at the census tract level, where urban land use densities have been achieved or increased over time compared to suburban or exurban land use densities. This analysis focuses on residential development because it is a significant determinant of urban form, especially at the urban fringes of rapidly growing regions. Evidence that substantial amounts of Florida's growth have resulted in urban land use densities is one indication that state, counties, and metropolitan areas have been successful at encouraging compact development.

Having implemented innovative land use regulation, growth management, and urban containment approaches, the State of Florida is seen as one of the growth management leaders among US states. Yet, little quantitative evidence exists to gauge the comparative success of Florida's strategies at the state or metropolitan scale that provides support for this assertion. This analysis intends to address this gap in the supportive literature by estimating the extent of urban development occurring within the State of Florida as well as its counties and metropolitan areas. To achieve this objective, the evaluation uses spatial analysis techniques within a geographic information system (GIS) to assess the location and extent of urban expansion (see Nelson and Sanchez, 2005). Using 1970, 1980, 1990, and 2000 census tract data, population density classifications were used to show changes in spatial patterns of urban, suburban, exurban, and rural settlement. The estimates for Florida MSAs are also compared with other selected US metropolitan areas to look at Florida's growth from a national perspective. The results are presented in both quantitative and graphic form. The following describes the methods used to generate the estimates of land use change.

Methods

We estimated land use change differently than prior work that measured changes in development patterns based on counties. The evaluation here requires a finer grain of geographic resolution; after all, many US counties contain a range of development densities from urban to suburban, exurban, and rural development. Our solution is to measure change in census tract population density over time, particularly change in urban classification status. To do this, we first classified all census tracts in Florida as urban, suburban, exurban, or rural based on certain residential density ranges. Density classifications were used to show patterns of urban (3,000+ persons/sq. mi.), suburban (1,000 to 3,000 persons/sq. mi.), exurban (300 to 1,000 persons/sq. mi.), and rural (<300 persons/sq. mi.). Based on prior conceptual work by Lang (1986) and Nelson (1992a, 1992b), classifications are relatively consistent with census criteria and practical observation. For example, we classify exurban census tracts as those with a residential density ranging from 300 to 1000 persons per square mile. At 2.5 persons per household, this implies 120 homes per square mile or an average of slightly more than 5 acres per home—clearly consistent with views on what constitutes urban-oriented rural residential densities (see also Daniels, 1999).

In order to model a realistic representation of urban form, the analysis interpolates population density information from census tracts, producing continuous value surfaces using GIS. There are several interpolation techniques available to do this within a GIS. To determine the most appropriate interpolation technique, we compared the three standard methods provided within ArcGIS 9.1 (Inverse Distance Weighting (IDW), Kriging, and Spline). Among these, the spline interpolation methodology[1] more effectively predicted changes in several test cases. This method was then used to predict the population density surface for the entire state of Florida and provide a descriptive analysis of land use change from 1970 to 2000. National or state parks, wetlands, or other protected lands were excluded from the land area and population density calculations. These areas were primarily in the Florida Managed Areas (FMA) program that the Florida Natural Areas Inventory (FNAI) has identified as having particular natural resource value and requiring protection or management for conservation purposes.[2] These areas represent approximately 20 percent of the total land area within Florida and have a significant impact on the amount of buildable land near urbanized areas in several counties.

Results

Using census tract data for 1970, 1980, 1990, and 2000, the analysis maps the locations of urban, suburban, exurban, and rural population densities and tabulates the square mileage for each of the four density categories for the entire state of Florida, the 19 metropolitan areas, and each of the 67 counties. Each section summarizes

---

1 More specifically, a spline-tension model was used. This surface produces a coarser surface with a better fit to data control points.

2 See the Florida Natural Areas Inventory web site at: http://www.fnai.org/ for more details on the program.

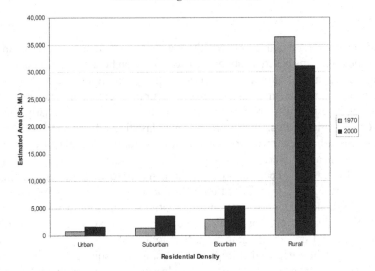

Figure 6.1   Florida land use/urban form change, 1970–2000

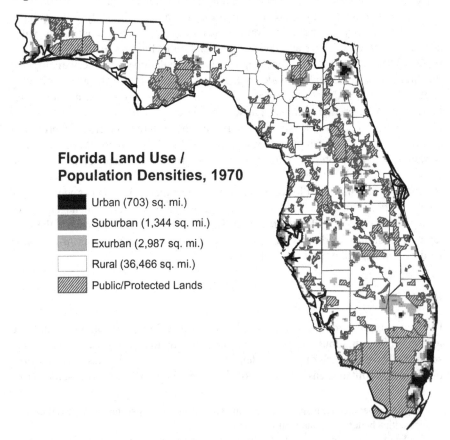

Figure 6.2   Florida population density, 1970

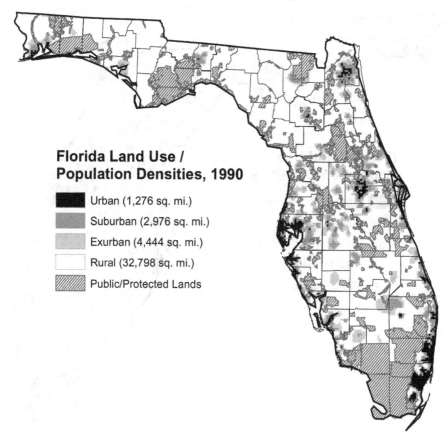

**Florida Land Use /
Population Densities, 1990**

- Urban (1,276 sq. mi.)
- Suburban (2,976 sq. mi.)
- Exurban (4,444 sq. mi.)
- Rural (32,798 sq. mi.)
- Public/Protected Lands

Figure 6.3   Florida population density, 1990

the estimates of land use change to report trends and also illustrates variable growth rates across the state.

Summary of Statewide Trends

Florida experienced very rapid population growth between 1970 and 2000. According to the US Census, the state had a population of 6,789,437 in 1970 and 15,982,824 in 2000 (a 135 percent increase), while the nation grew from 203,302,000 in 1970 to 281,422,000 in 2000 (a 38 percent increase). Five counties had population growth of over 500 percent during the 30 year period: They were Flagler (1018 percent), Hernando (669 percent), Osceola (583 percent), Collier (561 percent), and Citrus (515 percent). The slowest growing counties in the state were Franklin (39 percent), Madison (39 percent), Jackson (36 percent), and Gadsden (15 percent), which are all in the northern part of the state and were at or below the growth rate experienced by the United States. Overall the state has also experienced rapid growth in racial/ethnic minorities as well as relatively high rates of growth in persons of retirement age.

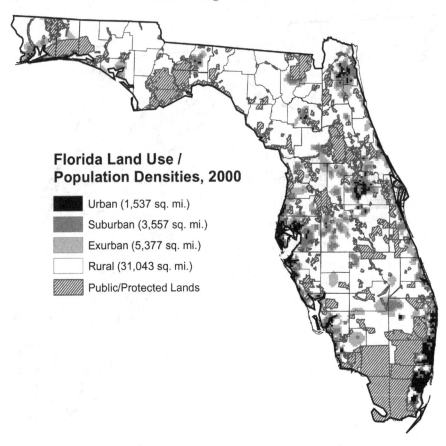

**Florida Land Use /
Population Densities, 2000**

■ Urban (1,537 sq. mi.)

■ Suburban (3,557 sq. mi.)

■ Exurban (5,377 sq. mi.)

□ Rural (31,043 sq. mi.)

▨ Public/Protected Lands

Figure 6.4   Florida population density, 2000

To accommodate this population growth, the amount of land at urban, suburban, and exurban densities increased 119 percent (834 square miles), 165 percent (2,213 square miles), and 80 percent (2,389 square miles), respectively. At the same time, Florida lost approximately 5,423 square miles (or 15 percent) of land at rural population densities.

For the purposes of this summary, the changes that occurred between 1970 and 1990 and between 1990 and 2000 were compared. The extent and location of high population densities (urban) and low densities (suburban and exurban combined) were compared. The period from 1970 to 1990 generally represents a pre-growth management (GM) urban containment stage and 1990 to 2000 represents a post-GM stage. High density land uses in the state grew by 4.1 percent annually[3] from 1970 to 1990 compared to 2.0 percent from 1990 to 2000. Low density areas grew by 3.6 percent annually from 1970 to 1990 compared to 2.0 percent from 1990 to 2000 (see Figures 6.2, 6.3, and 6.4). Significant growth occurred throughout Central Florida as well as along the southwest and southeast coastal areas, which are separated by the

---

3   Calculated as an average annual rate.

Table 6.1   High density land use rankings

| *Top 10 1970 to 1990 Annual High Density Change* | | | | *Top 10 1990 to 2000 Annual High Density Change* | | | | |
COUNTY	70–90 High	70–90 Rank	90–00 High	90–00 Rank	COUNTY	70–90 High	70–90 Rank	90–00 High	90–00 Rank
Sarasota	39.8%	1	0.0%	20	Leon	0.0%	15	20.0%	1
Okaloosa	21.5%	2	1.6%	16	Osceola	0.0%	15	15.0%	2
Manatee	20.4%	3	1.0%	18	Collier	0.0%	15	13.6%	3
Lee	13.6%	4	5.5%	8	Charlotte	0.0%	15	10.0%	4
Pinellas	9.6%	5	1.2%	17	Martin	0.0%	15	9.2%	5
Palm Beach	7.7%	6	7.2%	7	St. Lucie	0.0%	15	8.7%	6
Orange	5.7%	7	2.4%	12	Palm Beach	7.7%	6	7.2%	7
Polk	5.0%	8	0.0%	20	Lee	13.6%	4	5.5%	8
Broward	4.9%	9	2.2%	14	Volusia	3.1%	11	4.5%	9
Hillsborough	4.5%	10	2.3%	13	Seminole	0.0%	15	3.8%	10

Everglades National Park, the Big Cypress National Preserve, and the Everglades and Francis S. Taylor Wildlife Management Area.

Summary of County Trends

From 1970 to 1990, 15 of 67 counties added high density development more rapidly than they added low density development. By comparison, slightly more counties (18) added high density areas more rapidly than they added low density areas from 1990 to 2000. Nearly two-thirds of the counties had little estimated change in urban land uses in either period. The performance of counties between the periods from 1970 to 1990 and 1990 to 2000 was generally mixed: Counties that added high and low density land use at the fastest rates from 1970 to 1990 did not all continue the trend through 2000 (see Tables 6.1 and 6.2). Only Lee and Palm Beach counties

Table 6.2   Low density land use rankings

| *Top 10 1970 to 1990 Annual Low Density Change* | | | | *Top 10 1990 to 2000 Annual Low Density Change* | | | | |
COUNTY	70–90 Low	70–90 Rank	90–00 Low	90–00 Rank	COUNTY	70–90 Low	70–90 Rank	90–00 Low	90–00 Rank
Okeechobee	250.1%	1	0.6%	31	Walton	-4.9%	41	1732.4%	1
Monroe	135.5%	2	0.0%	34	Hardee	6.0%	17	144.5%	2
Hernando	84.4%	3	1.6%	25	Flagler	0.0%	33	30.9%	3
Collier	36.8%	4	2.7%	18	Wakulla	0.0%	33	24.2%	4
Pasco	20.0%	5	3.0%	15	Nassau	0.0%	33	12.7%	5
Martin	18.4%	6	0.3%	33	Sumter	3.5%	22	10.9%	6
St. Lucie	18.0%	7	1.6%	25	Alachua	7.5%	14	7.1%	7
Lee	15.7%	8	2.9%	16	Marion	13.6%	9	6.3%	8
Marion	13.6%	9	6.3%	8	Hendry	-1.3%	37	6.1%	9
Charlotte	10.3%	10	3.1%	14	St. Johns	7.9%	12	4.9%	10

Table 6.3   **Florida metro areas—1970 to 1990 high density change**

METRO	70–90 High	90–00 High	Regional	Subregional	SA*	SR*
			Containment Type			
Sarasota-Bradenton	29.9%	0.7%	X		X	
Fort Walton Beach	15.5%	1.4%		X		X
Fort Myers-Cape Coral	14.1%	5.8%		X	X	
Orlando	10.8%	3.4%		X	X	
Tampa-St. Petersburg-Clearwater	9.8%	1.4%		X	X	
West Palm Beach-Boca Raton	6.8%	6.6%	X		X	
Lakeland-Winter Haven	5.0%	0.0%				
Fort Lauderdale	4.6%	2.2%	X		X	
Daytona Beach	3.0%	4.4%				
Gainesville	2.6%	3.4%	X		X	
Miami	1.4%	0.1%	X		X	
Melbourne-Titusville-Palm Bay	0.9%	-1.2%	X		X	
Jacksonville	0.0%	1.7%		X	X	
Tallahassee	0.0%	20.0%		X	X	
Fort Pierce-Port St. Lucie	0.0%	9.9%	X		X	
Naples	0.0%	12.5%	X		X	
Ocala	0.0%	0.0%	X		X	
Punta Gorda	0.0%	10.0%				
Pensacola	0.0%	-1.7%			X	
ALL	4.1%	2.1%				

SA = strong accommodating, SR = strong restrictive (see Nelson and Dawkins, 2004)

were among the top counties for increasing high density land area during the periods from 1970 to 1990 and 1990 to 2000. On the other hand, only Marion County was among the leaders for increases in low density land uses during both time periods.

Summary of Metro Area Trends

From 1970 to 1990, eight of 19 metropolitan areas in Florida experienced higher rates of high density development compared to low density development (see Tables 6.3 and 6.4). On average, Florida metros added more high density area annually compared to low density development from 1970 to 1990 (4.1 percent versus 3.6 percent). From 1990 to 2000, high and low density development occurred at approximately the same annual rates (2.1 percent and 2.0 percent, respectively). This suggests that development during the 30-year period (1970 to 2000) was only slightly more likely to occur at urban densities, but also that the rates of high and low density development declined by about one-half from 1970 to 1990 and 1990 to 2000.

There was no evident relationship between the rates and types of growth and the types of existing growth management policies for each of the metros. We compared urban containment programs in terms of having regional (i.e., metro-wide) or subregional (i.e., county or local) focus. Regional programs are intended to

Table 6.4   Florida metro areas—1970 to 1990 low density change

METRO	70–90 Low	90–00 Low	Containment Type Regional	Subregional	SA*	SR*
Naples	35.1%	2.8%	X		X	
Fort Pierce-Port St. Lucie	17.9%	1.2%	X		X	
Fort Myers-Cape Coral	16.2%	2.9%		X	X	
Ocala	13.4%	6.1%	X		X	
Punta Gorda	10.8%	2.7%				
Daytona Beach	10.5%	2.3%				
Gainesville	7.6%	7.3%	X		X	
Tallahassee	7.3%	1.3%		X	X	
Tampa-St. Petersburg-Clearwater	5.6%	1.9%		X	X	
Jacksonville	5.5%	1.7%		X	X	
Melbourne-Titusville-Palm Bay	4.5%	1.7%	X		X	
Lakeland-Winter Haven	3.5%	1.5%				
Sarasota-Bradenton	2.8%	2.4%	X		X	
Orlando	2.0%	0.8%		X	X	
Pensacola	1.3%	2.7%				
West Palm Beach-Boca Raton	0.9%	1.2%	X		X	
Fort Walton Beach	0.4%	4.5%		X		X
Fort Lauderdale	-0.9%	2.2%	X		X	
Miami	-1.4%	0.0%	X		X	
ALL	3.6%	2.0%				

SA = strong accommodating, SR = strong restrictive (see Nelson and Dawkins, 2004)

be more geographically comprehensive compared to more localized programs (see Nelson and Dawkins, 2004, for more detail on program types). Neither regional nor subregional policies distinguished metro growth patterns. In addition, the perceived strength of urban containment programs (accommodating versus restrictive) also did not distinguish metros, as there was only one metro (Fort Walton Beach) with a strong-restrictive program. That metropolitan area, however, experienced relatively higher rates of high density growth and almost the lowest rates of low density growth, suggesting the impact of a strong-restrictive program on promoting higher density development and discouraging low density growth.

The rates and types of growth for the 19 Florida metros were also compared to the patterns of growth experienced by 46 large metros across the United States. As might be expected, several Florida metros experienced rapid urban development equivalent to some of the fastest growing large metros around the nation. In fact, the Naples metro ranked second behind booming Las Vegas in terms of urban development and land consumption. Others ranking very high included Fort Myers-Cape Coral, Sarasota-Bradenton, Fort Pierce-Port St. Lucie, and Tallahassee. On the opposite end of the spectrum, Miami and Pensacola ranked at the bottom with Northeastern metros such as Rochester, Providence, Hartford, Buffalo, and Pittsburgh (see Table 6.5).

Table 6.5   Florida metro comparisons to other US metros (average annual change)

Metropolitan Area*	1970–1990	1990–2000	1970–1990	1990–2000
	High Density		Low Density	
Las Vegas-Paradise, NV	15.7%	48.7%	9.1%	6.5%
NAPLES, FL	**0.0%**	**12.5%**	**35.1%**	**2.8%**
FORT MYERS-CAPE CORAL, FL	**14.1%**	**5.8%**	**16.2%**	**2.9%**
Riverside-San Bernardino-Ontario, CA	32.9%	4.1%	1.6%	0.0%
SARASOTA-BRADENTON, FL	**29.9%**	**0.7%**	**2.8%**	**2.4%**
FORT PIERCE-PORT ST. LUCIE, FL	**0.0%**	**9.9%**	**17.9%**	**1.2%**
TALLAHASSEE, FL	**0.0%**	**20.0%**	**7.3%**	**1.3%**
Phoenix-Mesa-Scottsdale, AZ	6.7%	6.6%	6.9%	7.4%
PUNTA GORDA, FL	**0.0%**	**10.0%**	**10.8%**	**2.7%**
Sacramento--Arden-Arcade--Roseville, CA	18.8%	2.8%	1.5%	0.2%
FORT WALTON BEACH, FL	**15.5%**	**1.4%**	**0.4%**	**4.5%**
GAINESVILLE, FL	**2.6%**	**3.4%**	**7.6%**	**7.3%**
Denver-Aurora, CO	4.8%	4.8%	8.1%	3.1%
DAYTONA BEACH, FL	**3.0%**	**4.4%**	**10.5%**	**2.3%**
OCALA, FL	**0.0%**	**0.0%**	**13.4%**	**6.1%**
TAMPA-ST. PETE-CLEARWATER, FL	**9.8%**	**1.4%**	**5.6%**	**1.9%**
Austin-Round Rock, TX	6.0%	4.5%	4.7%	2.9%
Atlanta-Sandy Springs-Marietta, GA	1.4%	5.6%	7.6%	3.6%
ORLANDO, FL	**10.8%**	**3.4%**	**2.0%**	**0.8%**
WEST PALM BCH-BOCA RATON, FL	**6.8%**	**6.6%**	**0.9%**	**1.2%**
Seattle-Tacoma-Bellevue, WA	3.6%	3.2%	4.4%	3.3%
San Diego-Carlsbad-San Marcos, CA	7.6%	2.4%	2.6%	-0.3%
Portland-Vancouver-Beaverton, OR-WA	3.3%	5.6%	2.0%	1.0%
San Antonio, TX	3.1%	3.9%	3.0%	1.8%
Dallas-Fort Worth-Arlington, TX	4.3%	3.6%	2.3%	1.3%
Houston-Baytown-Sugar Land, TX	3.2%	3.1%	4.0%	0.7%
Salt Lake City, UT	5.9%	4.1%	0.6%	0.1%
Washington-Arlgtn-Alxnd, DC-VA-MD-WV	2.8%	2.7%	3.7%	1.0%
LAKELAND-WINTER HAVEN, FL	**5.0%**	**0.0%**	**3.5%**	**1.5%**
San Jose-Sunnyvale-Santa Clara, CA	5.2%	3.7%	0.9%	-0.7%
JACKSONVILLE, FL	**0.0%**	**1.7%**	**5.5%**	**1.7%**
FORT LAUDERDALE, FL	**4.6%**	**2.2%**	**-0.9%**	**2.2%**
Nashville-Davidson--Murfreesboro, TN	0.5%	2.6%	2.9%	1.8%
Charlotte-Gastonia-Concord, NC-SC	0.4%	2.8%	2.6%	1.9%
New Orleans-Metairie-Kenner, LA	3.8%	0.4%	1.7%	0.7%
Minneapolis-St. Paul-Bloomington, MN-WI	1.2%	0.7%	2.7%	1.6%
Virginia Bch-Nrflk-Newport News, VA-NC	2.3%	1.7%	1.6%	0.5%
MELBOURNE-TITUS.-PALM BAY, FL	**0.9%**	**-1.2%**	**4.5%**	**1.7%**
Baltimore-Towson, MD	1.6%	1.5%	1.7%	0.6%
Richmond, VA	0.3%	1.0%	2.5%	1.7%
Los Angeles-Long Beach-Santa Ana, CA	1.6%	1.3%	2.0%	0.0%
Oklahoma City, OK	1.0%	2.3%	1.2%	0.3%
Kansas City, MO-KS	-0.3%	0.7%	2.6%	1.6%
Columbus, OH	1.6%	1.2%	0.9%	0.7%
Detroit-Warren-Livonia, MI	0.6%	0.4%	2.0%	1.0%
Memphis, TN-MS-AR	-0.9%	-1.6%	3.4%	2.5%
Chicago-Naperville-Joliet, IL-IN-WI	2.4%	-0.1%	0.1%	0.8%

Table 6.5   Continued

Metropolitan Area*	1970–1990 High Density	1990–2000	1970–1990 Low Density	1990–2000
Cincinnati-Middletown, OH-KY-IN	0.5%	0.0%	1.4%	0.9%
San Francisco-Oakland-Fremont, CA	2.1%	1.7%	-0.3%	-0.8%
Indianapolis, IN	0.1%	1.4%	0.9%	0.5%
Boston-Cambridge-Quincy, MA-NH	0.2%	0.3%	1.7%	0.3%
PENSACOLA, FL	**0.0%**	**-1.7%**	**1.3%**	**2.7%**
Louisville, KY-IN	-0.3%	0.1%	1.2%	1.3%
New York-Newark-Edison, NY-NJ-PA	0.4%	0.6%	0.9%	0.1%
St. Louis, MO-IL	-0.5%	-0.4%	1.5%	1.1%
Milwaukee-Waukesha-West Allis, WI	0.0%	0.5%	0.8%	0.3%
Birmingham-Hoover, AL	-1.9%	-1.6%	3.3%	1.7%
Cleveland-Elyria-Mentor, OH	-0.2%	0.0%	0.9%	0.6%
Phil-Camden-Wilmington, PA-NJ-DE-MD	-0.1%	0.2%	0.6%	0.3%
Rochester, NY	0.5%	-0.4%	0.6%	0.0%
Providence-New Bedford-Fall River, RI-MA	0.0%	-0.2%	0.5%	0.3%
MIAMI, FL	**1.4%**	**0.1%**	**-1.4%**	**0.0%**
Hartford-West Hartford-East Hartford, CT	-0.6%	-0.6%	0.9%	0.3%
Buffalo-Cheektowaga-Tonawanda, NY	-0.4%	-0.8%	0.1%	-0.2%
Pittsburgh, PA	-1.6%	-1.2%	0.2%	0.1%

* Ranked by overall growth from 1970 to 2000.

Conclusions and Implications

This chapter briefly summarizes urban development trends from 1970 to 2000 for the state of Florida. Although the analysis is primarily descriptive, it does highlight the trends in population density and land consumption patterns as a test for compact development outcomes. Urban, suburban, exurban, and rural density classifications were generated using surface interpolation methods within GIS. Estimates of geographic area within each class were compared over time to show the location and extent of development, with estimates broken out by county and metropolitan area to provide additional detail at sub-state level geography. While modeling errors are inevitable, it is suspected that errors were consistent across the state, thus making county and metro level comparisons reasonable. Further empirical testing is needed to find the best fitting surface models, which could produce more accurate estimates of historic land use consumption patterns. However, it is expected that the overall results and trends reported here will likely be unchanged.

Significant amounts of population growth from 1970 to 2000 were expected to be reflected in land consumption patterns across Florida counties. Strong, positive correlations between proportional increases in population sizes and low density land use development should indicate steady outward expansion. Conversely, negative (or no) correlations might indicate densification in urban or suburban areas, rather than in exurban or rural densities. From 1970 to 1990, the correlations (Pearson coefficients) between percent population change for counties and percent high density residential development and between percent population change and percent low density development were not statistically significant. From 1990 to 2000, the correlation for percent population change for counties and percent high density development

Table 6.6   Population and development change

County	Pop. Change 1970–1990	Pop. Change 1990–2000	High Density 1970–1990	High Density 1990–2000	Low Density 1970–1990	Low Density 1990–2000
Alachua	3.7%	2.0%	2.6%	3.4%	7.5%	7.1%
Baker	5.0%	2.0%	0.0%	0.0%	-1.3%	-1.4%
Bay	3.4%	1.7%	0.0%	0.0%	3.0%	2.2%
Bradford	2.7%	1.6%	0.0%	0.0%	0.0%	0.0%
Brevard	3.7%	1.9%	1.2%	-1.3%	4.4%	1.6%
Broward	5.1%	2.9%	4.9%	2.2%	-0.9%	2.3%
Calhoun	2.2%	1.8%	0.0%	0.0%	0.0%	0.0%
Charlotte	15.1%	2.8%	0.0%	10.0%	10.3%	3.1%
Citrus	19.4%	2.6%	0.0%	0.0%	0.0%	2.3%
Clay	11.5%	3.3%	0.0%	0.0%	6.8%	2.8%
Collier	15.0%	6.5%	0.0%	13.6%	36.8%	2.7%
Columbia	3.4%	3.3%	-5.0%	0.0%	-0.1%	0.6%
DeSoto	4.1%	3.5%	1.4%	0.1%	-1.4%	0.0%
Dixie	4.7%	3.1%	0.0%	0.0%	7.6%	1.9%
Duval	1.4%	1.6%	0.0%	0.0%	0.0%	0.0%
Escambia	1.4%	1.2%	0.0%	1.7%	3.9%	-0.3%
Flagler	27.2%	7.4%	0.0%	-1.5%	1.6%	3.0%
Franklin	1.4%	1.0%	0.0%	0.0%	0.0%	30.9%
Gadsden	0.3%	1.0%	0.0%	0.0%	0.0%	0.0%
Gilchrist	8.6%	4.9%	0.0%	0.0%	0.0%	0.0%
Glades	5.3%	3.9%	0.0%	0.0%	0.0%	0.0%
Gulf	0.7%	2.7%	0.0%	0.0%	2.5%	3.7%
Hamilton	2.0%	2.2%	0.0%	0.0%	0.0%	0.0%
Hardee	1.6%	3.8%	0.0%	0.0%	0.0%	0.0%
Hendry	5.9%	4.1%	0.0%	0.0%	6.0%	144.5%
Hernando	24.7%	2.9%	0.0%	0.0%	-1.3%	6.1%
Highlands	6.6%	2.8%	0.0%	0.0%	84.4%	1.6%
Hillsborough	3.5%	2.0%	0.0%	0.0%	1.7%	1.0%
Holmes	2.4%	1.8%	4.5%	2.3%	4.5%	1.7%
Indian River	7.5%	2.5%	0.0%	0.0%	0.0%	0.0%
Jackson	1.0%	1.3%	0.0%	0.0%	6.0%	3.7%
Jefferson	1.4%	1.4%	0.0%	0.0%	0.0%	0.0%
Lafayette	4.7%	2.6%	0.0%	0.0%	0.0%	0.0%
Lake	6.0%	3.8%	0.0%	0.0%	0.0%	0.0%
Lee	10.9%	3.2%	0.0%	0.0%	3.8%	1.8%
Leon	4.3%	2.4%	13.6%	5.5%	15.7%	2.9%
Levy	5.2%	3.3%	0.0%	20.0%	7.3%	1.3%
Liberty	3.2%	2.6%	0.0%	0.0%	0.0%	0.0%
Madison	1.1%	1.3%	0.0%	0.0%	0.0%	0.0%
Manatee	5.9%	2.5%	0.0%	0.0%	0.0%	0.0%
Marion	9.1%	3.3%	20.4%	1.0%	4.4%	3.5%
Martin	13.0%	2.6%	0.0%	0.0%	13.6%	6.3%
Miami-Dade	2.6%	1.6%	0.0%	9.2%	18.4%	0.3%
Monroe	2.4%	0.2%	0.0%	0.0%	135.5%	0.0%
Nassau	5.7%	3.1%	0.0%	0.0%	0.0%	12.7%
Okaloosa	3.2%	1.9%	21.5%	1.6%	0.4%	4.6%
Okeechobee	8.2%	2.1%	0.0%	0.0%	250.1%	0.6%
Orange	4.8%	3.2%	5.7%	2.4%	2.2%	-0.1%

Table 6.6   Continued

County	Pop. Change		High Density		Low Density	
	1970–1990	1990–2000	1970–1990	1990–2000	1970–1990	1990–2000
Osceola	16.3%	6.0%	0.0%	15.0%	-0.8%	0.6%
Palm Beach	7.4%	3.1%	7.7%	7.2%	0.9%	1.2%
Pasco	13.5%	2.3%	0.0%	0.0%	20.0%	3.0%
Pinellas	3.2%	0.8%	9.6%	1.2%	-1.6%	0.3%
Polk	3.9%	1.9%	5.0%	0.0%	3.5%	1.5%
Putnam	4.0%	0.8%	0.0%	0.0%	0.4%	-0.7%
Santa Rosa	5.8%	4.4%	0.0%	0.0%	1.4%	2.5%
Sarasota	6.5%	1.7%	39.8%	0.0%	1.8%	1.8%
Seminole	12.2%	2.7%	0.0%	3.8%	2.8%	1.1%
St. Johns	8.6%	4.7%	-5.0%	0.0%	7.9%	4.9%
St. Lucie	9.8%	2.8%	0.0%	8.7%	18.0%	1.6%
Sumter	5.6%	6.9%	0.0%	0.0%	3.5%	10.9%
Suwannee	3.6%	3.0%	0.0%	0.0%	0.0%	0.0%
Taylor	1.3%	1.3%	0.0%	0.0%	0.0%	0.0%
Union	1.3%	3.1%	0.0%	0.0%	-1.9%	-0.2%
Volusia	5.9%	2.0%	3.1%	4.5%	9.6%	0.6%
Wakulla	6.3%	6.1%	0.0%	0.0%	0.0%	24.3%
Walton	3.6%	4.6%	0.0%	0.0%	-4.9%	1732.4%
Washington	2.4%	2.4%	0.0%	0.0%	0.0%	0.0%

was 0.20 and that for percent population change and percent low density growth was not significant (see Table 6.6). This suggests that despite high rates of population growth from 1970 to 2000, there were no clear trends in how development within Florida counties was absorbed. Only a weak, positive correlation existed between the amount of high density development and the rate of population change. Because there was no detectable trend between high density development and population change from 1970 to 1990, this significant correlation for the post-GM period could suggest that more growth is occurring at urban densities compared to the pre-GM period.

So what does this tell us about the Florida growth management experiment? Is there any evidence that growth management has succeeded in limiting sprawl or promoting compact urban development? A simple comparison of Florida metropolitan areas with other selected US metros (see Table 6.7) shows some interesting results. For high density land use change in Florida metro areas, the annual average rate of increase rose from 2.8 percent in the pre-GM period to 4.2 percent in the post-GM era. Although this makes it appear that the advent of growth management had increased the pace of higher density development, Table 6.7 further shows that non-Florida metropolitan areas also experienced an increase in the rate of high density land use change, from 1.5 percent to 2.7 percent. We cannot conclude, therefore, that the faster pace of urban land growth in Florida was attributable to the 1985 GMA.

On the other hand, it appears that the Growth Management Act may have slowed the rate at which low density land use increased. Table 6.7 shows that the average annual increase in low density land use in Florida metropolitan areas slowed from 3.8 percent in the pre-GM period to 2.5 percent in the post-GM period. At the same

Table 6.7   Florida metro comparisons to other US metros (average annual
              percentage change)

	High Density Land Use Change		Low Density Land Use Change	
	1970–1990	1990–2000	1970–1990	1990–2000
Florida Metros	2.8%	4.2%	3.8%	2.5%
Non-Florida Metros	1.5%	2.7%	1.2%	1.2%

time, there was virtually no change in the rate of increase in low density land use in non-Florida metropolitan areas.

Overall, it appears that growth management has not encouraged Florida's metropolitan areas to increase the rate of development at densities of greater than 3,000 persons per square mile, but that growth management may have been the cause of the decline in the rate of development densities between 300 and 3,000 persons per square mile. At the same time, however, Table 6.7 reveals that Florida's rate of increase in low density land is significantly higher than the average for the 46 non-Florida metropolitan areas (2.5 percent versus 1.2 percent average annual increase, 1990–2000). These findings suggest, therefore, that while Florida's growth management laws may have contributed to a decrease in the rate of low density developments often associated with sprawl, the pace of growth in such densities is still much higher in Florida than in non-Florida metropolitan areas.

Growth management in Florida appears to have had an impact on sprawl, but nevertheless sprawl continues to be of greater significance in Florida's metropolitan areas than in non-Florida metropolitan areas. The growth of sprawl appears slowed, but not stopped, by growth management.

References

Ben-Zadok, E. (2005). Consistency, concurrency and compact development: Three faces of growth management implementation in Florida. *Urban Studies, 42*(12), 2167–2190.

Daniels, T. (1999). *Holding our ground.* Washington, DC: Island Press.

DeGrove, J. M. (1992). *Planning and growth management.* Cambridge, MA: Lincoln Land Institute.

Easley, G. (1992). *Staying inside the lines.* Chicago: American Planning Association.

Lang, M. (1986). Redefining urban and rural for the US census of population: Assessing the need and alternatives approaches. *Urban Geography, 2*, 118–134.

Nelson, A. C. (1992a). Characterizing exurbia. *Journal of Planning Literature, 6*(4), 350–368.

Nelson, A. C. (1992b). Preserving prime farmland in the face of urbanization. *Journal of the American Planning Association, 58*(4), 467–488.

Nelson, A. C. (1999). Comparing states with and without growth management: Analysis based on indicators with policy implications. *Land Use Policy, 16*, 121–127.

Nelson, A. C., and Dawkins, C. J. (2002). *Urban containment—American style(s).* Washington, DC: Brookings Institution.

Nelson, A. C., and Dawkins, C. J. (2004). *Urban containment in the United States.* Chicago: American Planning Association.

Nelson, A. C., and Duncan, J. B. (1995). *Growth management principles and practices.* Chicago: American Planning Association.

Nelson, A. C., and Sanchez, T. W. (2005). The effectiveness of urban containment regimes in reducing exurban sprawl. *NSL Network City and Landscape, 160,* 42–47.

Chapter 7

Growth Management and the Spatial Outcome of Regional Development in Florida, 1982–1997

John I. Carruthers, Marlon G. Boarnet and Ralph B. McLaughlin

This chapter evaluates growth management in Florida by examining its relationship to spatial patterns of economic development in the Atlantic Southeast region between 1982 and 1997. Specifically, the analysis uses a regional adjustment model—a dynamic, two-equation structural model that accounts for interaction between population and employment in the growth process—to project equilibrium densities of people and jobs county-by-county in Alabama, Florida, Georgia, Mississippi, North Carolina, South Carolina, and Tennessee in 1987, 1992, and 1997. The density forecasts, which express the direction(s) in which the space economy was "pushing" land use change in the region during the preceding five-year timeframes, are compared to actual outcomes as an *ex post* method of identifying the impacts of state policy. The study period covers three critical stages in the evolution of Florida's growth management program: the trailing years of the original Local Government Comprehensive Planning Act of 1975, the adoption of the 1985 Growth Management Act (GMA), and an ensuing 12 years of implementation and revision of that legislation. Each of these stages fits neatly within the timeframe of the analysis, allowing a number of tentative connections to be made between the state's planning framework and growth patterns. How do actual outcomes compare to those projected by regional adjustment models? Is there evidence of incremental change over the three five-year segments that may be attributed to policy differences? And, finally, what do the results imply about the effectiveness of Florida's growth management program?

A key point of departure for addressing these questions is Boarnet's (1994a) finding that local land use regulations may disrupt the equilibrium process that allows the spatial distribution of population and employment growth to be jointly determined. At issue in the present analysis is the possibility that, by interfering with the market tendencies that attract people and jobs to one another, centralized land use planning may inadvertently prevent the optimal organization of residential and nonresidential activities from being achieved. This framework explicitly recognizes growth management for what it is: a constraint on the land market that may produce more desirable outcomes by correcting for market failures (Brueckner, 2000) or less desirable outcomes, such as urban sprawl, by contributing to the proliferation of regulatory failure (Ulfarsson and Carruthers, 2006). Alternatively, centralized planning may work with the equilibrium process, producing outcomes that are more consistent with market tendencies by evening out the underlying regulatory landscape. Ideally, if it has been effective, Florida's GMA will be observed to promote higher densities (less urban sprawl), while at the same time coordinating development in a way that it stays on its projected equilibrium path.

A direct emphasis is placed on the interaction between population and employment because, as a practical matter, knowing the relative distribution of the two is at least as important to policy makers as knowing the overall outcome of land development. Moreover, the collocation of people and jobs is fundamental to contemporary regional development (Carlino and Mills, 1987; Clark and Murphy, 1996; Mulligan, Vias, and Glavac, 1999), meaning policies that interfere with that process must ultimately be judged counterproductive. In sum, from both practical and theoretical perspectives, it is appropriate to evaluate Florida's GMA on the basis of how it affects the nature of the growth process itself, along with land development patterns that it produces.

The remainder of the chapter is organized as follows. First, a background discussion explains the concept of spatial equilibrium, describes how regional adjustment models emulate it, and locates the analysis within a timeline of the evolution of growth management in Florida. Second, the empirical analysis uses a land use based regional adjustment model (Carruthers and Mulligan, 2006, in press; Carruthers and Vias, 2005) to project the spatial outcome of economic development in the Atlantic Southeast region over the 1982–1987, 1987–1992, and 1992–1997 timeframes. The estimation results are then compared to actual outcomes and relevant contextual information to evaluate how Florida's land use planning framework may have influenced the growth process. Last, the summary and conclusion section briefly revisits the research findings, outlines their policy implications, and suggests directions for future work on the relationship between state-based land use planning and the product of regional development.

Contemporary Regional Development, Spatial Equilibrium, and Regional Adjustment Models

Population and employment interact in the growth process as a result of people following jobs and jobs following people into and within regions, creating a so-called chicken-or-egg system of economic development (Muth, 1971). Generally speaking, population (employment) change between two points in time is a function of the level of employment (population) at the end of the time period, the level of population (employment) at the beginning of the time period, and other relevant initial conditions. Although there is some evidence that this type of growth process was at least sporadically at work in the United States as early as the 1950s—the process was first identified by Borts and Stein (1964)—it has more recently evolved into a dominant mechanism as a result of the far-reaching economic and demographic shifts that have occurred since that time. While people still move from place to place in search of jobs, it is increasingly common for residential consumer preferences, or quality of life, to play a role (see Frey, 1993). The result is a development process that is driven by a combination of factors affecting: (a) where firms want to locate and (b) where people want to live. The interaction between the two mechanisms stems from the fact that, just as people commonly move to places offering the best possible job market, firms increasingly move to places offering the best possible labor pool. On the latter point, it is instructive to note that one of the best ways for some companies to attract and retain talented workers is to locate in an area offering a high quality of life, either in the form of natural or locally-produced amenities (Carruthers and Mulligan, 2006). In this sense, economic development programs, which tend to focus heavily on the creation of work, may address only half of the process; and other forms of policy aimed at making and/or keeping regions desirable places to live, such as growth management, may be necessary to ensure sustained development.

The interaction between population and employment growth is best viewed from a perspective that characterizes the space economy as perpetually adjusting toward some unknown, optimal distribution of economic activity. If this steady state, or point of spatial equilibrium, were ever reached, both sets of economic actors, individuals and firms, would be indifferent among locations. More specifically, utility-maximizing individuals would be distributed in a way that allowed them to optimize on their consumption of goods and services, proximity to work, and access to non-market amenities, such as environmental attractions. At the same time, profit-maximizing firms would be distributed in a way that allowed them to optimize on agglomeration economies, wage differentials, labor supplies, regional comparative advantage, and other pertinent factors. Obviously, while this situation is attainable theoretically, it is not in practice—particularly because any change in the system, no matter how large or small, alters the end, "targeted" state. Thus many researchers view the space economy as being in a state of flux, continually adjusting toward the optimal organization of population and employment. Although it certainly involves theoretical abstraction, this approach provides a powerful analytical framework for understanding how contemporary regional development works.

Regional adjustment models are tools for simulating and, ultimately, empirically observing the growth process described in the preceding paragraphs. In applied terms, they are composed of a system of two autoregressive simultaneous equations that rely on the assumption, as outlined above, that population (employment) growth depends on a set of initial conditions, including its past level, and present employment (population). The core relationships that regional adjustment models emulate are as follows:

$$\Delta pd_i = f(pd_{it\text{-}}, ed_{it}, \boldsymbol{ic}_{pd}, \boldsymbol{u}_{pdit}), \tag{1a}$$
$$\Delta ed_i = f(pd_{it}, ed_{it\text{-}}, \boldsymbol{ic}_{ed}, \boldsymbol{u}_{edit}). \tag{1b}$$

In these equations, Δpd_i and Δed_i represent changes in population and employment densities in area i over a specified time period; pd and ed represent population and employment densities at the beginning ($t\text{-}$) and end (t) of that time period; $\boldsymbol{ic}_{pd}$ and $\boldsymbol{ic}_{ed}$ represent vectors of initial conditions (at $t\text{-}$) that explain changes in population density and employment density, respectively; and $\boldsymbol{u}_{it}$ represents a vector of unobserved effects that is different for each equation. Densities, rather than levels, are used to correct for heteroskedasticity, which, in this context, arises from inevitable variation in the size of the spatial units involved. Different versions of this modeling framework have been used for analyses of growth at the metropolitan (Boarnet 1994a, 1994b; Boarnet, Chalermpong, and Geho, in press; Bollinger and Ihlanfeldt, 1997; Steinnes and Fisher, 1974), sub-national (Carruthers and Vias, 2005; Duffy-Deno 1998; Henry, Barkley, and Bao, 1997; Henry, Schmitt, Kristensen, Barkley, and Bao, 1999; Henry, Schmitt, and Piguet, 2001; Steinnes, 1977; Vias, 1999; Vias and Mulligan, 1999), and national (Carlino and Mills, 1987; Carruthers and Mulligan, 2006, in press; Clark and Murphy, 1996; Deller, Tsai, Marcouiller, and English, 2001; Leichenko, 2001; Mills and Lubuele, 1995; Mulligan et al., 1999) scales. Although each of these studies has a different purpose, as a group, they broadly confirm that the spatial distribution of population and employment growth is jointly determined.[1]

Before moving on, it is necessary to frame the timeline over which Florida's growth management program has evolved with respect to the present analysis. As discussed in the introduction, the timeframe of the research presented in this chapter, 1982 to 1997, covers the final years of the original 1975 Local Government Comprehensive Planning Act, the adoption of the 1985 GMA, and an additional 12 years after the more rigorous revised program was put into effect. Despite the extensive effort put toward growth management in Florida, previous evaluations have revealed mixed results. For example, in a comparative analysis involving 14 states, including five with state-based land use policy frameworks, Carruthers (2002) found that Florida's planning mandate may actually contribute to increased land consumption. This result was consistent with earlier research pointing to Florida's concurrency requirements, combined with a failure to increase road capacity, as a cause of leapfrog development and urban sprawl (Blanco, 1998; Nicholas and Steiner, 2000). It might also be a consequence of the widespread variation in the degree to which local governments follow state mandated guidelines—and, ultimately, of selective use of appropriate coercive mechanisms (Deyle and Smith, 1998). Last, the middle five years of the timeframe, 1987 to 1992, were really a period of plan development, adoption, and review and not necessarily plan execution (Ben-Zadok, 2005). For these reasons, the questions motivating the present analysis remain wide open.

Empirical Analysis

Model Specification and Estimation Results

Equation set (1) was operationalized for the purposes of the present analysis via an econometric model with the following functional form:

$$\ln(pd_{it}/pd_{i-5}) = \Delta_0 + \Delta_1 \ln(pd_{it-5}) + \Delta_2 \ln(ed_{it}) + \Delta_3 ic_{pd} + \Delta_{pdit}, \quad (2a)$$

$$\ln(ed_{it}/ed_{i-5}) = \Delta_0 + \Delta_1 \ln(pd_{it}) + \Delta_2 \ln(ed_{it-5}) + \Delta_3 ic_{ed} + \Delta_{edit}. \quad (2b)$$

---

1 For detailed derivations of various types of regional adjustment models, see Steinnes and Fisher (1974), Carlino and Mills (1987), Boarnet (1994a, 1994b), Clark and Murphy (1996), Mulligan et al. (1999), Deller et al. (2001), and Carruthers and Mulligan (2006).

Changes in the dependent variables were expressed as ratios, rather than first differences, so that their natural log (ln) could be taken. Here, $t-5$ and t denoted the beginning and end, respectively, of the 1982–1987, 1987–1992, and 1992–1997 time periods; pd and ed represented the average population and employment per acre of developed land in county i; ic represented a vector of initial conditions, including fixed effects describing county typology (urban, suburban, exurban, and rural) and the base year of the observation (1982, 1987, and 1992), plus additional continuous variables that were different in each equation; Δ_0 and Δ_0 represented the equations' intercepts; Δ_1, Δ_2, Δ_1, and Δ_2 represented estimable parameters on the endogenous (pd_{it} and ed_{it}) and lagged (pd_{it-5} and ed_{it-5}) adjustment variables; Δ_3 and Δ_3 represented vectors of estimable parameters on the initial conditions; and $\Delta \sim (N, \Delta^2)$ represented the stochastic error term. Note that the initial conditions that are continuous variables were in natural log form and that urban and 1982 were excluded from the fixed effects groups in order to avoid perfect multicollinearity with the overall intercepts. The result was a land use based regional adjustment model similar to the ones developed by Carruthers and Mulligan (2006, in press) and Carruthers and Vias (2005).

The empirical model was specified with data collected for the entire Atlantic Southeast region, which contains Alabama, Florida, Georgia, Mississippi, North Carolina, South Carolina, and Tennessee, during appropriate years. This data set formed a panel of 616 counties observed at three points in time for a total sample size of 1,848. Because the model contained endogenous variables (pd_{it} and ed_{it}), ordinary least squares (OLS) was not a viable estimator. Instead, the model was estimated with two-stage generalized least squares (2SGLS), an estimator that handled the endogeneity and took advantage of the data set's panel structure to allow a different residual variance for each cross-section, or county. This latter feature did not affect the parameter estimates but, because it corrects the standard errors for cross-section specific heteroskedasticity, it significantly boosted the t-statistics.

The data used to specify the model came from several sources. First, population, employment, and income all come from the Bureau of Economic Analysis' *Regional Economic Information System* (available online at *http://bea.gov/bea/regional/docs/reis2004dvd.asp*). Second, the land use data used in the denominator of the density calculations came from the USDA's *National Resources Inventory*, which provides estimates of the number of acres of land in major land use categories throughout the continental United States by county in 1982, 1987, 1992, and 1997; here, the "urban" and "rural transportation" categories are summed to yield a single "developed" category, so the area measured approximates the area actually occupied by economic activity and, therefore, changes through time as counties grow. The same data were also used to calculate the percent of each county's total land area that was developed. Finally, the natural amenity index, produced by the USDA's *Economic Research Service*, is a composite index that accounts for such factors as the January temperature and sunlight, July temperature and humidity, topography, and water in each county of the continental United States (available online at *http://www.ers.usda.gov/Data/Natural/Amenities*). Specifically these data consist of (which is) z-scores, which express the counties' relative amenity levels. The same data set also includes the Beale code, used to form the county typology (see Carruthers and Vias,

Table 7.1   2SGLS estimates of Atlantic Southeast regional adjustment model

	ln Δ Population Density		ln Δ Employment Density	
	β	t-statistic	γ	t-statistic
Constant	0.6409	6.6458	-0.1233	-10.1974
Adjustment Variables				
ln Population Density t-5	-0.1923	-15.6821	-	-
ln Employment Density t	0.1061	10.8316	-	-
ln Employment Density t-5	-	-	-0.3099	-16.5895
ln Population Density t	-	-	0.3974	14.8798
Initial Conditions				
ln Per Capita Income t-5	-0.0567	-5.9300	-	-
ln Amenity Score[2]	0.0027	5.9125	-	-
ln % Developed t-5	-	-	0.0836	15.2788
County Type				
Suburban	-0.0470	-13.9417	0.0576	7.7096
Exurban	-0.0505	-14.0377	0.0704	8.9882
Rural	-0.0493	-9.4404	0.0711	6.9458
Temporal Effects				
1992	-0.0097	-3.1675	-0.0127	-3.3224
1997	-0.0201	-6.9604	-0.0117	-3.5970
n	1,848		1,848	
Adjusted R^2	0.57		0.51	
λ_1			0.93646	
λ_2			0.88908	
Estimated Steady State Fractions—Population : Employment			2.69 : 1.00	

2005). These three data sources provided all of the information necessary to estimate equation set (2).

The estimation results are shown in Table 7.1. All of the variables were significant at well over a 99 percent confidence interval, and the adjusted R^2s (0.57 and 0.51 for the population and employment equations, respectively) show that this simple model explains over half of the place-to-place variation in the density of people and jobs in the Atlantic Southeast. All four of the adjustment variables carried their expected signs, confirming that population and employment growth drive one another in the region. The parameters on the lagged variables were negative, because denser regions realize smaller relative changes. Further, the two tests documented at the bottom of the table illustrate that the model is dynamically stable and fractionally reasonable. The system's characteristic roots, the Δs, were less than one and, at the implied equilibrium, the ratio of population to employment density was 2.69:1, meaning that about 40 percent of the total population is employed in the steady state. Both of these conditions are necessary to the viability of regional adjustment models because, if they do not settle on a steady state, they produce multiple and, ultimately, nonsensical solutions; similarly, the implied population-to-employment has to be reasonable for their results to be believable (Mulligan et al., 1999). In short, all indications are that the model shown in Table 7.1 does a good job of revealing how the adjustment process described in the background section plays out in the Atlantic Southeast region. The actual spatial outcomes projected by the model are examined in detail below.

Working through the remaining explanatory variables in the population equation: (a) the two initial conditions, per capita income and the natural amenity index, show that, other things being equal, higher incomes led to more land consumption and desirable environmental features led to spatial clustering; (b) the county typology shows that suburban, exurban, and rural counties all realized lesser changes in density relative to urban counties, but not much variation relative to one another; and (c) the temporal fixed effects show that population densities declined by increasingly greater margins in the Atlantic Southeast over the 15-year timeframe. In the employment equation: (a) the initial condition, percent developed (selected on the basis of Carruthers and Mulligan, in press), shows that counties closer to being built-out realized greater increases in density; (b) the county typology shows that suburban, exurban, and rural counties realized greater changes in density relative to urban counties, but not much variation relative to one another; and (c) the temporal fixed effects show that employment densities have declined in the Atlantic Southeast since 1982.

As a set, the estimation results illustrate that growth in the Atlantic Southeast is a function of both opportunity and preference, meaning that it occurs due to economic forces and people's quality of life choices. The latter aspect, which results in jobs following people into and/or within the region, is important to recognize because it suggests that measures (like growth management) aimed at preserving local allure may be fundamental to sustained economic development. This point is reinforced by the spatial clustering, or increase in population density, associated with the USDA's natural amenity index. Specifically, even within the microcosm of this uniformly warm, lushly vegetated region, place-to-place variation in local conditions matters— so policy makers should carefully consider the benefits of protecting natural features such as Florida's Everglades. Finally, the results associated with the county typology reveal (in general) that, even as urban areas have experienced gains in *population* density relative to their suburban, exurban, and rural counterparts, *employment* density has gone the other way. That is, suburban, exurban, and rural counties have experienced increases in employment density relative to urban counties, probably due to the continuing decentralization of jobs that has swept the United States since the 1970s. Looking forward, the remainder of the chapter focuses on how the projected population and employment densities (at equilibrium) compare to actual outcomes.

Analysis of Spatial Outcomes

Equilibrium population and employment densities were calculated by reading the estimated adjustment speeds, Δ_1 and Δ_2, and applying them to observed densities at times t and t-5.[2] Results are shown for the entire Atlantic Southeast region

---

2   See Mulligan et al. (1999) or Carruthers and Mulligan (2006) for step-by-step explanations of how to do this.

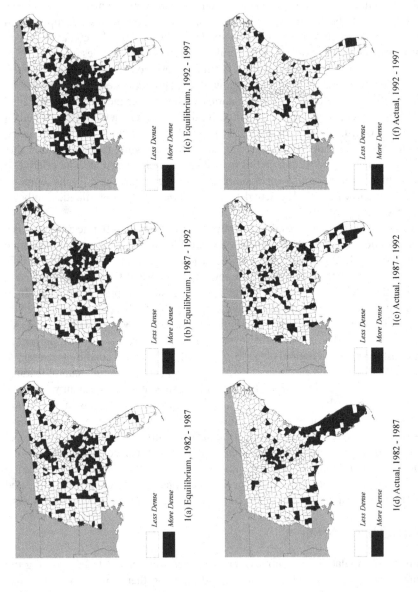

Figure 7.1   Population density change: Equilibrium vs. actual

1(a) Equilibrium, 1982 - 1987

1(b) Equilibrium, 1987 - 1992

1(c) Equilibrium, 1992 - 1997

1(d) Actual, 1982 - 1987

1(e) Actual, 1987 - 1992

1(f) Actual, 1992 - 1997

Less Dense

More Dense

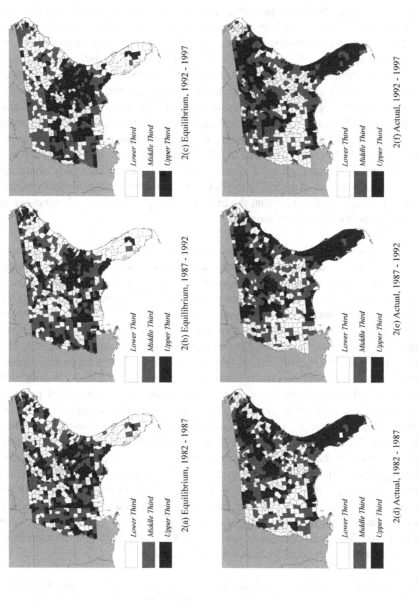

Figure 7.2   Population change: Equilibrium vs. actual

because it is the frame of reference for understanding what happened in Florida. The population results are mapped in Figures 7.1 and 7.2, which show projected (equilibrium) outcomes at the end of the three timeframes across the top, and actual outcomes across the bottom. The latter of the two figures deal with levels, which, in the case of equilibrium projections, were calculated from estimated densities. Figure 7.1 shows counties that were projected to (top row: a–c) and/or did (bottom row: d–f) realize increased densities in dark grey and those that were projected to and/or did realize decreased densities in white. Figure 7.2 shows population change by county, ranked by the lower (white), middle (grey), and upper (dark grey) thirds of the overall distribution; counties that lost population fall in the lower third. Figures 7.3 and 7.4 provide comparable illustrations of projected and actual employment outcomes. Finally, Figure 7.5 compares the top and bottom rows of Figures 7.1 and 7.3, showing whether or not actual population and employment densities moved in the same direction (dark grey) as the equilibrium estimates. The top rows of Figures 7.1 and 7.3 are interpreted as the direction in which the space economy was "pushing" densities, so Figure 7.5 speaks to the extent to which reality complied with market tendencies. Addressing the three questions motivating this analysis, the following paragraphs explore what the five figures imply about the effectiveness of Florida's growth management program.

How do actual outcomes compare to those projected by regional adjustment models? Throughout the Atlantic Southeast, market forces, which likely include the effects of the kind of market failures identified by Brueckner (2000), were pushing population densities in both directions over the 1982–1997 timeframe. The top row of Figure 7.1 exhibits a great deal of spatial variation in the projected direction of land use change, but the bottom row shows just the opposite: Despite being pushed in different directions, the region's population largely grew more spread out. The same is true for employment, except that, by and large, market tendencies veered toward increased density across the entire 15 years. Figures 7.2 and 7.4 show that population and employment levels, meanwhile, followed their projected paths in rural areas of the Atlantic Southeast, but went in essentially the opposite direction in many urban areas. In plain terms, these parts of the region grew much faster than they were projected to, particularly in Florida. Whether population and employment densities moved in the same (converging gap) or opposite (diverging gap) direction as the equilibrium state is shown in Figure 7.5. In the first case, population densities changed as forecast in most of the region, except for Florida and Georgia from 1982–1987 and continued to deviate from the projected path, which was mostly toward greater density, in the latter state over the next ten years. A more even pattern is visible in the employment map, which shows counties that deviated from equilibrium tendencies spread across the entire region.

Is there evidence of incremental change over the three five-year segments that can be attributed to policy differences? The critical turn in Florida's growth management program came in 1985, when the State Comprehensive Plan and the GMA were adopted, but that was followed by a period of gradual implementation in the late 1980s and early 1990s. Perhaps for that reason, Figures 7.1 and 7.3 reveal no easily

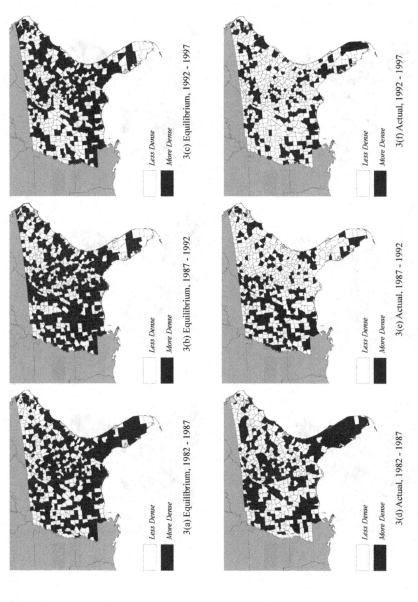

Less Dense
More Dense

3(c) Equilibrium, 1992 - 1997

Less Dense
More Dense

3(f) Actual, 1992 - 1997

Less Dense
More Dense

3(b) Equilibrium, 1987 - 1992

Less Dense
More Dense

3(e) Actual, 1987 - 1992

Less Dense
More Dense

3(a) Equilibrium, 1982 - 1987

Less Dense
More Dense

3(d) Actual, 1982 - 1987

Figure 7.3   Employment density change: Equilibrium vs. actual

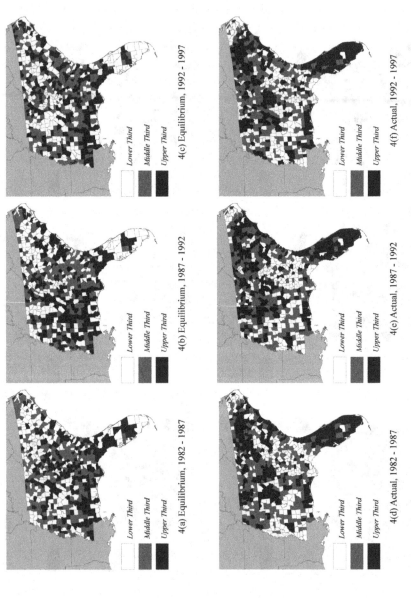

Figure 7.4   Employment change: Equilibrium vs. actual

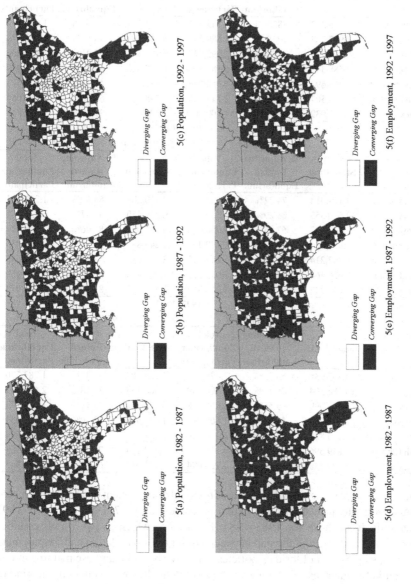

Diverging Gap

Converging Gap

5(c) Population, 1992 - 1997

Diverging Gap

Converging Gap

5(f) Employment, 1992 - 1997

Diverging Gap

Converging Gap

5(b) Population, 1987 - 1992

Diverging Gap

Converging Gap

5(e) Employment, 1987 - 1992

Diverging Gap

Converging Gap

5(a) Population, 1982 - 1987

Diverging Gap

Converging Gap

5(d) Employment, 1982 - 1987

Figure 7.5   Actual vs. equilibrium tendencies

Table 7.2   Equilibrium participants in the Atlantic Southeast

| | 1982–1987 | | | | | |
| | 1987 Population | | | 1987 Employment | | |
	State Total	% Adjusting Toward Equilibrium	Total Equilibrium Participants	State Total	% Adjusting Toward Equilibrium	Total Equilibrium Participants
Alabama	4,015,262	78.28%	3,143,305	1,922,545	56.02%	1,076,972
Florida	11,997,282	29.92%	3,589,193	6,140,169	46.05%	2,827,473
Georgia	6,208,479	33.71%	2,092,921	3,454,605	52.91%	1,827,860
Mississippi	2,588,545	61.15%	1,582,985	1,147,320	60.70%	696,419
North Carolina	6,403,693	76.61%	4,905,763	3,631,326	47.41%	1,721,524
South Carolina	3,380,507	80.92%	2,735,394	1,748,441	50.68%	886,054
Tennessee	4,782,939	88.02%	4,209,807	2,593,048	39.80%	1,031,948
	1987–1992					
	1992 Population			1992 Employment		
	State Total	% Adjusting Toward Equilibrium	Total Equilibrium Participants	State Total	% Adjusting Toward Equilibrium	Total Equilibrium Participants
Alabama	4,154,014	74.22%	3,082,994	2,110,288	58.69%	1,238,615
Florida	13,650,553	83.29%	11,369,765	6,820,450	91.37%	6,231,785
Georgia	6,817,203	57.95%	3,950,458	3,722,082	87.06%	3,240,621
Mississippi	2,623,734	67.82%	1,779,476	1,240,887	75.66%	938,883
North Carolina	6,897,214	72.65%	5,010,492	3,989,062	67.42%	2,689,543
South Carolina	3,620,464	83.48%	3,022,527	1,909,893	80.68%	1,540,969
Tennessee	5,049,742	84.38%	4,260,917	2,854,596	81.49%	2,326,073
	1992–1997					
	1997 Population			1997 Employment		
	State Total	% Adjusting Toward Equilibrium	Total Equilibrium Participants	State Total	% Adjusting Toward Equilibrium	Total Equilibrium Participants
Alabama	4,367,935	56.73%	2,478,020	2,335,282	76.38%	1,783,770
Florida	15,186,304	75.11%	11,406,439	8,068,158	33.18%	2,677,301
Georgia	7,685,099	49.56%	3,808,728	4,476,741	44.19%	1,978,173
Mississippi	2,777,004	57.90%	1,607,998	1,423,918	72.10%	1,026,637
North Carolina	7,656,825	66.83%	5,116,786	4,631,453	50.65%	2,345,694
South Carolina	3,859,696	81.85%	3,159,077	2,154,114	79.46%	1,711,764
Tennessee	5,499,233	89.54%	4,924,094	3,287,912	69.52%	2,285,645

visible difference in the spatial outcome of regional development that can be attributed to this shift: The projections and actual outcomes for 1987–1992 and 1992–1997 differ little from those for 1982–1987. Figure 7.5, which shows which counties were closing and opening the gap between equilibrium population and employment densities, suggests that Florida's policies may have brought land use patterns in line with market tendencies. If so, they would have done this by evening out the kind of patchwork regulatory landscape associated with a purely local planning framework (see Carruthers, 2002). In terms of what the data show, the pattern on the map comparing the projected and actual outcomes of population growth indicates that greater areas of the state have moved toward spatial equilibrium in recent years. The

Table 7.3   Equilibrium participants in the Atlantic Southeast vs. Florida

	Population			Employment		
	Mean	**Median**	**Florida**	**Mean**	**Median**	**Florida**
1982–1987	64.09%	76.61%	29.92%	50.51%	50.68%	46.05%
1987–1992	74.83%	74.22%	83.29%	77.48%	80.68%	91.37%
1992–1997	68.22%	66.83%	75.11%	60.78%	69.52%	33.18%

employment picture is somewhat less encouraging: At least in terms of land area, greater portions of Florida were following the equilibrium path at the beginning of the study period, before the State Comprehensive Plan and the GMA were adopted. Taken together, the two maps suggest that there was in fact a break from past trends in the middle of the 1980s, raising the possibility that growth management had a substantive impact on the state's economic development process.

A deeper look into the situation is provided via Table 7.2, which lists the total population and employment of each of the seven Atlantic Southeast states, the proportions that were adjusting toward equilibrium (measured as the total number of people and jobs in the "converging gap" (dark grey) counties, divided by the state totals), and the total number of equilibrium participants. In Florida, the proportion of population participating in the adjustment process increased from about 30 percent in the first time segment to 83 percent and 75 percent in the latter two; at the same time, the proportion of employment participating in the adjustment process rose from 46 percent to 91 percent between the first and second time segments but fell to 33 percent in the last time segment (see Table 7.3). Note that the erratic movement of the employment side of the system may be a symptom of the recession that occurred in the early 1990s. Some research has suggested that the equilibrium process that regional adjustment models emulate is sensitive to business cycles (Mulligan et al., 1999), so it may be that the shock of the recession is responsible for the difference between the projected and actual outcomes in Florida.

What do the results imply about the effectiveness of Florida's growth management program? On the upside, the implication of the data shown in Table 7.2 is that the shift in growth management policy that occurred in 1985 *may* have had a significant leveling effect on Florida's regulatory landscape. The previous growth policy, enacted in 1975, placed little emphasis on substance (DeGrove, 1992) so it is possible that it did more harm than good. Carruthers' (2002) evaluation of state policy frameworks found that simply requiring local governments to plan, without strict enforcement of appropriate standards and guidelines, can be counterproductive. What evolves from this type of policy is a patchwork regulatory landscape, which alters the land market in a way that creates a cycle of urban sprawl as a result of growth being directed to the urban fringe (see Carruthers, 2003; Ulfarsson and Carruthers, 2006). The situation on the employment side of the growth process notwithstanding, the possibility that Florida's GMA may have helped to bring economic actors (individuals and firms, or people and jobs) in line with market tendencies by alleviating local regulatory failures is beneficial. Theoretically, the space economy pushes the organization of residential and nonresidential activity toward an optimum configuration (given initial conditions), so it is appropriate to say that Florida's policy framework has had a positive effect by evening out place-to-place variation in the mechanisms, objectives, and process of land use regulation.

On the downside, Figures 7.1 and 7.3 illustrate that urban sprawl has continued to be the dominant mode of growth in Florida— an outcome that may be the result of concurrency requirements. As previous research has found, it does no good to require appropriate levels of infrastructure to be in place before development can proceed if adequate funding does not back the requirement. In Florida, growth occurs in a leapfrog fashion because development is forced into areas with adequate road capacity, meaning the concurrency requirement, combined with a failure to increase capacity, is a direct cause of sprawl (Blanco, 1998; Carruthers, 2002; Nicholas and Steiner, 2000). Finally, it is important to explain the implications that the adjustment process itself has for growth management policy. As already noted, there is evidence that, in addition to the traditional process of people following jobs, *jobs follow people into and/or within Florida and the Atlantic Southeast as a whole.* The point deserves particular emphasis because this type of growth is largely driven by quality of life factors: The process involves people moving in response to the region's desirability as a place to live, and job growth happening in the wake. In order for economic development to proceed in this way, people have to continue to want to move to Florida for its high quality of life. In other words, given the nature of the contemporary growth process, sustained economic development in the state may, in the end, depend upon the preservation of its environmental assets. What this means for growth management is that, to the extent that centralized land use policy involves a tradeoff between the day-to-day controversy, difficulty, and expense of implementation and long-term stewardship of the economy, the benefits almost certainly outweigh the costs.

Summary and Conclusion

This chapter examined the impacts of growth management in Florida by using a land use based regional adjustment model to project equilibrium densities of people and jobs county-by-county in seven Southeastern states in 1987, 1992, and 1997. The density forecasts, which express the direction(s) in which the space economy was pushing land use change in the region during the preceding five-year timeframes, were compared to actual outcomes as a means of identifying the impacts of state policy. The findings suggest that Florida's 1985 GMA may have helped to bring growth patterns in line with equilibrium tendencies by evening out place-to-place differences in local regulatory activity. Moreover, because population and employment growth are jointly determined in Florida (and the Atlantic Southeast as a whole), the long-term sustainability of economic development there may depend on policies that preserve the high quality of life that it has to offer. The implications of this research for Florida's growth management program are straightforward. First, the evidence tentatively suggests that it has had a beneficial effect by alleviating the kind of regulatory failure associated with a purely local approach to land use planning. Second, although the GMA does not appear to have had a meaningful impact on the spatial outcome of growth, this may have more to do with ancillary factors, such as infrastructure funding, than the legislation itself. Finally, because of the strong interdependence between population and employment growth in the Florida, future research on the costs and benefits of its planning efforts should carefully weigh the consequences of not having a policy framework directed at preserving its desirability as a place to live.

References

Ben-Zadok, E. (2005). Consistency, concurrency, and compact development: The faces of growth management implementation in Florida. *Urban Studies, 42,* 2167–2190.

Blanco, H. (1998, October). *The effectiveness of policies to contain urban sprawl and their evolution in Florida, Oregon, and Vermont.* Paper presented at the meeting of the American Collegiate Schools of Planning, Pasadena, CA.

Boarnet, M.G. (1994a). An empirical model of intrametropolitan population and employment growth. *Papers in Regional Science, 73,* 135–152.

Boarnet, M.G. (1994b). The monocentric model and employment location. *Journal of Urban Economics, 36,* 79–97.

Boarnet, M.G., Chalermpong, S., and Geho, E. (in press). Specification issues in models of population and employment growth. *Papers in Regional Science.*

Bollinger, C.R., and Ihlanfeldt, K.R. (1997). The impact of rapid rail transit on economic development: The case of Atlanta's MARTA. *Journal of Urban Economics, 42,* 179–204.

Borts, G.H., and Stein, J.L. (1964). *Economic growth in a free market.* New York: Columbia University Press.

Brueckner, J. (2000). Urban sprawl: Diagnosis and remedies. *International Regional Science Review, 23,* 160–171.

Bureau of Economic Analysis. (2004). *Regional Economic Information System* [Data file]. Available from Bureau of Economic Analysis Web site, http://www. bea.gov/bea/regional/docs/reis2004dvd.asp.

Carlino, G.A., and Mills, E.S. (1987). The determinants of county growth. *Journal of Regional Science, 27,* 39–54.

Carruthers, J.I. (2002). The impact of state growth management programs: A comparative analysis. *Urban Studies, 39,* 1959–1982.

Carruthers, J.I. (2003). Growth at the fringe: The influence of political fragmentation in United States metropolitan areas. *Papers in Regional Science, 82,* 472–499.

Carruthers, J.I., and Mulligan, G.F. (2006). *Human capital, quality of life, and the adjustment process in American metropolitan areas* (Working Paper No. REP 06-04). Washington, DC: US Department of Housing and Urban Development.

Carruthers, J.I., and Mulligan, G.F. (in press). Economic growth and land absorption in United States metropolitan areas. *Geographical Analysis.*

Carruthers, J.I., and Vias, A.C. (2005). Urban, suburban, and exurban sprawl in the Rocky Mountain west: Evidence from regional adjustment models. *Journal of Regional Science, 45,* 21–48.

Clark, D.E., and Murphy, C.A. (1996). Countywide employment and population growth: An analysis of the 1980s. *Journal of Regional Science, 36,* 235–256.

DeGrove, J. (1992). *The new frontier for land policy: Planning and growth management in the states.* Cambridge: Lincoln Institute of Land Policy.

Deller, S.C., Tsai, T.-H., Marcouiller, D.W., and English, D.B.K. (2001). The role of amenities and quality of life in rural economic growth. *American Journal of Agricultural Economics, 83,* 352–365.

Deyle, R., and Smith, R. (1998). Local government compliance with state planning mandates: The effects of state implementation in Florida. *Journal of the American Planning Association, 64,* 457–469.

Duffy-Deno, K.T. (1998). The effect of federal wilderness land on county growth in the intermountain western United States. *Journal of Regional Science, 38,* 109–136.

Frey, W.H. (1993). The new urban revival in the United States. *Urban Studies, 30,* 741–774.

Henry, M.S., Barkley, D.L., and Bao, S. (1997). The hinterlands' stake in metropolitan growth: Evidence from selected southern regions. *Journal of Regional Science, 37,* 479–501.

Henry, M.S., Schmitt, B., Kristensen, K., Barkley, D.L., and Bao, S. (1999). Extending Carlino-Mills models to examine urban size and growth impacts on proximate rural areas. *Growth and Change, 30,* 526–548.

Henry, M.S., Schmitt, B., and Piguet, V. (2001). Spatial econometric models for simultaneous equations systems: Application to rural community growth in France. *International Regional Science Review, 24,* 171–193.

Leichenko, R.M. (2001).Growth and change in US cities and suburbs. *Growth and Change, 32,* 326–354.

Mills, E.S., and Lubuele, L.S. (1995). Projecting growth in metropolitan areas. *Journal of Urban Economics, 37,* 244–360.

Mulligan, G.F., Vias, A.C., and Glavac, S.M. (1999). Initial diagnostics of a regional adjustment model. *Environment and Planning A, 31,* 855–876.

Muth, R. F. (1971) Migration: Chicken or egg? *Southern Economic Journal 37,* 295–306.

Nicholas, J.C., and Steiner, R.L. (2000). Growth management and smart growth in Florida. *Wake Forest Law Review, 35,* 645–670.

Steinnes, D.N. (1977).Causality and intraurban location. *Journal of Urban Economics, 4,* 69–79.

Steinnes, D.N., and Fisher, W.D. (1974). An econometric model of interurban location. *Journal of Regional Science, 14,* 65–80.

Ulfarsson, G.F., and Carruthers, J.I. (2006). The cycle of fragmentation and sprawl: A conceptual framework and empirical model. *Environment and Planning B, 33,* 767–788.

Vias, A.C. (1999). Jobs follow people in the rural Rocky Mountain west. *Rural Development Perspectives, 14,* 14–23.

Vias, A.C., and Mulligan, G.F. (1999). Integrating economic base theory with regional adjustment models: The nonmetropolitan Rocky Mountain west. *Growth and Change, 30,* 507–525.

Chapter 8

Growth Management or Growth Unabated? Economic Development in Florida Since 1990

Timothy S. Chapin

With the passage of the 1985 Growth Management Act (GMA), Florida's new growth management approach was met with both hope and apprehension by economic development practitioners, public officials, and state boosters. On the positive side, the state's approach offered hope to the cities in the state because of a set of requirements that all local governments develop goals, objectives, and policies that both directly and indirectly promote urban revitalization. With redevelopment promoted from the inside, through anti-blight policies, and development limited at the edges, through anti-sprawl policies, Florida's growth management approach could, in theory, provide a model that would help to revitalize the state's cities.

With the passage of local comprehensive plans and approval of these plans by the state in the late 1980s and early 1990s, there was great concern about the impact that state-mandated growth controls would have on the state's economy (DeGrove and Miness, 1992). Some predicted that the state's new growth controls would hinder the growth of the state's economy and, by extension, the revitalization of the central cities (Holcombe, 1990). In particular, the state's concurrency mandate raised some real concerns for the state economy and the health of Florida's heavily urbanized areas. Concurrency explicitly prohibited development in areas where roads were congested, parks and recreation facilities were inadequate, and stormwater facilities were insufficient—conditions found in almost all of the state's largest cities in 1990. A related concern was that concurrency would adversely affect two of the state's largest growth industries, construction and real estate, thereby limiting growth and expansion of the state economy. Although concurrency has been fine-tuned over the course of the decade, particularly through the relaxation of transportation concurrency requirements for urbanized areas (Steiner, 2001), the impact of this policy on the state's cities has remained largely undetermined.

Through an analysis of changes in economic indicators, this chapter assesses the impact of Florida's growth management approach on economic development outcomes in the state. More specifically, this work is aimed primarily at two questions:

1. How did Florida's economy fare during the initial implementation of the state's growth management mandate, 1990–2000?
2. Similarly, how did the state's largest cities fare during this period? Did they fare better or worse than their peers during this period?

To answer these questions, data from the US Census Bureau and the Brookings Institute were analyzed. Three separate indicator analyses were completed and are presented. First, Florida's economic performance were compared to the nation's during the period 1980–2000. Second, economic indicators for the state's largest cities were compared to those of the state. Finally, economic indicators for the state's largest cities were compared to peer cities. These analyses provide insights into the effects of the 1985 GMA on the Florida economy, as well as changes in the relative economic health of Florida's largest cities during this period.

The chapter is organized as follows. In the first section a brief overview of the nexus between growth management and economic development in the Florida planning process is provided. The next section outlines the data and methodology used to assess the impact of the 1985 GMA on the economies of the state and its largest cities. The following sections present the findings from these analyses. A concluding section offers an assessment of the impact of Florida's growth management system on the state economy.

The Growth Management–Economic Development Linkage in Florida

As detailed in Chapters 2 and 3, in 1985 Florida passed one of the most innovative and rigorous growth management programs this country has seen (Ben-Zadok and Gale, 2001; Gale, 1992; Porter, 1998). Most relevant to this chapter is the linkage between Florida's growth management approach and economic development efforts in the state. Although economic development has not featured prominently in the state's growth management system, DeGrove (2005, 56) describes it as one of the "weak links" in the state's system, there are some linkages to be found.

Several sections of the State Comprehensive Plan, which is supposed to guide the actions of state agencies and inform local and regional comprehensive plans, focus upon the economic health of the state. One of the twenty-six goals of the plan speaks to improving the economy. Other state goals direct state agencies and local governments to protect key industries (tourism and agriculture) and promote urban redevelopment, in the process increasing job opportunities for Floridians. However, while this language is found in the State Comprehensive Plan, it was never translated into a coherent state economic development policy (DeGrove, 2005). In addition, this plan has been ignored by the Legislature and sitting governors since its adoption in 1986. In 2001, Governor Bush appointed a Growth Management Study Commission, which recommended that the state scrap its comprehensive plan in favor of a short vision statement that makes a healthy, vibrant economy the highest priority (Florida Growth Management Study Commission, 2001).

In contrast to the rather cumbersome language for economic development at the state level, substantial linkages have been forged between growth management and urban revitalization at the local comprehensive plan level. Beyond inclusion of an "urban revitalization" goal in the State Comprehensive Plan, these linkages were established in the state statutes written and passed by the Legislature in 1985, as well as in the implementing language which was developed by the Florida Department of Community Affairs (DCA) in the late 1980s (Chapter 9J-5 of the Florida Administrative Code (FAC).

First, the Legislature established urban revitalization as a required part of the land use element of local comprehensive plans. The 1985 GMA specifically directs governments to identify blighted areas within their jurisdictions and to identify policies for promoting redevelopment within these areas. Chapter 9J-5 further requires that all comprehensive plans have at least one goal and objective that encourages the redevelopment and renewal of blighted areas. Urban revitalization is a required element of the long-term land use vision for a community, with specific policies aimed at achieving this outcome.

Second, Chapter 9J-5 also indirectly promotes urban revitalization through a mandate that all comprehensive plans include policies that limit urban sprawl. DCA provided a menu of options for "development controls", including urban service areas and urban growth boundaries, all of which were intended to "discourage urban sprawl". Chapter 9J-5 also identifies thirteen indicators of urban sprawl that will be viewed by DCA as not conforming with the state statute. Three of these indicators of sprawl speak directly to urban revitalization, as any policy that results in the following is *not* to be included in local comprehensive plans:

- Promotes, allows or designates significant amounts of urban development to occur in rural areas at substantial distances from existing urban areas while leaping over undeveloped lands which are available and suitable for development.
- Promotes, allows or designates urban development in radial, strip, isolated or ribbon patterns generally emanating from existing urban developments.
- Discourages or inhibits infill development or the redevelopment of existing neighborhoods and communities.

The Research Approach

While the GMA was passed in 1985, it is important to recognize that most local governments did not implement this legislative mandate until the late 1980s, at the earliest. DCA was tasked with reviewing and ultimately approving comprehensive plans produced by local governments. DCA began this work in 1988 and most local governments had their plans approved in the period 1989–1991. After several years of legal challenges and a period of ramping up for local government and state DCA planning staff, communities in the state began to implement their comprehensive plans in the early 1990s. These plans influenced government expenditures on infrastructure, led to the purchase of environmentally sensitive lands, dictated where development could and could not proceed (and at what intensity), and more generally shaped the form of growth in the state.

The decade of the 1990s, then, represents the appropriate time period for an assessment of the impacts of the state's growth management approach. Consequently, the analyses that follow generally utilize two time periods: 1) pre-growth management (1980–1990) and 2) post-growth management (1990–2000). To assess the impact of the state's growth management approach on the state economy and on the state's largest cities, I rely upon data from the US Census Bureau. The Bureau's County Business Patterns dataset was employed to examine employment trends for Florida versus the United States during the period 1980–2000. In addition, data on poverty levels and median household incomes were obtained from the Census. Using growth rate calculations and share of growth analyses, insights into Florida's economic performance pre-growth management and post-growth management were gleaned.

In addition, data made available by the Brookings Institution to researchers, policy makers, and elected officials were employed (Brookings Institution Living City Census Series overview, 2004). The Living Cities dataset brings together a variety of indicators for the 100 largest cities in the United States in the year 2000, including variables on population, income/poverty, race/ethnicity, education, and work. The data were captured from the last three Decennial Census datasets, for the years 1980, 1990, and 2000.

These data provide the basis for comparisons between Florida's largest cities and peer cities throughout the nation. In this peer analysis, I compared the five Florida cities in the Living Cities dataset (Miami, Jacksonville, Tampa, St. Petersburg, and Hialeah) with a set of peer cities. For Miami, Tampa, St. Petersburg, and Hialeah, peer groups were identified on the basis of population size in 1990; for each city, the seven cities with the next highest population sizes and the seven cities with the next lowest population sizes were included in its peer group. In the case of Jacksonville, its peer group was identified as other large cities that have also undergone city-county consolidation. Jacksonville and Duval County consolidated after a voter referendum in 1967, establishing the area as one of the first to undertake consolidation. Under this model the city and county are now effectively one jurisdiction. Of the nation's largest cities, the following cities have in place some form of city-county consolidation (National League of Cities, 2004) and therefore represent Jacksonville's peer group: Augusta (GA), Boston, Denver, Honolulu, Indianapolis, Lexington (KY), Nashville, New Orleans, Philadelphia, and San Francisco.

After each Florida city's peer group was identified, key indicators of economic performance were compared for the Florida cities and their peers, including standardized median household income and the percentage of people living in poverty. In addition, I calculated each city's city-MSA income ratio for both 1989 and 1999. To generate these income ratios each city's median household income was divided by the MSA's median household income and then multiplied by 100. A value less than 100 indicates that a city's median household income in that year was less than that of the MSA in which it was located. For example, Tampa's city-MSA income ratio was 87.5 in 1989, indicating that the city's median household income was roughly 87.5 per cent of the MSA's median household income in that year. These city-MSA income ratios were then compared to determine whether or not each city was gaining or losing ground on its metropolitan region in terms of median income over the period of study.

Table 8.1   US and Florida population and employment growth, 1980–2000

Changes in Population, 1980–2000					
	Pop 1980	Pop 1990	Pop 2000	Change 80–90	Change 90–00
US	226,545,805	248,709,873	281,421,906	9.8%	13.2%
Florida	9,746,324	12,937,926	15,982,378	32.7%	23.5%
Changes in Employment, 1980–2000					
	Emp 1980	Emp 1990	Emp 2000	Change 80–90	Change 90–00
US	74,835,525	93,476,087	114,064,976	24.9%	22.0%
Florida	2,975,177	4,607,247	6,217,386	54.9%	34.9%

Florida's Economic Performance 1980–2000

The period 1980–2000 was generally a period of strong economic growth for the United States and the state of Florida. Despite economic slumps in the early 1980s and early 1990s, private employment soared during these decades. Table 8.1 compares population and employment growth for Florida and the United States. Between 1980 and 2000 the nation's population grew by roughly 55 million, while the economy grew by almost 40 million private sector jobs, representing growth rates of 24 percent and 52 percent, respectively. A major beneficiary of this ongoing growth has been the state of Florida. Between 1980 and 2000, the state added 6.2 million residents, a 64 percent increase. In part driven by this population growth, Florida's economy added jobs at an even greater rate than the nation's, with 3.2 million more private sector jobs in 2000 than 1980, a 109 percent increase.

When these indicators are analyzed by decade, the data suggest that Florida's growth continued during the period when growth management was being implemented in the state. A simple share of growth analysis reveals that during the 1980s, Florida grew by 3.2 million people and 1.6 million jobs, capturing 14.4 percent and 8.8 percent of the nation's growth in population and employment, respectively. In the post-growth management period of the 1990s, Florida's population and economic growth continued largely unchecked. The state added another 3 million residents and 1.6 million jobs, although its share of the nation's growth fell to 9.9 percent of new residents and 7.8 percent of new jobs. At an aggregate level, these figures indicate that, even after the onset of growth management, the growth machine that is Florida continued to add people and jobs at rates the envy of most states.

Supporting this finding of continued growth in the Florida economy are annual employment growth rates for the period 1980–2000 (presented in Figure 8.1). These data illustrate how closely the Florida economy mirrors that of the national economy, as both the state and nation experienced significant employment hits during the recessions of the early 1980s and 1990s. However, in every year Florida has experienced greater rates of job growth than the nation, even with the coming of growth management in the 1990s. While some public officials in Florida expressed concerns about the influence of growth management on the state's economy in the early 1990s (DeGrove, 2005), pushing for revisions or outright repeal of the GMA, these data illustrate that the slowed economy of the early 1990s was a national

Table 8.2   Florida economic indicators, 1980–2000

	1980	1990	2000	1980 State Ranking	1990 State Ranking	2000 State Ranking
Total Population	9,746,324	12,937,926	15,982,378	7	4	4
FL's Share of the National Population	4.3%	5.2%	5.7%	--	--	--
Total Private Employment	2,975,177	4,607,247	6,217,386	7	5	4
FL's Share of National Private Emp.	4.0%	4.9%	5.5%	--	--	--
Total Private Payroll (in millions) (2)	$35,362,189	$89,563,372	$177,378,971	9	7	5
FL's Share of National Private Payroll	3.4%	4.3%	4.6%	--	--	--
Median Household Income (1,2)	$31,395	$35,856	$38,819	39	29	34
Percent Poverty (1,3)	13.5%	12.7%	12.5%	36	29	33

Source: US Census Bureau, Decennial Censuses and County Business Patterns

1 Household Income and Percent Poverty statistics are for 1979, 1989, and 1999.
2 Median Household Income and Private Payroll in standardized 1999 and 2000 dollars, respectively.
3 Rankings for Percent Poverty scaled from best to worst (i.e., Florida had the 29th best Poverty rate in 1990).

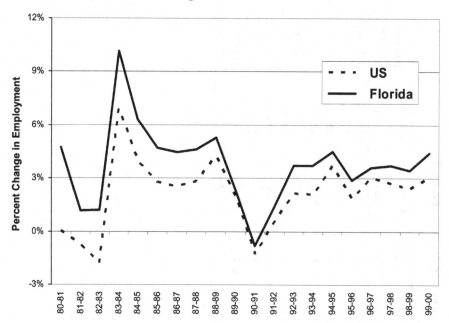

Figure 8.1   US and Florida annual employment growth rates, 1980–2000

problem. Since 1993, the Florida economy has returned to its pattern of easily outdistancing the nation's annual employment growth rate.

Despite this growth, other traditional measures of economic development suggest that all is not well with Florida's economy. Table 8.2 reveals that while Florida was adding jobs, these were not always high paying. The state's median household income levels and poverty rates, as measured in the decennial censuses of 1980, 1990, and 2000, rank in the bottom half for all states. During the 1990s, Florida's median household income actually fell relative to other states in the nation, from 29th to 34th in the state rankings. Similarly, during the 1990s, as Florida was adding jobs and payroll, the state's percentage of people living in poverty changed very little. During this decade, Florida went from having a percent living in poverty below that of the nation (12.7 percent for Florida vs. 13.1 percent for the US in 1989) to having a rate higher than that of the nation (12.5 percent vs. 12.4 percent in 1999).

What is remarkable about this finding is that these relative declines in economic performance occurred in the face of continued growth in the state economy. Florida continues to add jobs and payroll at rates well above those of most states, yet along key income and poverty measures it continues to lag behind. At the heart of this incongruence is a Florida economy that remains exceedingly tied to its staples: tourism and entertainment. While these industries have been a major part of the state's economic growth in the last several decades, they traditionally pay less and offer fewer benefits than many other industries.

Overall, then, the post-growth management decade of the 1990s was in many ways a continuation of both the good and the bad of the Florida economy. The state experienced tremendous increases in population and employment, typical measures

Table 8.3 Summary economic indicators for the state of Florida and its largest cities, 1990 and 2000

	Population 1990	Population 2000	Percent Change	Median HH Inc, 1989	Median HH Inc, 1999	Percent Change	Percent Poverty, 1989	Percent Poverty, 1999	Percent Change
Florida	12,937,926	15,982,378	23.5%	$35,670	$38,819	8.8%	12.7%	12.5%	-1.6%
Jacksonville	635,230	735,617	15.8%	$36,736	$40,316	9.7%	13.0%	12.2%	-6.1%
Miami	358,548	362,470	1.1%	$21,966	$23,483	6.9%	31.2%	28.5%	-8.7%
Tampa	280,015	303,447	8.4%	$29,555	$34,415	16.4%	19.4%	18.1%	-6.8%
St Petersburg	238,629	248,232	4.0%	$30,600	$34,597	13.1%	13.6%	13.3%	-2.2%
Hialeah	188,004	226,419	20.4%	$30,426	$29,492	-3.1%	18.2%	18.6%	2.3%
Orlando	164,693	185,984	12.9%	$33,889	$35,732	5.4%	15.8%	15.9%	0.5%
Ft Lauderdale	149,377	152,125	1.8%	$35,353	$37,887	7.2%	17.1%	17.7%	3.7%
Tallahassee	124,773	150,581	20.7%	$30,439	$30,571	0.4%	22.3%	24.7%	10.8%
Hollywood	121,697	139,261	14.4%	$35,500	$36,714	3.4%	11.0%	13.2%	20.2%

of economic growth. In both the pre- and post-growth management periods, the state experienced very similar levels of growth in population and private sector employment. However, indicators of economic development signal some troubling aspects to Florida's growth. More employment hasn't translated into income growth over that of other states. More troubling is the very minor decline in poverty during a decade of tremendous prosperity for the state. It is within this context that an analysis of the relative economic performance of the state's largest cities is undertaken.

Comparing the Economies of Florida and its Largest Cities, 1990–2000

This section presents a comparison of changes in the state economy with those of the state's largest cities during the period of growth management implementation, 1990–2000. Table 8.3 summarizes key indicators for Florida and the nine cities in the state greater than 100,000 in population in 1990. As discussed in the previous section, by 1990 Florida was by most indicators an economically healthy state, with growth rates the envy of most states. The decade of the 1990s continued this general trend of economic growth, as the state added another 3 million people and saw its median household income grow by 9 percent, and its poverty rate decline.

Table 8.3 hints at a different story for Florida's largest cities. At the outset of the decade, Florida's cities generally lagged behind the state on these indicators. In 1990, all but one city had a median household income below that of the state and all but one had a poverty rate higher than the state's. Further, in terms of population growth, during the 1980s only Hialeah, Orlando, and Tallahassee grew at rates comparable to the state, while the remaining cities captured very little of the state's residential growth (numbers not presented). Similar to the situation in most states, as of 1990 Florida's largest cities were generally growing at a slower rate than the state as a whole, with populations that earned less money and, consequently, were more likely to live in poverty. Given these conditions, it is understandable why urban revitalization was one component of the 1985 GMA. The inclusion of these urban-focused policies was also due, in part, to the sheer number of residents living within the state's urban areas; in 1990, roughly one in six Floridians lived in one of these nine cities.

When changes in these indicators during the period 1990–2000 are analyzed, the data yield a few key findings. First, Table 8.3 shows that the state's largest cities still generally lag behind the state in terms of population growth rates, 1999 median household income levels, and 1999 percentages of people living in poverty. Only consolidated Jacksonville-Duval County has better income and poverty statistics than the state. Overall, in 2000 Florida's cities still remained far behind state as a whole.

However, Table 8.3 also provides some evidence of economic progress during the 1990s for the state's four largest cities (Jacksonville, Miami, Tampa, and St. Petersburg). Each of these cities saw gains in their median household income very close to or above that of the state as a whole. In addition, each city experienced greater declines in their percent poverty than the state over the course of the 1990s. While the state's poverty rate remained largely unchanged between 1989 and 1999

Table 8.4   Summary of the Jacksonville peer city analysis

Cities	Pop Change 1990–2000	Rank	Pct Change in Med HH Inc, 1989–1999	Rank	Pct Change in Inc Ratios, 1989–1999	Rank	Change in Pct Poverty 1989–1999	Rank
Jacksonville, FL	15.8%	2	9.7%	4	-0.9%	6	-6.1%	5
Augusta, GA	4.4%	7	0.6%	9	-2.4%	7	7.1%	9
Boston, MA	2.6%	8	4.6%	8	-0.3%	4	4.4%	8
Denver, CO	18.6%	1	21.2%	2	1.0%	3	-16.6%	1
Honolulu, HI	1.7%	9	-6.5%	11	-5.2%	10	40.1%	11
Indianapolis, IN	6.9%	6	6.4%	7	-4.0%	9	-5.4%	6
Lexington, KY	15.6%	3	9.3%	5	-0.6%	5	-8.9%	4
Nashville, TN	11.7%	4	8.7%	6	-3.6%	8	-0.9%	7
New Orleans, LA	-2.5%	10	13.1%	3	1.6%	2	-11.6%	2
Philadelphia, PA	-4.3%	11	-3.7%	10	-6.8%	11	12.9%	10
San Francisco, CA	7.3%	5	27.3%	1	5.7%	1	-10.6%	3
Averages	7.1%		8.3%		-1.4%		0.4%	

(down 1.6 percent), poverty rates in three cities declined by 6 percent or more during the decade.

In contrast, the other five cities in Table 8.3 struggled in the 1990s. These cities had less growth in their median household incomes than the state as a whole, with Hialeah even experiencing a decline in this indicator. Similarly, each of these cities saw increases in their percentage of people living in poverty between 1989 and 1999, a period of substantial economic growth for the state economy. This finding is partly attributable to a "Miami effect," one that saw Miami's poor, largely Hispanic immigrant population spread into neighboring jurisdictions in the 1990s. Three of these cities, Hialeah, Ft Lauderdale, and Hollywood, are located in the greater Miami consolidated metropolitan statistical area. During the 1990s, these cities saw an increase in their Hispanic populations, as well as their minority populations. Tallahassee's remarkable poverty rate is in part attributable to a substantial and growing college-student population; the city's three higher learning institutions are home to over 50,000 students.

Taken together, these data suggest a mixed bag of outcomes for the state's largest cities in the 1990s. These cities generally experienced positive changes to their economic indicators during the decade in which growth management was implemented. Median household incomes were generally up and poverty rates are down, with the greatest improvements made in the state's four biggest cities. However, despite policies aimed at urban containment and urban revitalization, Florida's cities are still not capturing large shares of the state's population growth and they still substantially lag the state (and nation) along key indicators of economic health.

Peer City Analyses

A peer city analysis was completed for Florida's five most populous cities. For each city, a table is presented that summarizes changes in key economic indicators and ranks these changes for all cities in the peer group.

The Jacksonville Peer Group Analysis

Jacksonville's peer group includes other cities with some form of city-county consolidated government. As such this group is much more diverse in population size than any of the other peer groups. This group includes large cities, like Philadelphia, Indianapolis, and San Francisco, as well as much smaller cities like Lexington (KY) and Augusta (GA). Table 8.4 provides a summary of key indicators of economic development for the period 1990–2000.

An analysis of economic changes in the 1990s finds that Jacksonville continues to be a relatively well-off city. During the decade Jacksonville continued to grow, adding almost 100,000 residents—the most of this peer group—and establishing the city as the nation's 14th largest city in 2000. The median household income of Jacksonville also grew in the 1990s, by roughly $3,600, while the city's poverty rate declined by three-quarters of a percentage point. The city's income ratio did

Table 8.5 Summary of the Miami peer city analysis

Cities	Pop Change 1990–2000	Rank	Pct Change in Med HH Inc, 1989–1999	Rank	Pct Change in Inc Ratios, 1989–1999	Rank	Change in Pct Poverty 1989–1999	Rank
Miami, FL	1.1%	11	6.9%	7	3.8%	1	-8.7%	4
Buffalo, NY	-10.8%	15	2.3%	10	-3.4%	11	3.8%	11
Cincinnati, OH	-9.0%	13	8.2%	6	-2.6%	8	-10.0%	3
Fresno, CA	20.7%	2	-0.3%	13	-2.4%	7	8.9%	12
Honolulu, HI	1.7%	10	-6.5%	15	-5.2%	14	40.1%	15
Mesa, AZ	37.6%	1	9.0%	5	-2.7%	9	-6.2%	7
Minneapolis, MN	3.9%	9	15.5%	1	1.0%	3	-8.7%	5
Oakland, CA	7.3%	7	13.9%	3	1.2%	2	3.0%	10
Omaha, NE	16.1%	3	14.5%	2	0.2%	5	-10.3%	1
Pittsburgh, PA	-9.5%	14	6.2%	8	-1.8%	6	-4.9%	9
Sacramento, CA	10.2%	6	1.3%	11	-6.7%	15	16.2%	14
Santa Ana, CA	15.1%	4	-4.9%	14	-3.6%	12	9.4%	13
Toledo, OH	-5.8%	12	1.0%	12	-4.3%	13	-6.3%	6
Tulsa, OK	7.0%	8	5.8%	9	-3.1%	10	-5.8%	8
Wichita, KS	13.2%	5	9.8%	4	0.8%	4	-10.2%	2
Averages	6.6%		5.5%		-1.9%		0.7%	

fall slightly during the decade, although it remained the second highest of the peer group in 2000.

When changes in Jacksonville's indicators during the period 1990 to 2000 are compared to those of the peer group, I find that the city fared reasonably well during the decade. The city's income change ranked 4th in this peer group and 35th out of the entire database. Similarly, the city's decline in its poverty rate ranked 5th in its peer group and 38th overall. While Jacksonville's income ratio actually declined, from 95.9 in 1989 to 95.0 in 1999, a relative change of -0.9 percent, this very modest decline still placed the city in the top half of all cities in the dataset on this criterion (a rank of 40th).

Taken together, changes in these indicators reflect a Jacksonville economy that generally improved during the 1990s. The city's median household income continued to climb at a rate greater than the state as a whole and its poverty rate fell below that of the state. The peer city analysis reveals that Jacksonville generally fared better than its peer group along these indicators.

The Miami Peer Group Analysis

As of 1990, the city of Miami was a very diverse, very poor city of roughly 360,000 persons. Miami's population is among the most diverse in the nation, roughly a quarter African-American and three-fifths Hispanic. Along key measures of economic performance in 1990 Miami showed evidence of tremendous economic problems, ranking last in median household income and third from the bottom in percent poverty for all ninety-nine cities. Included in Miami's peer group are cities ranging in population size from about 290,000 to 370,000 residents. This peer group includes older, industrial cities, such as Pittsburgh, Toledo, and Cincinnati, as well as growing Sunbelt cities like Mesa and Tulsa. On almost all indicators, Miami lagged far behind its peer cities in 1990.

During the 1990–2000 period Miami experienced very little population growth, increasing by only 3,900 people (see Table 8.5). Similarly, the city remains far behind most of its peers in terms of median household income levels and poverty rates, although both of these indicators improved during the decade. Miami's closest peers in this group are the cities of Buffalo, Cincinnati, and Pittsburgh, cities typically found on the list of America's most troubled cities.

However, Table 8.5 illustrates that despite Miami's very troubled baseline economic conditions, the city experienced changes along some of these dimensions that are the envy of most cities in its peer group. Improvements in Miami's city-MSA income ratio and its percent poverty were among the highest of these fifteen cities. Miami's income ratio experienced a percentage increase of 3.8 percent during the decade, which suggests that Miami's incomes are slowly approaching those of the entire MSA. This increase in the city-MSA income ratio ranked 1st out of these fifteen cities and 8th out of the entire set of ninety-nine cities. Similarly, Miami's roughly 9 percent decrease in percent poverty during the period ranked 4th in its peer group and 24th out of all cities.

Overall, while Miami still combats tremendous poverty, ranking dead last in poverty for large cities in the 2000 Census; and while the city added very few

Table 8.6 Summary of the Tampa peer city analysis

Cities	Pop Change 1990–2000	Rank	Pct Change in Med HH Inc, 1989–1999	Rank	Pct Change in Inc Ratios, 1989–1999	Rank	Change in Pct Poverty 1989–1999	Rank
		8		2		1		4
Tampa, FL	8.4%		16.4%		5.2%		-6.8%	
Anaheim, CA	23.1%	4	-8.4%	15	-7.1%	13	32.5%	15
Arlington, TX	27.2%	3	4.7%	9	-5.0%	12	20.1%	14
Birmingham, AL	-8.7%	13	7.3%	8	-7.3%	14	-0.5%	9
Buffalo, NY	-10.8%	15	2.3%	11	-3.4%	9	3.8%	11
Col. Springs, CO	28.4%	2	20.1%	1	-1.5%	5	-20.0%	1
Louisville, KY	-4.8%	11	10.3%	5	-3.2%	8	-4.5%	8
Mesa, AZ	37.6%	1	9.0%	7	-2.7%	7	-6.2%	7
Newark, NJ	-0.6%	10	-4.2%	13	-7.6%	15	7.9%	12
Norfolk, VA	-10.3%	14	4.0%	10	-1.9%	6	0.4%	10
Omaha, NE	16.1%	5	14.5%	3	0.2%	3	-10.3%	2
Santa Ana, CA	15.1%	6	-4.9%	14	-3.6%	10	9.4%	13
St. Paul, MN	5.5%	9	12.7%	4	-1.5%	4	-6.7%	5
Toledo, OH	-5.8%	12	1.0%	12	-4.3%	11	-6.3%	6
Wichita, KS	13.2%	7	9.8%	6	0.8%	2	-10.2%	3
Averages	8.9%		6.3%		-2.9%		0.2%	

residents during the decade, there is some evidence that the city's economy improved in the 1990s relative to its peers. The city's income ratio increased at a rate greater than all but a handful of America's largest cities. In addition, the city's decline in percent poverty was among the larger declines of all cities. Unlike many of its peers, Miami's economy experienced some improvement during the 1990s and it began to close the income gap with the MSA.

The Tampa Peer Group Analysis

In 1990, Tampa's economic indicators reflect the troubled status of many of Florida's cities at the outset of the decade. Tampa grew very little during the 1980s, at just over 3 percent, while the city's 1989 median household income and poverty rate placed the city well within the bottom half of the rankings for the nation's largest cities. Tampa's peer group contains a mix of cities, including other cities located in metropolitan areas with more than one large city (St. Paul, Anaheim, and Arlington, TX). This peer group also runs the gamut from booming city on the rise (Arlington and Colorado Springs) to older cities with uncertain futures (Newark, Norfolk, and Buffalo). Its peer group ranges in population from 260,000 to 336,000.

During the 1990s Tampa experienced positive change along all indicators. The city grew by over 20,000 residents, standardized median household income grew by almost $5,000, and the city's poverty rate declined. In addition, the city's income ratio improved substantially during this period, growing from 87.5 in 1989 to 92.0 in 1999. Table 8.6 summarizes changes for Tampa and its peer cities along these indicators. The peer analysis indicates that Tampa fared by far the best of its peer group. Most impressively, Tampa's median household income improved by over 16 percent during the 1990s, ranking 2nd amongst its peers, while the city-MSA income ratio improved by over 5 percent in the same period, ranking 1st in the peer group and 6th out of the entire set of ninety-nine cities. Tampa was one of only three cities in its peer group to experience a positive increase in its income ratio and was the only city with a positive increase above 1 percent. The change in poverty also ranked Tampa high amongst its peer cities (4th). All of these gains were achieved with only modest population growth during the decade (8.4 percent, or 8th in the peer group).

Overall, there is every indication that the 1990s were a strong decade for the Tampa economy. For each economic indicator, Tampa performed at or near the top when compared to its peers. Most impressive has been the city's income rise relative to the MSA's. The city's incomes are approaching that of the region, indicating some success in attracting or retaining businesses and families in the urban center. While the city still has a substantial proportion of its population living in poverty, on most other measures Tampa experienced very positive urban economic development outcomes during the 1990s.

The St. Petersburg Peer Group Analysis

At the outset of the 1990s, the city of St. Petersburg was a relatively healthy central city of 240,000. Unlike most Florida cities, St. Petersburg experienced no population growth in the 1980s, although the region's population grew substantially during

Table 8.7   Summary of the St. Petersburg peer city analysis

Cities	Pop Change 1990–2000	Rank	Pct Change in Med HH Inc, 1989–1999	Rank	Pct Change in Inc Ratios, 1989–1999	Rank	Change in Pct Poverty 1989–1999	Rank
St. Petersburg, FL	4.0%	9	13.1%	1	2.1%	2	-2.2%	6
Akron, OH	-2.7%	11	10.1%	4	-2.0%	7	-14.8%	1
Anaheim, CA	23.1%	4	-8.4%	15	-7.1%	12	32.5%	15
Arlington, TX	27.2%	2	4.7%	10	-5.0%	10	20.1%	13
Aurora, CO	24.4%	3	7.9%	7	-10.1%	15	19.4%	12
Baton Rouge, LA	3.8%	10	6.9%	9	-2.9%	8	-8.4%	4
Birmingham, AL	-8.7%	14	7.3%	8	-7.3%	13	-0.5%	8
Corpus Christi, TX	7.8%	7	8.9%	6	-1.4%	5	-12.2%	2
Jersey City, NJ	5.0%	8	0.4%	12	0.0%	3	-1.5%	7
Las Vegas, NV	85.2%	1	11.0%	2	4.3%	1	3.8%	10
Lexington, KY	15.6%	5	9.3%	5	-0.6%	4	-8.9%	3
Louisville, KY	-4.8%	12	10.3%	3	-3.2%	9	-4.5%	5
Norfolk, VA	-10.3%	15	4.0%	11	-1.9%	6	0.4%	9
Riverside, CA	12.7%	6	-7.8%	13	-6.1%	11	31.9%	14
Rochester, NY	-5.1%	13	-8.3%	14	-7.3%	14	10.2%	11
Averages	11.8%		4.6%		-3.2%		4.4%	

this decade. Along many measures, St. Petersburg was one of the state's healthier central cities, although its poverty rate and income levels lagged those of the state and nation. St. Petersburg's peer group contains cities with a population of between roughly 220,000 and 270,000. Included in this list are some long established cities, like Birmingham, Baton Rouge, and Rochester (NY), as well as newer, booming cities like Las Vegas, Arlington (TX), and Aurora (CO).

When the relative changes for these indicators are reviewed, shown in Table 8.7, I find that St. Petersburg fared as well if not better than any of its peer group. While Las Vegas boomed in population and income, its poverty rate also increased substantially during this period. St. Petersburg's change in income ranked 1st in its peer group and 21st out of all cities. Similarly, St. Petersburg and Las Vegas were the only cities in the peer group to see a positive change in the city-MSA income ratio between 1989 and 1999. The city's 2.1 percent increase in its income ratio ranked 14th out of all cities in the database. While St. Petersburg's percent poverty saw a relative decline of only 2.2 percent during the decade, this ranked 6th in its peer group. Given that the entire state's percent poverty remained stable between 1989 and 1999, this relative decline for St. Petersburg is all the more impressive.

Like sister city Tampa, the decade of the 1990s was a good one for St. Petersburg. The city improved along all measures reviewed, even outpacing booming cities like Las Vegas, Aurora, and Arlington on key economic indicators. It is worth noting that these economic changes occurred largely in the absence of population growth, as St. Petersburg grew by only 4 percent during the decade. As with Tampa, the experience of St. Petersburg offers additional evidence that Florida's cities experienced positive economic development outcomes above those of their peer cities during the 1990s.

The Hialeah Peer Group Analysis

The city of Hialeah is located in Miami-Dade County, to the north of Miami. Hialeah is a city largely defined by its Hispanic population, with a 1990 population of 188,000 that was 87.6 percent Hispanic. It is important to note that no other city in the entire ninety-nine city database is as dominated by a single racial or ethnic minority as Hialeah. Although better off than the largest city in the MSA (Miami), Hialeah's indicators suggest a city with a very troubled economy in 1990; the city's poverty rate topped 18 percent and its income ratio was well below 90.0. In 1990 Hialeah was generally worse off than its peer group along all economic indicators; its percent poverty was among the highest and its income level and city-MSA income ratio were among the lowest of this group. What is particularly striking is that the city ranked so low, even given a peer group that included other smaller southern cities with struggling economies, such as Lubbock, Mobile, Montgomery, and Augusta.

The 1990s were a period in which Hialeah's economy showed no improvement. Although the city grew by almost 40,000 during the decade, this was the only sign of improvement for the city. The city's median household income actually declined, from $30,436 to $29,492, and its percent poverty rose almost half a point. Of the ninety-nine cities in the dataset, only sixteen saw declines in their median household incomes, placing Hialeah in the company of Philadelphia, Baltimore, and Newark. Further, the city's income ratio experienced a 5.1 point drop during the decade,

Table 8.8   Summary of the Hialeah peer city analysis

Cities	Pop Change 1990–2000	Rank	Pct Change in Med HH Inc, 1989–1999	Rank	Pct Change in Inc Ratios, 1989–1999	Rank	Change in Pct Poverty 1989–1999	Rank
Hialeah, FL	20.4%	2	-3.1%	13	-5.9%	13	2.3%	10
Augusta, GA	4.4%	11	0.6%	12	-2.4%	7	7.1%	12
Des Moines, IA	2.8%	13	10.8%	3	-3.9%	9	-11.9%	1
Garland, TX	19.4%	3	1.6%	11	-9.3%	15	13.3%	14
Glendale, CA	8.3%	7	-6.3%	15	0.8%	2	7.9%	13
Grand Rapids, MI	4.6%	10	7.0%	6	0.9%	1	-2.5%	8
Greensboro, NC	22.0%	1	4.7%	8	-2.8%	8	5.9%	11
Lincoln, NE	17.5%	4	11.5%	2	0.0%	4	-10.4%	2
Lubbock, TX	7.2%	9	1.7%	10	-0.3%	5	-5.9%	6
Madison, WI	8.8%	6	9.8%	4	-5.3%	11	-6.6%	5
Mobile, AL	1.3%	14	7.9%	5	-7.0%	14	-5.7%	7
Montgomery, AL	7.7%	8	4.3%	9	-3.9%	10	-1.9%	9
Shreveport, LA	0.8%	15	6.5%	7	-0.5%	6	-9.8%	3
Spokane, WA	10.4%	5	12.1%	1	0.4%	3	-8.2%	4
Yonkers, NY	4.3%	12	-5.4%	14	-5.3%	12	40.8%	15
Averages	9.3%		4.3%		-3.0%		1.0%	

suggesting that as the MSA economy grew, the residents of Hialeah were being left behind (see Table 8.8).

Overall, Hialeah's experience diverges greatly from that of Florida's other large cities. Whereas these other cities experienced very positive economic changes in the 1990s, Hialeah's plight only worsened. As noted earlier, part of this explanation is the city of Miami's desperate economic problems, problems that have spread into surrounding cities in the MSA. In addition, Hialeah's huge and growing Hispanic population, which represented roughly 90 per cent of the city's population in 2000, also helps to explain these economic declines. As Hialeah becomes home to even more immigrant Hispanics, their economic indicators are likely to continue to substantially lag those of the region and the state.

Conclusion

This chapter has summarized changes in indicators for the Florida economy and in the economies of the state's largest cities between 1990 and 2000. If, as some feared, the state's system for managing growth was harmful for the state economy, then these indicators would reveal an economy that has slowed since implementation of the 1985 GMA by local governments in the early 1990s. To date, surprisingly little research has been completed on this topic. One early analysis of the impact of the Florida GMA offered some preliminary support for the view that growth management would harm the state's economy (Feiock, 1994), whereas another study found no negative economic impacts attributable to comprehensive plans prepared pursuant to the GMA (Denslow, O'Dell, Shermyen, and Audirac, 1994).

The evidence presented in this chapter suggests that while growth management in Florida may have managed growth, it did not limit it. Quite the contrary, the state's population and employment growth continued largely unabated during the 1990s. Florida's economy continued to boom, adding over 3 million residents and 1.6 million jobs, levels that were very similar to the pre-growth management decade of the 1980s. Further, when the annual growth rates for private sector jobs were reviewed, there is no evidence that growth management slowed the state's remarkable job creation rate. Even in an era of mandated comprehensive planning and aggressive state and local planning, the Florida economy continued to create thousands of new jobs every year, with the state often leading the nation in new job creation.

As for the second research question guiding this work, the state-city comparisons indicate that the economies of Florida's four largest cities, Jacksonville, Miami, Tampa, and St. Petersburg, generally performed well during the post-growth management implementation period of the 1990s. Each of these cities saw greater declines in their poverty rates than did the state between 1989 and 1999, while three of the four experienced increases in their median household income greater than that of the state during the same period. What is particularly noteworthy about these economic gains is that they occurred despite only lukewarm population growth, with only Jacksonville experiencing a double-digit population rate increase during the 1990s (although even this rate was only two-thirds that of the state's). In addition, the improvement in the poverty rates of these cities is remarkable given that the state's overall poverty rate declined very little between 1989 and 1999.

The peer city analyses found that these cities performed better than their peers during the 1990s. Tampa and St. Petersburg fared very well when compared to their peer cities, while Jacksonville and Miami also experienced changes in their economic indicators that placed them near the top of their peer group. Of perhaps most interest was the finding that improvements to the city-MSA income ratios of Miami, Tampa, and St. Petersburg were at the top of their peer groups, and near the top for all cities in the database (Miami's income ratio increase ranked 6th, Tampa's ranked 8th, and St. Petersburg's ranked 14th out of ninety-nine cities). Although Jacksonville saw a slight decrease in its city-MSA income ratio, this change still placed the city 40th in the overall rankings. In contrast to these four cities, Hialeah's economic performance during the 1990s was among the worst of the nation's largest cities. Hialeah lagged both the state and its peer group on almost all indicators and by 2000 Hialeah was among the poorest and most impoverished cities in the nation. As noted earlier, a major explanatory factor in Hialeah's economic performance is the city's huge (and increasing) immigrant Hispanic population.

When viewed together, what do these results tell us about the performance of Florida's economy and the performance of the state's largest cities during the period when state-mandated growth management was implemented in the state? First and foremost, the evidence undermines the view that the coming of growth management would spell trouble for the Florida economy. During the 1990s Florida's economy continued its boom of recent decades, adding over 1.6 million private sector jobs and almost doubling the state's private sector payroll. While Florida's ongoing population boom has been well documented, growth in private sector employment actually outpaced population growth over the decade.

Second, these findings also undermine the view that growth management might harm the state's largest cities, causing them to lag even further behind the state once comprehensive plans were adopted and concurrency began to be enforced. By all measures, the state's most populous cities participated in the state's growth, seeing income increases and declines in poverty rates that generally bettered those of the state during the decade. Similarly, the peer analysis supports this conclusion, as the state's largest cities generally outperformed their peers over the decade.

Overall, then, the evidence presented in this chapter yields the conclusion that Florida's state-mandated growth management approach appears to have had a benign impact on the state economy and the economies of the state's largest cities. Further, there is some limited evidence that the approach may have contributed to the economic revitalization of the most populous cities in the state. This work complements findings from other studies of the impacts of growth management on urban economies. In the two best studies on this topic, Nelson and Peterman (2000) and Dawkins and Nelson (2003) both find that growth management programs promote economic development for mid-sized cities (Nelson and Peterman) and large central cities (Dawkins and Nelson). While neither study focused specifically upon Florida, they did include Florida cities in their analyses. The findings of this indicator analysis offer some additional support for establishing a positive relationship between growth management and economic development outcomes.

Before closing, three caveats are worth noting. First, it is important to underscore that the state of Florida's growth management approach is, in the terms of Dawkins and Nelson (2004), best understood as an "accommodating" one. While the state requires local governments to develop comprehensive plans and to develop a future land use map that illustrates the location and general intensity of growth, the state also requires local governments to accommodate all growth that they are projected to experience. Local governments in Florida cannot "say no" to growth; they must plan for it and, in the long run, provide capital facilities and services to support it. Dawkins and Nelson (2004) contrast the accommodating approach with a "restrictive" system in which local governments can make the decision to close their door to growth. Florida's "growth management" approach is aptly named, as the state encourages local governments to manage growth, but not limit it.

Second, while Florida's largest cities did indeed experience economic improvements during the 1990s, they still generally lag behind the state and many of their peer cities on baseline indicators of economic health. Despite improvements to their economies in the 1990s, Miami, Tampa and Hialeah, in particular, still have significant proportions of their populations living in poverty and income levels in these cities remain in the bottom third of all large cities in the nation. While urban economic development was achieved in the 1990s, there is still much work to be done to improve conditions in the state's largest cities.

Third, while Florida's growth management approach does include requirements that directly and indirectly support urban economic development, it is important to note that these requirements by no means add up to a coherent state urban policy. While some commentators, such as David Rusk, have argued for a coherent and focused national urban policy, the same criticism could be leveled at the state of Florida. While other states have growth management policies that explicitly and more comprehensively support urban economic development— Maryland's smart growth approach is an excellent example (DeGrove 2005)—Florida's growth management approach only indirectly speaks to urban (re)development. If Florida is indeed interested in using growth management to promote urban economic development, then the state needs to go further in establishing existing urban centers as a priority. For example, as part of its growth management approach Florida could, like Maryland, prioritize infrastructure projects in existing urban areas when disbursing state funds.

In conclusion, all indicators reveal that Florida's economy continued to boom during the 1990s and that the state's largest cities shared in this economic growth. Taken together, these indicators suggest that during the first decade of growth management in Florida, the state's approach to managing growth did not stop the flood of new residents, immigrants, and businesses. A look back reveals that the 1990s were yet another decade of unabated growth in a state still coming to terms with the 3 million net new residents that had arrived each of the previous three decades.

References

Ben-Zadok, E., and Gale, D. (2001). Innovation and reform, intentional inaction, and tactical breakdown: The implementation record of the Florida concurrency policy. *Urban Affairs Review, 36*(6), 836–871.

Brookings Institution Living Cities Census Series. (2004). *Living Cities Databook Series* [Data file]. Available from Brookings Institution Web site, http://www.brookings.edu/es/urban/livingcities.htm.

Dawkins, C., and Nelson, A. (2003). State growth management programs and central city revitalization. *Journal of the American Planning Association, 69*(4), 381–396.

Dawkins, C., and Nelson, A. (2004). Urban containment in the US. (*APA Planning Advisory Service Report No. 520*). Chicago: Planners Press.

DeGrove, J., and Miness, D. (1992). *The new frontier for land policy: Planning and growth management in the states.* Cambridge, MA: Lincoln Institute of Land Policy.

DeGrove, J. (2005). *Planning policy and politics.* Cambridge, MA: Lincoln Institute of Land Policy.

Denslow, D., O'Dell, W., Shermyen, A., and Audirac, I. (1994). The economic impact of local comprehensive plans. *Bureau of Economic and Business Research Monographs* (No. 8). Gainesville, FL: University of Florida.

Feiock, R. (1994). The political economy of growth management. *American Politics Quarterly 22*(2), 208–220.

Florida Growth Management Study Commission. (2001, February). *A liveable Florida for today and tomorrow.* Retrieved October 2001, from http://www.floridagrowth.org/pdf/gmsc.pdf.

Gale, D. (1992). Eight state-sponsored growth management programs: A comparative analysis. *Journal of the American Planning Association 58*(4), 425–439.

Holcombe, R. (1990). Growth Management and Land Use Planning in Florida. *Florida Policy Review 6*, 1. Tallahassee, FL: Florida State University.

National League of Cities (2004). About cities: Consolidated city-county governments. Retrieved December 13, 2004, from http://www.nlc.org/about_cities/cities_101/150.cfm.

Nelson, A., and Peterman, D. (2000). Does growth management matter?: The effect of growth management on economic performance. *Journal of Planning Education and Research 19*, 3: 277–85.

Porter, D. (1998). The states: Growing smarter? In Urban Land Institute (Ed.), *Smart growth: Economy, community, environment* (pp. 28–35). Washington, DC: Urban Land Institute.

Steiner, R. (2001). Florida's transportation concurrency: Are the current tools adequate to meet the need for coordinated land use and transportation planning? *Florida Journal of Law and Public Policy 12*, 269–297.

Chapter 9

Compact Urban Form or Business as Usual? An Examination of Urban Form in Orange County, Florida

Gerrit-Jan Knaap and Yan Song

Orange County is the most populous county in Central Florida, with nearly a million people living in an area of more than a thousand square miles. Without including the visitors who can inflate the county's population by 120,000 or more per day, the county has more residents than Alaska, Delaware, Montana, North and South Dakota, Vermont, and Wyoming. The county is one of the fastest growing in the state, adding about 30,000 new residents per year.

The Orlando/Orange County Convention and Visitors Bureau reports 95 theme parks and attractions in the area, but the three most widely known are Walt Disney World, Sea World, and Universal Studios, each of which draws both national and international visitors. The economic impact of Walt Disney World alone is immense. Disney attracts more than forty million visitors per year, pays hundreds of millions of dollars in taxes, and is one of the state's largest employers (Wetherell, 2002, p. 2). Disney has also played an important part in the growth of the Orlando region, eventually converting the formerly small city into a huge metropolis with one of the busiest airports in the country.

In 1967, the Florida Legislature passed Chapter 67–764 which created the Reedy Creek Improvement District (the "District" or "RCID"). The District includes approximately 25,000 acres (39 square miles) of land, almost all of which is owned by Disney. The RCID has been granted nearly a full array of governmental powers by the state, including the power to construct, operate, and maintain public utilities and to issue bonds; and perhaps most important, the District and all development within it is specifically exempt from the zoning and other regulations of Orange and Osceola counties, in which it is located. Moreover, development within the RCID has been exempt from the Developments of Regional Impact process, the State's special planning and regulatory process for large developments that have multi-county impacts.

Development outside the RCID but within unincorporated Orange County is subject to county zoning and subdivision regulations and concurrency requirements. Unincorporated Orange County imposes minimum lot area requirements, as well as minimum setbacks and maximum building height requirements for development in each of its agricultural, residential, and commercial zones (Orange County Zoning Ordinances, § 38–1501 *et seq.*). Density bonuses are offered to developers within specified zoning districts in exchange for dedication of land for certain public purposes or contribution to the Orange County Parks and Recreation Department parks fund (Orange County Zoning Ordinances, §§ 38.558, 38–607).

Development within the numerous incorporated cities contained in Orange County—including Orlando, Apopka, Ocoee, and Winter Garden—is subject to local (city) land development regulations. In 1991, the City of Orlando enacted a more aggressive program for discouraging sprawl and promoting a compact urban form by including in its zoning code a minimum floor area ratio and density requirements for development on vacant land. Orlando has also established a comprehensive system of density (intensity) bonuses, the stated purpose of which is to "discourage the proliferation of urban sprawl, encourage a compact urban form, encourage the redevelopment and renewal of blighted areas, and provide incentives for infill development" (Orlando Illustrated Land Development Code, § 58.1001.

Despite the extensive growth management framework put in place by the state of Florida, and implemented through local government comprehensive plans, the extent to which development patterns in the Orlando metropolitan area have changed remains in dispute. In this chapter, we employ several measures of urban form to evaluate development patterns and trends in the Orlando metropolitan area. Our intent is not to isolate and analyze the effects of particular regulations or plans. Instead, we use Orange County as a case study site to compare patterns of urban development in Florida to those in other states and to analyze trends in development patterns over time. We also do not claim that the trends in Orange County are representative of those in other Florida metropolitan areas. But because Orange County is a rapidly growing area in the central part of the state, we feel our analysis provides some insights into larger Florida development trends.

Quantitative Analysis of Urban Form

In this chapter, we analyze the pattern of urban development in Orange County in three ways. First, we construct measures of urban form for five metropolitan areas across the United States: Orange County, Florida; Maricopa County, Arizona; Montgomery County, Maryland; Minneapolis-St. Paul, Minnesota; and Portland, Oregon. We then divide each of the metropolitan areas into neighborhoods—defined by traffic analysis zones (TAZs)[1]—and compute several measures of urban form for each neighborhood. This allows us to compare recently developed neighborhoods in Orange County to recently developed neighborhoods in other parts of the country.

In our second approach, we identify all the single-family homes constructed in Orange County in 2000 and measures the urban form characteristics of the neighborhoods (in this study, defined as quarter-mile buffers around the building sites), in which these new homes are located. Then, using cluster analysis, we identify

---

1   TAZs are geographic units designed for use in transportation planning and are roughly coincident with census block groups. In previous work we explored alternative definitions of neighborhood. Our analysis showed not that TAZs are necessarily the best geographic unit, but that they were useful for demonstrating differences in development patterns that were not inconsistent with alternative geographic units.

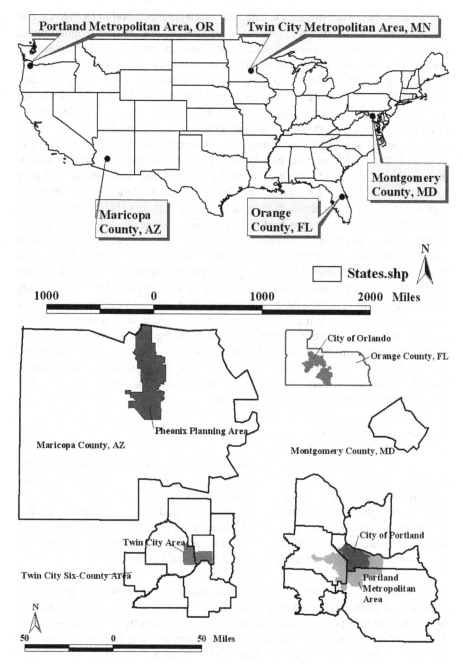

Figure 9.1   Location and relative size of the five study areas

specific neighborhood types and enumerate how many single-family homes were built in each type of neighborhood. This allows us to examine the kinds of neighborhoods in Orange County in which single-family homes are currently being built. Finally, in our third approach we estimate the date when each neighborhood in Orange County was built and examine trends in urban development patterns in these neighborhoods over time. This allows us to assess whether development trends in Orange County are becoming more or less consistent with the principles of smart growth.

We obtained data for the five study areas: Maricopa County; Orange County; Minneapolis-St. Paul; Montgomery County; and Portland. Site locations are illustrated in Figure 9.1. The sites were clearly not randomly selected, but still provide a range of locations, sizes, growth rates, and local regulatory environments.

Policy makers in each of these areas have taken unique approaches to the control of sprawl. Portland is perhaps best known for its urban growth boundary and its metropolitan-wide "2040" plan for controlling growth within that boundary. Orange County also has a growth boundary and, like all places in Florida, has an elaborate concurrency policy that limits development until adequate public facilities are in place. Minneapolis-St. Paul is best known for its tax-base sharing system and its metropolitan-wide urban service boundary. Besides an urban growth limit and an elaborate adequate public facilities ordinance, Montgomery County has a well-known program of transferable development rights. Metropolitan Phoenix is not known for its growth controls but faces limits to growth imposed by the availability of water. It is important to note that our sample includes four study areas with some of the most advanced urban growth management systems in the United States. Also, our analysis of development patterns includes the entire metropolitan areas for the Portland, Minneapolis-St. Paul, and Phoenix study areas, but only parts of the larger Washington, DC and Orlando metropolitan areas for Montgomery and Orange counties, respectively. This is especially noteworthy for Montgomery County, since the study area does not contain the central city.

Measures of Urban Form

To begin our analysis, we obtained GIS data from each of our five study sites. These data included: (1) attributes of parcels such as year built of the structure, land use type, lot size, and floor space; (2) street network centerlines; (3) bus routes; (4) political and planning boundaries, such as county and city boundaries and urban growth boundaries; (5) open space, and (6) aerial photographs. We then used these data to compute measures of urban form. Our measures fall into three categories: Street Network Design; Land Use Intensity; and Land Use Pattern.[2]

Street Network Design   We compute two measures of street network design: internal and external connectivity. Internal connectivity involves the number of nodes and intersections within the neighborhood, while external connectivity measures distance between points of access into and out of the neighborhood. Internal connectivity measures transportation route options within a neighborhood, while external connectivity measures route options between neighborhoods.

---

2   All the calculations were computed using ArcInfo and/or ArcView.

- *Int_Connectivity*—The number of intersections within the neighborhood, divided by the sum of its cul-de-sacs (or dead ends) and intersections. The higher this ratio, the greater the internal connectivity.
- *Ext_Connectivity*—The median distance between ingress/egress (access points) to and from the neighborhood, in feet. The greater the distance, the poorer the external connectivity.

Land Use Intensity   We offer two measures of development intensity: single family lot size and single family floor space.

- *Lot Size*—The median lot size in square feet of single-family dwelling units in the neighborhood. The smaller the lot size, the higher the intensity.
- *Floor space*—The median floor space in square feet of single-family dwelling units in the neighborhood. The larger the floor space, the higher the intensity.

Land Use Pattern   We offer one measure of land use mix and two measures of accessibility. Our measure of land use mix is based on the concept of entropy—a measure of variation, dispersion or diversity (Turner, Gardner, and O'Neill, 2001). Our measures of accessibility capture the distance of single-family homes from commercial uses and the percent of single-family homes that are within walking distance of a commercial use.

- *LU_Mix*—A diversity index $H_1 = \dfrac{-\sum_{i=1}^{s}(p_i)\ln(p_i)}{\ln(s)}$ where H_1 = diversity, p_i = proportions of each of the five land use types such as Single-Family Residential , Multifamily Residential, Industrial, Public, and Commercial uses, and s = the number of land uses (which, in this study, equals five). This index can range from zero, which connotes single-use property, to one. The higher the value, the more even the distribution of land uses.
- *Comdis*—The median straight-line distance from single-family dwelling units in the neighborhood to the nearest commercial-use property, in feet. The greater the distance, the lower the accessibility.
- *Ped_Com*—The percentage of Single-Family Residential units in the neighborhood within one-quarter mile of commercial-use property, along a road network. The greater the percentage, the greater the pedestrian accessibility.

Five Study Areas and Characteristics of Recently Developed Neighborhoods in Study Areas

To analyze patterns of recent developments in each study area, we computed the median value and the coefficient of variation of each measure of urban form for all neighborhoods developed after 1995. The results are presented in Table 9.1. The last column of Table 9.1 presents the results of F-tests which are used to test the equality of means across study areas. These test results indicate whether the measures of urban form vary significantly across study areas.

Table 9.1   Urban form median values and coefficients of variation for neighborhoods built after 1995

	Montgomery County	Orange County	Maricopa County	Portland	Twin Cities	F-test
Int_Connectivity	0.76	0.78	0.79	0.65	0.77	18.3*
	(0.16)	(0.11)	(0.18)	(0.22)	(0.13)	
Ext_Connectivity	989	631	937	392	389	21.9*
	(0.36)	(0.43)	(0.42)	(0.46)	(0.43)	
Lot_Size	8035	7695	8165	6838	—	not sig
	(0.32)	(0.45)	(0.45)	(0.36)		
Floor space	2900	2107	2047	1883	—	not sig
	(0.26)	(0.29)	(0.33)	(0.26)		
LU_Mix	0.36	0.39	0.45	0.48	0.42	13.8*
	(1.81)	(1.46)	(1.13)	(0.98)	(1.45)	
Comdis	2545	3653	1676	1851	965	11.9*
	(0.68)	(0.31)	(0.67)	(0.46)	(0.72)	
Ped_Com	0.11	0.28	0.19	0.30	0.22	13.3*
	(1.80)	(2.13)	(1.47)	(0.66)	(1.00)	

* $p < .05$. Coefficients of variation are provided in parentheses.

As shown, some aspects of development patterns vary within and across study areas, while others do not. The coefficients of variation indicate that land use mix and pedestrian accessibility vary most within metropolitan areas.[3] The variation in these values is particularly high in Montgomery and Orange counties. This implies that metropolitan areas have some areas that are characterized by a mixture of uses and commercial properties that are accessible by foot and some areas that are not— and that the intrametropolitan differences are statistically significant. In addition, internal and external connectivity, land use mix, distance to nearest commercial use, and pedestrian accessibility all vary significantly across study areas. Lot size and single-family floor space do not differ significantly across study areas.

Orange County falls in the middle of the study areas on most of these measures of urban form. In terms of internal connectivity, it is second best, though the variation between metropolitan areas is not large. In external connectivity, it ranks about 250 feet better than Montgomery County and about 250 worse than Portland. By lot size Orange County places in the middle of a fairly narrow range, whereas by floor space it ranks in the middle of a larger range. A relatively large proportion of single-family homes (28 percent) are within one-quarter mile of a commercial-use property, yet the median distance to commercial uses is the greatest of the five study areas.

These results can be interpreted in at least two ways. On one hand, growth management efforts have not produced results that distinguish Orange County from the other areas in this study. On the other hand, the other study areas are among the most actively involved in growth management. Thus, while development trends in

---

3   For these variables in most of the study areas, the standard deviation is greater than the mean value.

Table 9.2   Summary statistics for urban form variables

Variable	Unit of Measure	Mean	StDev	Minimum	Maximum
BlockArea	Acre	449	979	0	18,077
Nbr_BlockSize	Feet	17,557	19,758	1,256	253,429
BlockSize	Feet	41,961	58,998	292	320,719
Nbr_BlockArea	Acre	205	504	2	14,211
StreetLength	Feet	876	888	299	30,297
#Cul-de-sac	# of counts	4	3	0	24
#Intersection	# of counts	14	8	0	69
Nbr_FloorSpace	Square Feet	2,043	590	948	9,444
FloorSpace	Square Feet	2,197	800	998	12,959
#Lots	# of counts	189	86	1	554
Nbr_LotSize	Square Feet	10,012	22,123	3,899	1,341,171
LotSize	Square Feet	10,879	24,861	1,160	1,341,171
Commercial	Acre	2	7	0	110
Com_Dist	Feet	2,145	1,756	37	19,564
Bus_Dist	Feet	1,434	5,174	29	63,557
OpenArea	Acre	3	3	0	10

Orange County do not stand out when compared to this set of study areas, this is a good group from which to compare favorably.

Quantitative Analysis of Orange County Neighborhoods

Measuring the development characteristics of recently built neighborhoods, of course, provides an incomplete picture. If a number of homes are being built in older, existing neighborhoods, then measures of new neighborhoods fail to capture these trends. In Portland, for example, a very large number of homes are being built in inner city neighborhoods with traditional characteristics of urban form.

To explore this possibility we next identify the Orange County neighborhoods where single family homes were recently built. We first identified the 5,810 new or redeveloped single-family homes that were built in Orange County in the year 2000. We next computed a set of urban form measures for the neighborhoods around these new homes, based upon a quarter-mile buffer. Based upon our measures of urban form, we then classify these neighborhoods into specific types and enumerate the number of homes built in each neighborhood type. Summary statistics for all these measures are provided in Table 9.2.

Street Design Measures
- *#Intersection*—number of intersections in the buffer area of the parcel;
- *#Cul-de-sac*—number of cul-de-sacs in the buffer area;
- *StreetLength*—length of street miles in the buffer area;
- *BlockSize*—perimeter of the block where the parcel is located;
- *Nbr_BlockSize*—median perimeter of the blocks in the buffer area;

- *BlockArea*—area of the block where the parcel is located;
- *Nbr_BlockArea*—median area of the blocks in the buffer area;

Density Measures
- *LotSize*—lot size of the parcel;
- *Nbr_LotSize*—median lot size of single-family parcels in the buffer area;
- *#Lots*—number of single-family lots in the buffer area;
- *FloorSpace*—floor space of the single-family house on the lot;
- *Nbr_FloorSpace*—median floor space of all single-family houses in the buffer area;

Mixed Land Uses Measures
- *Commercial*—acres of commercial land use in the buffer area;
- *Com_Dist*—distance from the lot to the nearest commercial land;

Natural Environment Measures
- *OpenArea*—acres of open space per buffer;

Alternative Transportation Modes
- *Bus_Dist*—distance from the lot to the nearest bus route.

Critical Dimensions of Neighborhood Form

Some of these measures of physical neighborhood form are highly correlated. The distribution of cul-de-sacs, for example, is highly correlated with the distribution of large blocks. Therefore it is useful to condense these variables into a smaller set of variables that removes the correlations in the data. For this purpose we use factor analysis, a technique for data reduction, to help us understand the dimensional structure of our group of variables.

From the various aspects of physical neighborhood form, we use factor analysis to extract six dimensions (or factors). The results are presented in Table 9.3. The variables are listed in the order of the size of their factor loadings sequentially for each factor. The extracted factors reproduce about seventy per cent of the total variation among the cases. We used principal component analysis for the extraction, and Varimax with Kaiser Normalization as the rotation method, because this combination explained the most variation in the data.

Inspection of Table 9.3 reveals six dimensions of physical neighborhood form that emerged from the analysis. The last row of the table presents the percent of the total variation accounted for by each factor. The first factor reflects the dimension *Street Block Design*. Factor loadings indicate that smaller street blocks, both the block where the home locates and other blocks in the buffer area, contribute to a smaller value of factor 1. The density variables loaded highly on the next factor: Smaller lots (of each home and of other single-family homes in the buffer areas) and more lots contained in the buffer areas contribute to a smaller value of factor 2. The third factor relates to *House Size*, showing that larger houses contribute to a larger value of factor 3. The fourth factor reflects *Connectivity*: Fewer cul-de-sacs, more

Table 9.3   Factor analysis of physical neighborhood form dimensions

Variable	Factor 1 Street Block Design	Factor 2 Density	Factor 3 House Size	Factor 4 Connectivity	Factor 5 Commercial Uses	Factor 6 Natural Environment
BlockArea	**0.88**	0.06	0.17	0.09	0.02	-0.06
Nbr_BlockSize	**0.87**	0.08	0.06	0.20	0.16	-0.17
BlockSize	**0.81**	-0.05	0.01	0.02	-0.11	-0.17
Nbr_BlockArea	**0.80**	0.08	0.21	-0.11	0.20	-0.01
Nbr_LotSize	0.04	**0.87**	0.11	0.16	-0.27	-0.21
LotSize	0.06	**0.77**	0.01	0.20	0.09	-0.21
#Lots	-0.06	**-0.55**	0.06	0.16	0.06	-0.06
Nbr_FloorSpace	0.13	0.24	**0.88**	-0.05	-0.01	0.18
FloorSpace	-0.03	0.03	**0.86**	-0.27	0.20	-0.01
#Cul-de-sac	0.06	0.18	-0.05	**0.73**	-0.13	0.17
#Intersection	0.02	-0.15	0.18	**-0.61**	-0.11	-0.03
StreetLength	0.04	0.06	-0.07	**-0.50**	-0.05	0.26
Bus_Dist	0.07	0.20	-0.06	**0.44**	-0.15	-0.14
Commercial	-0.01	-0.10	-0.04	-0.10	**-0.66**	0.06
Com_Dist	0.03	0.26	0.06	0.01	**0.65**	0.15
OpenArea	0.15	-0.02	-0.11	0.02	-0.21	**-0.57**
% Variation	0.20	0.11	0.15	0.09	0.09	0.06

intersections, more street miles, and shorter distances to nearby bus stops contribute to a smaller value of factor 4. The fifth factor reflects the level of *Commercial Uses*: more commercial land uses and shorter distance to commercial units contribute to a smaller value of factor 5. Finally, the open area variable is the only one to load on the natural environment factor.

Classifying Neighborhood Types

We then carried out a cluster analysis to identify neighborhood characteristics shared among the homes in our study, no matter where they are located spatially. Specifically, we used a K-means cluster analysis[4] to classify all 5,810 homes into different neighborhood types (or clusters) on the basis of similarities and dissimilarities in the values of the six factors previously derived.[5] This technique reveals clusters

---

4   K-means clustering begins with a grouping of observations into a predefined number of clusters. It evaluates each observation and moves it into its nearest cluster. The nearest cluster is the one which has smallest Euclidean distance between the observation and the centroid of the cluster. When a cluster changes by losing or gaining an observation, the cluster centroid is recalculated. At the end, all observations are in their nearest cluster.

5   Cluster analysis procedures are affected by the magnitude of the variables included; that is to say, variables with large numbers have a greater impact on the outcome of the analysis than variables with small magnitudes. To control for this imbalance, scaling may be necessary to convert the original variable values to standard scores. Because the six factor scores derived from the factor analysis are used here in the cluster analysis, the magnitude of the variables is not a concern.

Table 9.4 Cluster centroid values for the neighborhood types

Physical Form Dimensions	Cluster 1 Urban Core	Cluster 2 Inner Ring	Cluster 3 Outer Ring	Cluster 4 Greenfields	Cluster 5 Rural Devs
Street Block Design	-0.5923	-0.2590	-0.5162	1.9308	7.2959
Density	-0.0941	0.3253	0.5298	-0.1429	20.9385
House Size	1.0988	0.2695	2.0988	-0.3622	3.4448
Connectivity	-0.5605	0.5898	0.6940	-0.1037	1.1857
Commercial Uses	-2.4736	0.3757	0.5384	0.2342	0.6438
Natural Environment	3.4455	4.6753	0.2625	0.5856	-0.3899
Counts	368	1532	3095	803	12
Percentage of All	6.33	26.37	53.27	13.82	0.21

Table 9.5 Cross-tabulation of neighborhood types by age of buffer

Neighborhood Type	Pre-1960	1960s	1970s	1980s	1990s	2000s	All
Urban Core	84	119	36	63	54	12	368
Inner Ring Suburbs	8	12	9	39	109	1351	1532
Outer Ring Suburbs	0	15	24	120	426	2510	3095
Greenfields	0	0	0	0	5	798	803
Rural Developments	0	0	1	0	3	8	12
All	92	146	70	222	597	4679	5810

such that each neighborhood type is internally as similar as possible but externally dissimilar to other neighborhood types.

The best clustering solution, based on the interpretability of the results and associated cluster statistics, was found to be a five-cluster solution. The five neighborhood types thus revealed are: the urban core, inner ring suburbs, outer ring suburbs, greenfield developments, and rural developments The values of the cluster centroids for each of the five neighborhood types are presented in Table 9.4. The centroid values of the individual clusters uncover the characteristics of the neighborhood types by scoring their performance on the six dimensions of physical neighborhood form. The last two rows of Table 9.4 reveal the distribution of homes by cluster. Table 9.5 provides additional information on the distribution of homes within each cluster by age of their immediate neighborhoods—determined by the median "year built" attribute of all single-family units contained in the quarter-mile buffer areas.

Characterization of Neighborhood Types

Based on the information provided in Tables 9.4 and 9.5, we now describe these different neighborhood types. Figure 9.2 (in the color plate section) illustrates locations of new homes by neighborhood type. Figure 9.3 illustrates the prototypical site design of each neighborhood type.

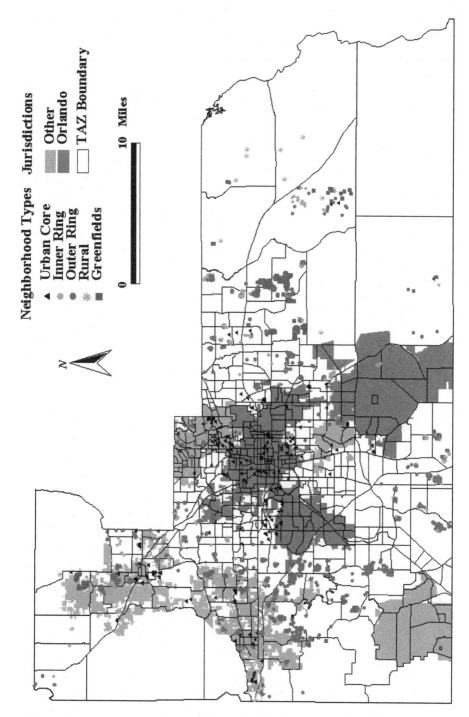

Figure 9.2   Distribution of neighborhood types in Orange County, Florida

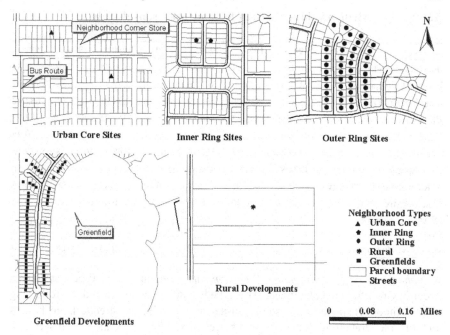

Figure 9.3   Site design of each neighborhood type

Neighborhood Type 1—Urban Core   Neighborhood Type 1 structures are located in central city areas or at the immediate inner ring suburbs. There are 368 new homes in this neighborhood type, and they make up 6 percent of our sample. Most of these neighborhoods have the following characteristics: grid street networks, small lots and houses, high density, abundant commercial stores, and accessible bus service. However, there is typically no open space nearby (see Figure 9.3). It is noteworthy that many of the Type 1 homes are built in older neighborhoods; 65 percent of them are built in areas developed before the 1980s (see Table 9.5).

Neighborhood Type 2—Inner Ring Suburbs   Neighborhood Type 2 can be classified as the inner ring suburban infill developments. Type 2 has 1,532 new homes, or 26 percent of our observations. These neighborhoods have small houses, access to public transit, and less access to nearby open space (see Table 9.4). Only about five percent of Type 2 neighborhoods were built before the 1990s (see Table 9.5).

Neighborhood Type 3—Outer Ring Suburbs   Neighborhood Type 3 can be classified as the outer ring suburban developments. This is the largest neighborhood type, with 3,095 new homes—fully 53 percent of our total. Type 3 neighborhoods are characterized by curvilinear street arrangements, some cul-de-sacs, wide streets, relatively large and clustered lots of uniform size and shape, predominantly detached single-family homes close to suburban malls, and some open space nearby (see Figure 9.3).

Table 9.6   Median values of urban form measures, by decade

	1940s	1950s	1960s	1970s	1980s	1990s	2000s
Internal Connectivity	0.94	0.91	0.87	0.82	0.78	0.79	0.83
External Connectivity	321.34	350.80	387.26	553.60	642.50	693.55	776.99
Length of Cul-De-Sac	339.98	334.06	416.49	463.25	438.91	375.61	682.42
Block Size	1774.50	1843.33	2246.18	2231.96	2365.71	2090.32	2487.77
Number of Blocks	0.66	0.46	0.48	0.97	0.42	0.28	0.45
Lot Size	16847.84	10000.28	17547.14	20169.25	19108.71	10753.73	8389.16
Floor Space	1284.81	1317.96	1519.53	1513.02	1666.95	1936.91	2057.80
SFR Dwelling Unit Density	4.87	4.30	3.81	3.46	3.42	3.06	4.15
Land use Mix	0.39	0.40	0.37	0.41	0.46	0.42	0.33
Distance to Commercial Unit	1578.66	1612.20	1866.76	2372.42	3401.19	3973.70	5945.11
Ped. Access to Commercial	0.82	0.59	0.32	0.19	0.15	0.08	0.10

Neighborhood Type 4—Greenfield Developments   Neighborhood Type 4 structures occur in Greenfield developments. This type has 803 new homes and the following neighborhood characteristics: some cul-de-sacs, curvilinear streets, large houses and lots, moderate to high density, and scarce mixed land uses and transit services. Type 4 neighborhoods are also distinct in their inviting open spaces (see Figure 9.3). Greenfield neighborhoods are very new: Almost all were built after 2000.

Neighborhood Type 5—Rural Developments   Neighborhood Type 5 structures, dispersed across the rural landscape at the periphery area (see Figure 9.2), can be characterized as rural developments. Only 12 homes in our study belong to this neighborhood type. These neighborhoods have a set of typical rural development features (see Figure 9.3): large blocks, extremely low density, extremely large lots and houses, no access to transit or mixed land uses, and dominant rural land uses with abundant open spaces.

In sum, our analysis of newly-constructed single-family homes in Orange County suggests that they have been built in five distinct neighborhood types that range in character from Urban Core to Rural Development. The majority of homes built in 2000 were built in outer ring suburbs, whereas very few were built in the urban core.

Changes in Orange County Neighborhoods over Time

Finally, to get a sense of how development patterns in Orange County have changed over time, we conducted a time series analysis. We began by date stamping neighborhoods based on the median "year built" attribute of the single-family houses they comprise. We then computed the median values of various urban form measures for every neighborhood built in each decade since 1940. The results are presented in Table 9.6.

As shown in this table, the proportion of cul-de-sacs in Orange County fell from the 1940s to the 1980s and then began rising , though the changes over time are relatively small. The distance between access points into neighborhoods, however, rose throughout the post-war period. Like internal connectivity, single-family lot sizes rose sporadically through the early post-war period but began falling in the 1970s. Single-family house sizes, measured by square feet of living space, have risen throughout the post-war period. Land use mix in single-family neighborhoods has fallen recently but displays no clear temporal trend. The median distance to a commercial use has risen since 1940 and most dramatically since 1990. Finally, the percent of homes within one-quarter mile of a commercial use has fallen steadily since 1940, with a slight uptick in 2000.

These trends paint a relatively clear picture of development trends in Orange County. Though lot sizes are falling, house sizes are rising. Neighborhoods are becoming slightly more internally connected but more regionally isolated. What's more, neighborhoods are becoming more homogeneous, and fewer homes are within walking distance of commercial uses. With the exception of falling lot size and slightly greater internal connectivity, the trends are clearly not in the direction of smarter growth.

Conclusions

Before we draw implications from this study, it is necessary to note several significant limitations of our approach. First, our classification of neighborhood types only focuses on physical neighborhood form. We did not consider any social or economic characteristics. Second, we examined the development patterns only for single-family homes. Due to data limitations, we were not able to examine trends in multifamily developments. Finally, though our results provide some interesting information about development trends in Orange County, it is impossible to generalize these results to all of Florida or even Central Florida.

Although the state of Florida is recognized as a national leader in growth management, its program has evolved in fits and starts and, with the exception of concurrency, lacks a consistent theme or framework for effectively managing urban growth. This lack of coherence and consistency is evident in Orange County, where intergovernmental complexity mixes with large-scale private developments and continuous rapid growth. Although every jurisdiction in the county has a plan, it is hard to characterize overall growth in the county as well planned.

To provide some assessment of growth management in Orange County, we examined patterns of urban form in three ways. First, we constructed measures of urban form and used them to characterize and compare neighborhoods built after 1995 in each of five study areas across the United States. Second, we identified all the single-family homes constructed in Orange County in 2000 and computed our measures of urban form for the neighborhoods in which these new homes were located. Then, using cluster analysis, we distinguished specific neighborhood types and enumerated how many single-family homes were built in each type of neighborhood. Finally, we estimated the dates when all neighborhoods in Orange County were built and examined trends in urban development patterns over time.

The results are mixed at best. In most measures of urban form, Orange County falls near the middle of the five study areas. That is, among these areas, Orange

County doesn't have a street network, house sizes or densities, or land use patterns that stand out, though its median distance from single-family homes to commercial uses is the highest. Perhaps given Orange County's rapid rate of growth and the national reputations of the comparison groups, these results are commendable.

Our analysis of the types of neighborhoods in which new houses are being built also presents a mixed picture. Over half of all new development in Orange County is taking place in outer ring neighborhoods, characterized by curvilinear streets, cul-de-sacs, wide streets, uniformly large lots, and predominantly single-family homes—typical suburban sprawl. Yet the second most common type of neighborhood in which new houses are being built is the inner ring neighborhood, with suburban infill and better transit access. Greenfield neighborhoods are the next most common, with Urban Core and Rural neighborhoods falling well below these three dominant neighborhood types. Without comparison to other study areas, it is difficult to judge these results; but it is also difficult to argue that these proportions in Orange County are ideal.

Our analysis of trends over time are perhaps least ambiguous. Over the postwar period in Orange County, lot sizes have fallen and house sizes have risen. Neighborhoods have become slightly more internally connected but less regionally connected. Further, neighborhoods have become more homogeneous, with fewer homes within walking distance of commercial uses. With the exception of falling lot size and slightly greater internal connectivity, these trends are not encouraging.

Though the results are not encouraging, they are perhaps not the best test of growth management success in Florida. Besides the difficulty of generalizing from Orange County to the rest of the state, it is not clear that measures of urban form are the best means to evaluate growth management success—especially in Florida. The focus of growth management in Florida has long been concurrency, that is, the timely provision of public services as new development occurs. Clearly our measures of urban form provide no insights into the success of that effort. Neither do they offer any insights into the degree that Florida has been successful in protecting its unique natural environment. In fact, most of our measures reflect perhaps more on the influence of local subdivision ordinances than on any other element of state or local comprehensive planning. To date, the state has not weighed in on these elements. Despite these caveats, we believe that our measures of urban form do provide insights into the kinds of urban environments currently under construction in the state of Florida. And to the extent that Orange County is exemplary, there remains considerable room for improvement.

References

Turner, M. G., Gardner, R. H., and O'Neill, R. V. (2001). *Landscape ecology in theory and practice: Pattern and process.* New York: Springer Verlag.

Wetherell, K. (2002). Florida law because of and according to Mickey: The "top 5" Florida cases and statutes involving Walt Disney World. *Florida Coastal Law Journal, 4, Fl. Coastal L. J.1.(Fall).*

Chapter 10

The Spillover Effects of Growth Management: Constraints on New Housing Construction

Yan Song

In response to a population explosion and its associated consequences—among them jetport construction in environmentally sensitive areas and encroachment of saltwater into freshwater reservoirs (Catlin, 1997; Anthony, 2003), Florida has adopted two waves of state-based growth management programs since 1972. The focus of the earlier growth management programs was on the protection of the state's physical environment. With concerns growing over the increase in sprawl, the state's Local Government Comprehensive Planning Act was passed in 1975. Flaws in this Act, which mandated that all incorporated municipalities and counties prepare and adopt comprehensive plans, soon were noted. Much has been written to present the mixed results that this earlier program has achieved. For example, DeGrove (1989) argued that the centralized approach to the planning process failed to ensure cogent implementation mechanisms at the local level. In addition, the lack of state financial resources has weakened the local governments' ability to prepare quality comprehensive plans (DeGrove, 1990).

To address the deficiencies in the earlier statewide growth management program, the Florida legislature adopted the 1985 Growth Management Act (GMA). This "second wave" of state growth management identified specific requirements for local comprehensive (or growth management and land use) plans and was intended to contain urban sprawl through coordinated land use among local governments (DeGrove, 1992). The underlying rationale behind this move was that fragmented local land use decisions made by individual communities acting in their own self-interests cannot be expected to contain urban sprawl. Consistent actions by the local agencies in the face of rapid population growth and land development were essential to the state and regional growth vision (Carruthers, 2002). Vertical consistency specified that each local comprehensive plan not conflict with the goals of the state and regional comprehensive plans. Horizontal consistency required that local comprehensive plans be compatible with each other (DeGrove, 1989).

It has been two decades since the adoption of the 1985 GMA, which mandated enforcement mechanisms to ensure uniform compliance among local jurisdictions. Despite these efforts, local inconsistencies in enforcement and implementation remain a major problem (Burby and May, 1997). Florida's growth management is still lacking uniform application at the local level (Carruthers, 2002). Deyle and Smith (1998) found substantial variation in the degree to which communities meet their planning requirements. Feiock (2004) identified tremendous variation in the complexity and restrictiveness of the regulatory process at the local level.

Many have argued that inconsistent growth management programs at the local level could result in the displacement of growth from more restrictive to less restrictive communities (Pollakowski and Wachter, 1990; Cho and Linneman, 1993). It is conjectured that the capricious enforcement style of implementing Florida's 1985 GMA could have led to redistribution of sprawling developments. Feiock (2004) argued that when different results in housing markets occur under alternative local institutional settings, households and land developers might attempt to select and build at locations with relatively less restrictive land use regulations.

However, the link between local land use policies and the redistribution of growth has received little formal analysis. This study explores whether the inconsistent enforcement of Florida's 1985 GMA at the local level has displaced new housing construction from more stringent to less stringent communities. A review of previous studies on the effects of local land use regulations on the housing market is described in Section 2. Section 3 describes the data and methods used to carry out this study, while Section 4 presents the empirical analysis. Finally, Section 5 concludes with policy implications.

Land Use Regulation at the Local Level and Its Impacts on Housing Markets

The literature on the impacts of local growth management actions on the redistribution of housing construction is surprisingly sparse, given the importance of growth controls in affecting the housing development process. Several studies have found that growth management actions at the local level could restrict the supply of housing (Atash, 1990; Evans, 1992; Mayer and Somerville, 2000). Atash (1990) argued analytically that strict standards imposed by local land use regulations, long regulatory delays that add to developers' carrying costs, and growth controls have all kept the levels of housing production far below potential demand. Atash categorized the impacts of zoning, subdivision, environmental and growth controls on housing costs into three groups: direct costs, indirect costs, and costs of unnecessary or excessive requirements and defined them as follows:

- Direct costs cover the administrative expense paid by a developer to meet all requirements associated with local land use regulations. These include application fees, review fees, inspection fees and permit fees.
- Indirect costs are due to delays and inefficiencies in the administrative process. They relate to such overhead expenditures as staff, property taxes, inflation and interest on loans which continue regardless of whether or not a project has been halted.
- Costs due to excessive or unnecessary requirements of local land use regulations emanate from regulations which specify a development standard above and beyond what is necessary for protection of the basic health, safety and welfare of the occupants of a housing unit (p. 232).

Mayer and Somerville (2000) examined the relationship between land use regulation and residential construction empirically. In their study, they considered two classes of regulations, those that impose explicit financial costs on builders, such as development or impact fees, and those that delay or lengthen the development process. They argued that:

> ... many types of regulations can be collapsed into these two categories. With sufficient time or money, various regulations such as minimum building standards or lot sizes can be overcome. In other cases, a sufficiently high payment or provision of amenities to affected individuals or communities can weaken or remove opposition to a project. Usually, if a developer is willing to fight long and hard enough, approvals will eventually be forthcoming (p. 642).

By using quarterly data from a panel of 44 US metropolitan areas between 1985 and 1996, Mayer and Somerville (2000) found that land use regulations decrease the level of new housing construction. The findings suggested that metropolitan areas with more extensive regulation could have up to 45 percent fewer housing starts. More importantly, they found differences by type of regulation: Development or impact fees have negligible impact on new housing construction, but regulations that lengthen the development process or otherwise constrain new development have larger and significant effects. Mayer and Somerville did not examine the impacts of local regulations on the spillover of housing construction.

Landis (1992) assessed the overall significance of local growth controls on housing price by focusing on seven California communities that had imposed stringent controls and six similar jurisdictions that had not. Landis found that much of the development spilled over into neighboring jurisdictions with less onerous growth control measures and thus concluded that ad hoc growth control actions could have supply-limiting effects. Cho and Linneman (1993)'s study addressed varying implementation of local plans and the consequent effect upon adjacent communities. Specifically, their research found that communities with a high proportion of residentially-zoned land which are adjacent to communities with a low proportion of residentially-zoned land tend to experience positive housing development spillover. Conversely, proportionally low residentially-zoned communities next to proportionally high residentially-zoned communities experience negative housing spillover. What this means is that residential construction moves from communities that are less welcoming, zoning-wise, to communities that are more welcoming with their zoning. Despite the fact that these communities' comprehensive plans are consistent with each other, one community implements its plan much differently than the other; and it is that difference in implementation that has an effect on the development patterns.

In sum, little empirical study has been performed on the effects of different levels of implementation of land uses on redistribution of housing construction. This study attempts to address this issue. The primary purposes of this study are, first, to examine differentiated implementation stringency of growth management actions at the local level and, second, to investigate empirically the nature of the interjurisdictional spillover effect. Specifically, data on comprehensive plan expenditures, plan quality, development impact fee, regulatory process, and other growth management actions such as Urban Growth Boundaries (UGB) or Urban Service Areas (USA) are collected from 67 counties in Florida. A Regulation Ruggedness Index (RRI) is further constructed to measure the interjurisdictional spillover effect caused by different levels of implementation of the GMA.

Methods and Data

I built a housing supply model in order to examine what factors affect the location of housing construction. From the housing supply model, I tried to test whether or not the UGB influences urban residential development patterns, after controlling other variables affecting housing supply such as housing value, household income, and accessibility and location factors. I built a standard regression model where housing units constructed by census-block group are a function of the housing market, neighborhood and location, and accessibility variables.

Housing market variables include median household income and mean housing value. Income and housing value are good indicators when developers select development sites. The model uses lag variables for housing value and household income, which are measured for four years prior to the new construction because it is rational to argue that the housing value and income where developers want to invest are available prior to investment decisions.

Methods and Variables

This section describes the method used to examine the impacts of differentiated implementation stringency of GMA at the local level on housing construction. A yearly new housing supply function for 67 counties in Florida for the years from 1993 to 2002 is used to explore the link between local land use regulation and interjurisdictional spillover of growth. Specifically, the amount of new housing construction Q_i in jurisdiction i is specified as follows:

$$Q_i = Q_i(\Delta P_i, \Delta C_i, X_i, R_i, RRI_i)$$

where ΔP_i and ΔC_i are the changes in housing prices and costs (Mayer and Somerville, 2000);[1] X_i is a vector of other supply-side control variables; R_i is a set of variables measuring implementation stringency of GMA by the local jurisdiction, and RRI_i—Regulation Ruggedness Index—measures the interjurisdictional spillover effects caused by different level of implementation stringency of GMA. Table 10.1 provides a list of variable definitions and data sources.

Dependent Variable   To estimate the supply function mentioned above, the number of new single family building permits is used as the measure of new housing construction. The use of number of building permits as a proxy of new housing

---

1   Mayer and Somerville (2000) suggested that the new housing construction should be modeled as a function of changes in housing prices and costs rather as a function of the levels of those variables.

Table 10.1 Definition of variables and data sources

Dependent Variable	Data Source
START: Yearly Housing Starts (Building Permits) 1993–2002	US HUD: www.huduser.org

Independent Variables	Data Source
PRICE: Price change rate measured by repeat sales index	Florida Statistical Abstract (1)
RPR: Change in real prime rate	Office of Federal Housing Enterprise Oversight (OFHEO)
COAST: Dummy variable, 1 if the county fronts the coast	GIS
DHWY: Distance, in kilometers, from the centroid of the county to the nearest interstate	GIS
MSA: Dummy variable, 1 if the county is part of a census-defined Metropolitan Statistical Area in 2000	US Census
POP90: Population in 1990	US Census
BUILT90: Percentage of the county's buildable land area that was considered "urbanized" by the Census Bureau in 1990 (2)	GIS;
EXPPC: Planning expenditures per capita	Florida Legislative Committee on Intergovernmental Relations: http://fcn.state.fl.us/lcir/cntyfiscal.html
#GOAL: Number of goals, objectives and policies in plan	Survey of plans
#IMPV: Number of land improvement standards	Survey of plans
AICP: Dummy variable, 1 if the highest planning official is certified AICP	Telephone survey
IMPFEE3: Average impact fee charged for a new 3-bedroom single family house (1847 square feet)	Data compiled by Sarasota County
#IMP: Number of categories for which impact fees are charged	Telephone survey
IMPFEE: Dummy variable, 1 if impact fees are charged	Telephone survey
DELAY: Number of months to receive subdivision approval	Telephone survey
APPOn: Dummy variable, 1 if applications for development are available online	Telephone survey
SUBOn: Dummy variable, 1 if developers can submit their development applications online or there are fast track options	Telephone survey
PLANNER: Dummy variable, 1 if the county has a "planner on call," customer service center or other established way for the public to easily get information and help with planning or development questions	Telephone survey
MAPOn: Dummy variable, 1 if zoning and/or future land use maps of the county are available online	Telephone survey
BOUND: Dummy variable, 1 if there is an urban growth boundary or urban service area	Telephone survey
RRI (Regulation Ruggedness Index)_I: Ratio of the factor score of all regulatory variables for own and adjacent jurisdictions	The above collected regulatory data and GIS
RRI (Regulation Ruggedness Index)_ II: Ratio of the length of time (months) to receive subdivision approval for own and adjacent jurisdictions	The above collected regulatory data and GIS

1. I thank Susan Floyd at the Bureau of Economic and Business Research (BEBR) at the University of Florida for providing this information.
2. Buildable land is land that is not water, or a designated national or state park, wildlife refuge, or forest.

construction builds upon Mayer and Somerville's study (2000) in which they find a reliable relationship between the housing starts in period *t* and permits in period *t* and *t − 1*.

Independent Variables Changes in housing price are measured by the repeat sales price indices (Mayer and Somerville, 2000). As a timely, accurate indicator of house price trends, the repeat sales price index measures the movement of single-family house prices.

Changes in the cost of construction are measured by the changes in the real prime rate. Mayer and Somerville (2000) suggested that this variable is appropriate to measure the changes in the cost of construction since most construction financing is based on adjustable interest rates that vary with the prime rate.

Other supply-side control variables are used to characterize the location, the urban primacy, and the population size of the local jurisdictions (McDonald and McMillen, 2000). A variable measuring the percentage of the jurisdiction's buildable land area that was considered "urbanized" by the Census Bureau in 1990 is included. Yearly dummies are also included.

Land use regulation variables are of the most interest in this study. To measure the level of implementation of the GMA, the Plan Expenditures variable (R_j) was developed. Local comprehensive planning expenditures are selected (Liou and Dicker, 1994) as a measure of local regulatory efforts as they indicate the complexities of the planning activities. Following Liou and Dicker (1994), the calculated per capita comprehensive planning expenditures of local governments are used.

Liou and Dicker (1994) pointed out that although expenditures for planning indicate local planning efforts, expenditures do not necessarily capture the effectiveness of managing growth. To measure level of enforcement of the GMA, quality of the comprehensive plan could be a good supplementary measure.

Comprehensive plans provide local governments a statement of goals and policy and a legal document for techniques of land use controls. Therefore, number of goals, objectives, and techniques specified in the plans is included as a measure on the complexity of regulatory requirements (Feiock, 2004). In addition, a variable measuring the number of land improvement standards specified in plans (such as right-of-way widths, pavement widths, sidewalk widths, curbs, and cul-de-sac diameter) is included to quantify the designing of land improvement standards and to measure the type and magnitude of required subdivision improvements. Finally, the expertise of the planner who is in charge of drafting the comprehensive plan might be a determining factor of the plan's quality; therefore, a dummy variable indicating whether the highest planning official in the planning department is certified by the American Institute of Certified Planners (AICP) is included.

Feiock (2004) suggested the use of development impact fees as a measure of the restrictiveness of the "second wave" growth management program. Several impact fee variables, variables measuring the average impact fee charged for a new 3-bedroom single family house (1,847 square feet), the number of categories for which impact fees are charged, and a dummy variable indicating if impact fees are charged, are developed and included respectively in the model to explore if there is any effect of impact fees on housing construction.

Atash (1990) and Mayer and Somerville (2000) suggested that complicated subdivision review and permitting processes could add long delays to the development process. The length of time needed to receive subdivision approval reflects the complexity of growth management actions adopted in local jurisdictions and is therefore included as a variable in this study. On the other hand, technical assistance from local planners provided to developers might expedite the development process; therefore, a set of dummy variables are included to see whether the jurisdiction has posted its application forms for development on a website, whether there is online submittal of permit applications (or other fast track permit options), whether the jurisdiction offers "planner on call" or similar dedicated customer service staff time, and whether the county has posted zoning and future land use maps on a website.

Finally, a dummy variable is included to see if the jurisdiction has adopted urban growth boundaries.

To model the effects of the relative implementation stringency of adjacent jurisdictions, the Regulation Ruggedness Index (RRI) is constructed. The first step is to develop a stringency index for each jurisdiction. Pollakowski and Wachter (1990) and Cho and Linneman (1993) suggested a method to develop a stringency index for each jurisdiction by combining the set of regulatory measures based on different weights. Acknowledging the arbitrary weights assigned to the regulatory variables by this method, and the possible correlation among the regulatory variables, two different approaches to constructing RRI are experimented with in this study. The first approach adopts factor analysis, a technique for data reduction, on a stand-alone basis to construct one single index from the above set of regulatory variables. This is done because the large number of variables precludes modeling all the measures individually. The second approach to developing a stringency index for each jurisdiction is simply to use the variable on the length of time for obtaining subdivision approval.

To develop the RRI, the second step is then to calculate the ratio of stringency indexes for own and adjacent jurisdictions. In this study, this ratio is adjusted by the extent that those jurisdictions share a boundary using GIS. By construction, the higher the jurisdiction's RRI, the higher the relative stringency of implementing the GMA in that jurisdiction.

Data Collection and Survey

Sixty-seven counties in Florida are used as unit of analysis for carrying out this study. It is essential to note that by sampling counties only, this study investigates the impacts of planning and regulations only in unincorporated areas. Spillover effects caused by regulation practiced by municipalities are not considered in the study.

For each county, the above mentioned variables are constructed based on survey data and secondary data (Table 10.1). A survey for Florida planning officials is designed in order to obtain critical information about the development process as well as ascertain attitudes toward the strictness of growth management regulations. Interviewees included planning directors or senior planners from 67 counties. The key factual question is the length of time necessary for a developer to obtain approval for a new subdivision. Additional factual data covered a range of technical growth

Table 10.2   Summary statistics for dependent and regulatory variables

Variable	Mean	Std Dev	Minimum	Maximum
Permit 93	1363	1877	10	9808
Permit 94	1437	2038	20	10665
Permit 95	1255	1776	15	8188
Permit 96	1359	1855	21	9584
Permit 97	1342	1674	10	7494
Permit 98	1455	1869	9	8753
Permit 99	1591	2022	12	8574
Permit 00	1589	2037	6	9160
Permit 01	1772	2257	20	8508
Permit 02	1921	2421	16	9437
EXPPC93	4.54	5.41	0	30.44
EXPPC94	4.60	6.31	0	35.23
EXPPC95	5.93	12.16	0	94.37
EXPPC96	6.01	6.50	0	31.3
EXPPC97	6.41	7.13	0	35.14
EXPPC98	6.43	6.01	0	28.36
EXPPC99	7.72	11.55	0	83.65
EXPPC00	6.45	7.41	0	45.57
EXPPC01	6.65	6.24	0	28.22
EXPPC02	8.45	17.17	0	138.15
#IMPV	1.67	3.54	0	5
AICP	0.51	0.50	0	1
IMPFEE3	3350	4474	0	17917
#IMP	3.79	2.07	1	7
IMPFEE	0.54	0.50	0	1
DELAY	6.00	8.47	1	30
APPOn	0.68	0.47	0	1
SUBOn	0.26	0.38	0	1
PLANNER	0.32	0.47	0	1
MAPOn	0.48	0.50	0	1
BOUND	0.57	0.50	0	1

management techniques, such as whether the county employed impact fees, urban growth boundaries, online submittal of permit applications, or fast track permit options. In addition to technical data, questions are also asked to assess how easily development can proceed. Interviewees were asked about the resources they offer to developers. Their responses are then translated into a series of dummy variables. These variables included whether the county offered "planner on call" or similar dedicated customer service staff time, whether the county had posted its application forms for development on a website, and whether the county has posted zoning and future land use maps on a website. These factors, when present, could substantially increase the ease of developing in a county, holding regulations constant.

Results

Table 10.2 presents the summary statistics for the dependent variable and the regulatory variables. They show that the number of new single-family building permits increased steadily every year since 1997. Per capita planning expenditures also increased over the study period, though with a slight drop in 2000 and 2001. Of the counties surveyed, 51 percent had planning directors certified by AICP, 54 percent had impact fees, 68 percent had online application forms for development, 26 percent had online submittal service or other fast track options, 32 percent offered technical assistance to the public on their planning or development questions, 48 percent had online future zoning or land use maps, and 57 percent had urban growth boundaries or urban service areas.

Note that the standard deviations of most of the regulatory variables indicate that there was a large variation in the adoption and implementation of land use regulations across the counties. For instance, the average impact fee charged for a new 3-bedroom single-family house ranged from $0 to $17,917, the number of categories for which impact fees are charged ranged from 1 to 7, the number of land improvement standards specified in plans (such as right-of-way widths, pavement widths, sidewalk widths, curbs, and cul-de-sac diameter) ranged from 0 to 5, and the number of months to receive subdivision approval ranged from 1 to 30.

Table 10.3 presents the results of various regressions from the yearly new housing supply function for 67 counties in Florida for the years from 1993 to 2002. Regression (1) is a simple OLS specification. The coefficients for price changes are statistically significant. The positive coefficients indicate that increase in housing prices could result in increase in new construction spread over the current and the previous year. Variables controlling for the location of the counties are also significant: Coastal counties and counties that were closer to interstate highways had more new housing construction. Variables controlling for the primacy of the counties have mixed results: whether the county is part of a census-defined Metropolitan Statistical Area did not influence new housing construction significantly, while a larger population size in 1990 and a smaller percentage of the county's buildable land area considered "urbanized" by the Census Bureau in 1990 contribute to more housing construction.

Of most interest in this study are the regulatory variables and the Regulation Ruggedness Index. Consistent with previous studies on the link between local land use regulations and housing supply, use of impact fees did not affect housing construction significantly. This is not surprising, given the fact that neither developers nor landowners bear the burden of paying impact fees. In addition, adoption of UGB or USA by the county did not have a significant impact on new housing construction. Local comprehensive planning expenditures, a measure of the complexities of the planning activities, affected new housing construction negatively: more complicated planning activities or growth management actions, as measured by greater expenditure levels on planning activities, were associated with fewer housing starts.

Two of the three variables used to examine the relationship between plan quality and housing construction—number of goals, objectives, and techniques specified in the comprehensive plans and whether the highest planning official in the planning

Table 10.3 Summary of the regression results

Variable	Regression (1)—OLS		Regression (2)—IV	
	Estimate	t Value	Estimate	t Value
Intercept	-3.19	-10.94	-3.04	-8.54
	(0.29)		(0.65)	
Price change rate (PRICE)	4.58	6.34	3.21	7.97
	(0.66)		(0.54)	
Price change rate (t-1)	4.97	5.53	4.54	6.98
	(0.64)		(0.48)	
Change in real prime rate (RPR)	0.0049	4.22	0.0016	4.83
	(0.0049)		(0.0037)	
Location: Coastal (COAST)	0.06	2.06	0.12	2.01
	(0.03)		(0.04)	
Location: Distance to Interstate highway (DHWY)	-0.0089	-6.86	-0.01	-4.83
	(0.0013)		(0.002)	
Urban primacy: MSA (MSA)	not sig.		not sig.	
Lg(Population in 1990) (POP90)	1.08	21.47	1.11	12.33
	(0.05)		(0.09)	
Buildable area in 1990 (BUILT90)	-1.38	-15.39	-1.53	-10.86
	(0.1)		(0.14)	
Planning expenditure per capita (EXPPC)	-0.0029	-3.87	-0.0108	-2.43
	(0.0015)		(0.0026)	
Number of goals, objectives and policies (#GOAL)	not sig.		not sig.	
Number of land improvement standards (#IMPV)	-2.04E-05	-6.27	-0.09	-1.99
	(0.00)		(0.05)	
AICP	not sig.		not sig.	
Impact fee variables (IMPFEE3, #IMP, or IMPFEE)	not sig.		not sig.	
Length of time (months) to receive subdivision approval (DELAY)	-0.19	-4.26	-0.13	-4.04
	(0.06)		(0.07)	
Application online (APPOn)	not sig.		not sig.	
Submit application online (SUBOn)	0.23	8.87	0.05	6.56
	(0.03)		(0.05)	
Planners on call (PLANNER)	0.11	3.96	0.36	4.37
	(0.03)		(0.15)	
Future zoning or land use maps online (MAPOn)	0.32	5.83	0.31	4.26
	(0.05)		(0.14)	
UGB or USA (BOUND)	not sig.		not sig.	
Regulation Ruggedness Index (RRI)	-0.26	-6.27	-0.22	-4.29
	(0.08)		(0.15)	
Adjusted R Square	0.8389			

Notes: Number of observations = 670. Numbers in parentheses are standard errors.

department is certified AICP—were not significant factors. Interestingly, the third variable, the number of land improvement standards specified in plans, a measure used to measure the type and magnitude of required subdivision improvements, was found to have a significant effect; a greater number of standards resulted in less housing construction. This is consistent with Atash (1990), who argued that excessive subdivision regulations, used to govern the range and quality of on-site and off-site improvements, may have delaying effects on the approval procedures and inflationary effects on the cost of housing. Overall, these results indicate that comprehensive plans, which are intended to be direction-setting documents, do not create regulatory authority and thus do not affect housing construction significantly. In contrast, subdivision regulations, which establish standards for land improvements, might have a more binding effect on the development process; a larger number of standards can dampen the residential development market.

Consistent with the findings of Mayer and Somerville (2000), the coefficient for the delay variable was significantly negative: As the number of months needed for approval increased, the number of single family housing permits decreased. On the other hand, instruments that could shorten delay (such as technical assistance provided by local planners, online zoning and future land use maps, and fast track permit options) were associated with increases in housing construction.

Finally, the findings on the estimates of the RRI are interesting. When this index was constructed by using the ratio of the factor score of all regulatory variables for own and adjacent jurisdictions, it was not significant. However, when developed by comparing the ratio of the length of time (months) to receive subdivision approval for own and adjacent jurisdictions, the RRI was significantly negative: Longer approval time in a county, compared to that in adjacent counties, was associated with a decrease in the number of single family housing permits. In other words, a delay of receiving subdivision approval could push housing construction away, inducing the displacement of growth.

A potential problem suggested by Mayer and Somerville (2000) is that the local implementation of the GMA may be endogenously determined with new housing construction. Faster growing communities are more likely to implement the GMA more strictly, to preserve the general health and amenities of their neighborhoods. Conversely, slow growing areas might implement the GMA less strictly, to foster local economic growth. To correct for this potential endogeneity, an instrumental variable (IV) estimation was used with the following instruments: 1980 per capita income in the counties, 1980 population, and whether the county has citizen referendums (following Mayer and Somerville, 2000). The results of this regression are also presented in Table 10.3. Consistent with the results of OLS estimation, regulatory variables remained significant. This indicates that even though the local implementation of the GMA may be endogenous, the interjurisdictional spillover effect caused by different levels of its implementation is still significant.

Conclusion

This study has explored if a spillover supply effect exists, whether an area's greater relative stringency in implementing growth management controls results in more housing construction in neighboring jurisdictions. The empirical analysis employed yearly single family housing permit data on a panel of 67 counties in Florida from 1993 to 2002. The findings suggest that land use regulations have significant effects on the distribution of new housing construction.

The results of this study indicate regulations that lengthen the development process have significant, negative effects on new housing construction. However, factors that shorten delays in the process (such as technical assistance provided by local planners, posting zoning and future land use maps online, and adopting fast track permit options) can mitigate these effects. Although the complexity of comprehensive plans (measured by the number of goals, objectives, and techniques) has little effect on housing construction, more sophisticated subdivision plans or regulations (measured by the number of land improvement standards specified) seem to result in fewer housing starts. This reinforces Mayer and Somerville (2000)'s finding that regulations which constrain new development and thus lengthen the development process lower the level of new housing construction. In addition, higher local comprehensive planning expenditures (which might indicate more complicated growth management actions) are associated with less housing construction. However, adoption of a UGB or USA and enactment of impact fees seems to have little impact on new construction.

Finally, the results show that a community with a longer delay in subdivision approval relative to adjacent communities tends to experience negative spillover. In other words, a longer approval time could drive housing construction away to other nearby communities.

In summary, this study provides strong evidence that developers are responsive to project delays, which typically take two forms: 1) delay in obtaining subdivision approvals and 2) delay caused by satisfying requirements related to the design of land improvement standards. According to Atash (1990, 235), each month of delay in getting approval can add 1 to 2 percent to the cost of each housing unit. Therefore, on average, the cost of delay in obtaining approvals can represent between 13 percent and 26 percent of the final home price. Lowering housing subdivision standards, however, could reduce housing costs by as much as 10 percent. This study also suggests that uneven stringency, marked by variance in the delay in development process, may cause development to be inefficiently dispersed in the region. Introducing multiple reviews in the development process by a community can influence not only its own housing construction but also that of nearby communities. Less "smart" developments, to avoid stringent reviews or subdivision design standards, might possibly flee to other communities which implement the GMA loosely.

Given these findings, the challenge for Florida and other pro-comprehensive planning states is to seek remedies for the interjurisdictional displacement of growth. Obviously, a comprehensive approach to the statewide growth management effort must include a component for effective and consistent implementation strategies at the local level. Experience suggests that both a 'top-down' approach and a 'bottom-up' approach must be realized. The first approach reflects the need for policy direction from the state government in order to achieve vertical consistency, while the second approach ensures that the implementation experience of local governments is evaluated for consideration as state policy evolves (Powell, 2000). The ability of the state to provide specific goals and determine the role that each jurisdiction should play in accommodating these goals can warrant a more integrated and coordinated planning system at the local level and avoid unplanned spillover of growth.

References

Anthony, J. (2003). The effects of Florida's Growth Management Act on housing affordability. *Journal of the American Planning Association 69*: 282–295.

Atash, F. (1990). Local land use regulations in the USA: A study of their impacts on housing cost. *Land Use Policy 7*(3), 231–242.

Burby, R., and May, R. (1997). *Making governments plan.* Baltimore: Johns Hopkins University Press.

Catlin, R. (1997). *Land Use Planning, Environment Protection and Growth Management.* Chelsea, Michigan: Sleeping Bear Press, Inc.

Cho, M., and Linneman, P. (1993). Interjurisdictional spillover effects on land use regulations. *Journal of Housing Research 4*(1), 131–163.

Carruthers, J. I. (2002). The impacts of state growth management programs: A comparative analysis. *Urban Studies 39* (11): 1959–1982.

DeGrove, J. M. (1989). Consistency, concurrence, compact: Florida's search for a rational growth management system. *The Political Chronicle 7*(September), 32–38.

DeGrove, J. M. (1990). The politics of planning a growth management system: The key ingredients for success. *Carolina Planning 16*(Spring), 26–34.

DeGrove, J. M. (1992). *The new frontier for land policy: Planning and growth management in the states.* Cambridge, MA: Lincoln Institute of Land Policy.

Deyle, R. and Smith, R. (1998). Local government compliance with state planning mandates: The effects of state implementation in Florida. *Journal of the American Planning Association 64*: 457–469.

Evans, A. (1992). Town planning and the supply of housing. In G. Keating, P. Warburton, et al. (Eds.), The state of the economy (pp. 81–93). London: Institute for Economic Affairs.

Feiock, R. C. (2004). Politics, institutions and local land-use regulation. *Urban Studies 41*(2), 363–375.

Landis, J. D. (1992). Do growth controls work? A new assessment. *Journal of the American Planning Association 58*(4), 489–508.

Liou, K. T., and Dicker, T. J. (1994). The effect of the Growth Management Act on local comprehensive planning expenditures: The south Florida experience. *Public Administration Review 54*(3), 239–244.

Mayer, C. J.,and Somerville, C. T. (2000). Land use regulation and new construction. *Regional Science and Urban Economics 30*, 639–662.

McDonald, J. F., and McMillen, D. P. (2004). Determinants of suburban development controls: A Fischel expedition." *Urban Studies 41*(2), 341–361.

Pollakowski, H. O.and Wachter, S. M. (1990). The effects of land-use constraints on housing prices. *Land Economics 66*(3), 315–324.

Powell, D. (2000, August). Growth management: Florida's past as prologue for the future. Paper presented to the Growth Management Study Commission, Tallahassee, FL.

Chapter 11

Are We Any Safer? An Evaluation of Florida's Hurricane Hazard Mitigation Planning Mandates

Robert E. Deyle, Timothy S. Chapin, and Earl J. Baker

Florida is the most hurricane-prone state in the United States (Blake, Jarrell, Rappaport, and Landsea, 2005). Forty percent of all hurricanes that made landfall on the US mainland between 1851 and 2004 struck Florida (110 out of 273). Florida also has experienced nearly twice as many major hurricanes (Category 3 or greater)[1] than the second-most-exposed state, Texas (35 versus 19). While hurricane landfall probabilities vary to some degree within the state, all areas of the coast are vulnerable.

Threats to public safety from hurricane flooding were one of the Florida Legislature's important concerns when it enacted the 1985 Growth Management Act (GMA) (DeGrove and Miness, 1992). The statute expresses the Legislature's intent that "local government comprehensive plans ... protect human life and limit public expenditures in areas that are subject to destruction by natural disaster" (*Fla. Stat.* §163.3178.1, 2006). State regulations that were subsequently promulgated by the State Department of Community Affairs (DCA) require coastal communities to include policies in their comprehensive plans that limit development in and direct populations away from "coastal high hazard areas" (CHHAs) (*Fla. Admin. Code* § 9J-5.012(3)(b)(6), 2006). Local governments also are required to adopt policies to maintain or reduce evacuation clearance times[2] within larger areas designated as "hurricane vulnerability zones" (HVZs) (*Fla. Admin. Code* §§ 9J-5.003(18) and 9J-5.012(3)(b)(7), 2006).

Florida is widely recognized as having the strongest state mandate for local governments to incorporate hazard mitigation policies in their comprehensive plans (Berke and French, 1994; Burby and Dalton, 1994; May et al., 1996; Burby and May, 1997). Because most local comprehensive plans adopted pursuant to the 1985

---

1 Tropical cyclone intensity in the United States is measured using the Saffir Simpson index which defines tropical storms and five categories of hurricanes based on maximum sustained wind speed and storm surge height. Major hurricanes (Category 3 or greater) have wind speeds in excess of 110 mph and storm surges greater than eight feet.

2 An evacuation clearance time is the time it takes for the last evacuating vehicle to pass a given point, such as the county line.

GMA have been in place since about 1990, Florida offers an excellent setting in which to assess the effectiveness of such a mandate. We report here the results of an evaluation of the extent to which land use changes subsequent to local comprehensive plan approval are consistent with the state's hazard mitigation policy mandates concerning development within CHHAs and evacuation times within HVZs.

In this chapter, we describe aggregate changes in several measures of residential exposure to hurricane flooding for a sample of 89 Florida coastal communities in 15 counties and extrapolate our findings to the entire coastal area of the state. In addition, we describe the impacts that these residential land use changes have had on hurricane evacuation clearance times and emergency shelter demand in a subset of our county sample. We begin by setting the context for this work with a discussion of Florida's growth management and comprehensive planning approach. We then present the results of our analyses followed by a discussion of hypotheses that might explain our observations.

Context

Florida's 1985 growth management legislation requires local governments in coastal areas to prepare coastal management elements as part of their comprehensive plans that address, among other goals, the state's intent to protect human life. The focus of the state regulations that detail the substantive content of the coastal element is on limiting development in and directing populations away from "coastal high hazard areas" and on maintaining or reducing hurricane evacuation clearance times within "hurricane vulnerability zones". The CHHA is defined as the evacuation zone for a Category 1 hurricane, while the HVZ is defined as the area that would be evacuated for a 100-year storm or a Category 3 hurricane (*Fla. Admin. Code* §§ 9J-5.03(17) and (57), 2006). In addition, local governments are specifically required to (*Fla. Admin. Code* §§ 9J-5.06(3) and 9J-5.012(2) and (3), 2006):

- include an objective to maintain or reduce hurricane evacuation times;
- inventory and analyze the effects of future land uses on hurricane evacuation, including the effects of anticipated population densities;
- analyze measures the local government could adopt to maintain or reduce hurricane evacuation times;
- include policies and regulatory or management techniques that address hurricane evacuation;
- include an objective in the future land use element of the comprehensive plan that coordinates coastal planning area densities with the applicable hurricane evacuation plan;
- inventory existing and proposed land uses in CHHAs and structures repetitively-damaged by coastal storms, and identify measures which can be used to reduce exposure to coastal flooding hazards including relocation and public acquisition;
- include an objective that directs population concentrations away from CHHAs; and

- include policies that limit development within CHHAs and that achieve general hazard mitigation including regulation of land use so as to reduce the exposure of people and property to natural hazards.

Elsewhere, in the definitions section, the regulations stipulate that "[w]hen preparing and implementing the hurricane evacuation or hazard mitigation requirements of the coastal management element, the coastal planning area shall be those portions of the local government's jurisdiction which lie in the hurricane vulnerability zone" (*Fla. Admin. Code* § 9J-5.003(18), 2006). As a result, DCA staff have interpreted the mandate concerning hurricane evacuation times as applying to the HVZ. State law also requires local governments to enact land development regulations and take other initiatives to implement these policies (*Fla. Stat.* § 163.3167, 2005).

The first comprehensive plans developed under Florida's 1985 local planning mandate were adopted by local governments and approved by the DCA between 1988 and 1991. Land development regulations and other growth management strategies implementing these policies were adopted by local governments to varying degrees in the years following approval of the plans. Data collected by state agencies suggest, however, that development has continued in many coastal areas of the state, with resultant increases in exposure and vulnerability and associated costs to local and state governments. County hurricane evacuation studies indicate that populations in category 1 hurricane storm surge zones[3] increased an average of 24 percent between 1990 and 1999 (Florida Department of Community Affairs, Coastal Management Program, 2000).

The State Division of Emergency Management recommends that all communities, with the exception of the Florida Keys, strive to achieve evacuation clearance times of 16 hours or less.[4] The most recently available data from the state (Florida Department of Community Affairs, Division of Community Planning, 2005) indicate that 14 of the 35 coastal counties in Florida (excepting Monroe County) would not be able to meet this threshold for a Category 5 hurricane. Of those 14 counties, 12 would not be able to complete an evacuation for a Category 4 hurricane within 16 hours, 8 would evacuate too slowly for a Category 3 storm, and 5 for a Category 2 hurricane. Additionally, even when people have sufficient notice to evacuate, they may not have any safe place to go. As of 2004, Florida was short more than 590,000 shelter spaces for a Category 4/5 hurricane (Florida Department of Community Affairs, Division of Emergency Management, 2004). A few coastal counties have

---

3 Storm surge zones are the areas likely to be inundated by a rise in the sea surface associated with the onshore movement of a tropical cyclone. Evacuation zones are based in part on storm surge zones.

4 Typically, the National Hurricane Center (NHC) issues hurricane warnings approximately 24 hours prior to anticipated landfall. On average, tropical storm-force winds reach land about 8 hours before the eye of a hurricane. Thus a window of about 16 hours is available for evacuation following receipt of an NHC hurricane warning. Earlier evacuation, in response to an NHC hurricane watch, is sometimes called for, but officials run a higher risk of unnecessary evacuation under such circumstances. This is the case in the Florida Keys, where physical constraints are generally viewed as precluding evacuation clearance times of less than about 24 hours.

surpluses, but most do not. Shortages range from 107 in Wakulla County to 93,527 in Pinellas County.

Increases in populations and property at risk also raise the public costs of hurricane disaster response and recovery. Between 1994 and 2003, hurricanes and tropical storms in Florida resulted in more than $1.17 billion in local government costs that were eligible for disaster assistance under the federal Public Assistance Program (Florida Department of Community Affairs, Division of Emergency Management, 2003). June 2006 estimates of approved Public Assistance expenditures for the eight hurricanes that struck Florida in 2004 and 2005 total approximately $2.7 billion (G. Freerksen, personal communication, June 2, 2006).

Deyle and Smith (1998) analyzed the extent to which local governments in Florida had incorporated required policies concerning hurricane hazards in their comprehensive plans. They found substantial variation among the 18 communities whose plans they reviewed, and they found evidence that this variation was due in part to the manner in which the DCA carried out its role in reviewing and approving the initial plans prepared by the local governments. There has not been, however, any systematic analysis of the extent to which the policies that are included in local comprehensive plans have been implemented by local governments.

In this chapter we examine the patterns of residential development that occurred within CHHAs and HVZs during the initial 11 to 12 years that these policies should have been in effect. These development trends offer some empirical evidence of the extent to which local comprehensive plan policies may have affected development patterns within hurricane hazard zones. In addition, we examine their associated impacts on shelter demand and evacuation clearance times. More formally, we address the following research questions:

1. How did residential land use patterns change in coastal communities in Florida subsequent to approval of local comprehensive plans by the DCA?
2. To what extent did land use changes differ within and outside of the communities' hurricane hazard zones?
3. How did land use changes within hurricane hazard zones affect the exposure of people and property to hurricane flooding?
4. How did the population increases associated with land use changes that occurred after plan approval affect county-level emergency shelter demand and evacuation clearance times?

Aggregate Changes in Exposure to Hurricane Flood Risks

Florida has 35 coastal counties and an additional 158 municipalities that are required to include coastal management elements in their comprehensive plans. We were unable to analyze all of these jurisdictions because of various data constraints.[5] Nonetheless,

---

5   Sample selection criteria include the following:
a.   coastal jurisdictions must include a coastal element in the local comprehensive plan;
b.   comprehensive plan approved between 1989 and 1991;
c.   2002 property appraiser tax roll data and parcel geometry available;

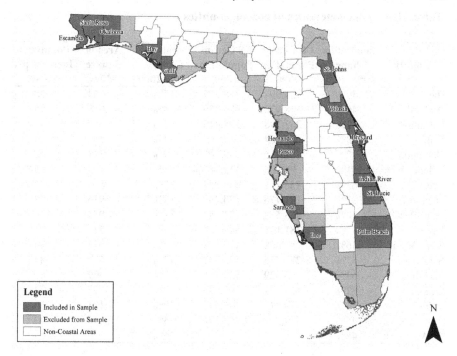

Figure 11.1 Florida coastal county sample

our ultimate sample of 89 jurisdictions (15 counties and 74 municipalities) provides good coverage of the range of geographic and socio-economic variation among Florida's coastal jurisdictions and thus can be viewed as a reasonable sample of the likely variation in both hazard exposure and planning responses (see Figure 11.1).

Table 11.1 summarizes key characteristics of the 15 counties from which our sample is drawn. Total 1990 populations range from a low of 11,504 in Gulf County, a rural Panhandle county with substantial amounts of undeveloped coastal land at that time, to a high of 863,518 in Palm Beach County, one of the densely-populated counties of southeast Florida. Our sample constitutes 36 percent of the total 2002 population of all 35 coastal counties in Florida and 41 percent of the area. The decadal (1990–2000) average growth rate for our sample of 15 counties (26 percent) is nearly identical to the average for all 35 coastal counties (27 percent). Growth in individual counties in our sample ranges from rates of 12 to 20 percent in some of the more built-out coastal areas in northeast Florida (Brevard and Volusia counties) and the central west coast (Sarasota County) and some of the slower-growing Panhandle counties (Bay, Escambia, and Okaloosa) to rates in excess of 40 percent in very fast growing areas (Santa Rosa and St. Johns counties). The densities of these population

---

d. parcel geometry in useable format, e.g., shapefile or ArcInfo coverage; not un-projected CAD files;

e. reliable land use coding data—suspect data field checked;

f. some vacant land in Category 1 or Category 3 zone in 1995; and

g. no exceptional circumstances.

Table 11.1   Characteristics of coastal counties

County	Population Change 1980–1990	US Census Population 1990	US Census Population 2000	Population Change 1990–2000	Area in Square Miles	Population Increase per Sq Mi
Bay	30%	126,994	148,217	17%	764	28
Brevard	46%	398,978	476,230	19%	1,018	76
Escambia	12%	262,798	294,410	12%	662	48
Gulf	8%	11,504	14,560	27%	555	6
Hernando	127%	101,115	130,802	29%	478	62
Indian River	51%	90,208	112,947	25%	503	45
Lee	63%	335,113	440,888	32%	804	132
Okaloosa	31%	143,776	170,498	19%	936	29
Palm Beach	50%	863,518	1,131,184	31%	1,974	136
Pasco	45%	281,131	344,765	23%	745	85
Santa Rosa	46%	81,608	117,743	44%	1,017	36
Sarasota	37%	277,776	325,957	17%	572	84
St. Johns	63%	83,829	123,135	47%	609	65
St. Lucie	72%	150,171	192,695	28%	572	74
Volusia	43%	370,712	443,343	20%	1,103	66
Totals:	45%	3,579,231	4,467,374	25%	12,312	72
Averages:	48%			26%		65

increases average 65 people per square mile, with a range from 6 people per square mile in Gulf County to highs of 132 and 136 in Lee and Palm Beach counties.

Our analysis of land use change patterns and resulting exposure of people and property to hurricane flood hazards required us to document the locations of hurricane hazard zones and land uses prior to and following approval of each jurisdiction's comprehensive plan. We also needed to determine the assessed value of property improvements and the numbers of permanent and seasonal residents associated with developed property. In the following sections we summarize the methods we employed to answer our first three research questions and then present the results of those analyses.[6]

How Did Residential Land Use Change Subsequent to Approval of Local Comprehensive Plans?

To answer this question, we needed to make the following determinations for each coastal jurisdiction:

- Where did residential development exist at the time the local comprehensive plan was approved by the DCA and how much vacant residential land was available at that time?
- Where did residential development exist at the end of the "post-plan" time period in 2002?

---

6   For full details of our methods, see Chapin, Deyle, and Baker (2006).

- Where did residential development exist, and how much vacant residential land was available, in the base year at the start of the "pre-plan" time period?

To make these determinations, we used 2002 county property appraiser tax roll data and property parcel geometry obtained through the Florida Department of Revenue. We deduced the development status of each parcel in years prior to 2002 from "actual year built" data that we obtained directly from the counties.

Communities are required to adopt land development regulations to implement the policies in their comprehensive plans within one year of plan approval by the DCA. We identified property records as having been developed "pre-plan" where the actual year built for improvements on the parcel was the year of plan approval or earlier. This approach accounts for the expected lag in plan implementation after the adoption and approval of the local plan. We designated property parcels with an actual year built after the year of plan approval as having been developed "post-plan."

We defined the base year of the pre-plan time period for each community so that the lengths of the post-plan and pre-plan time periods were the same. Thus, for example, if a community's comprehensive plan was approved in 1990, the length of the post-plan period would be 12 years (1990 to 2002). The base year for the corresponding pre-plan period would therefore be 1978.

We defined the 2002 developed land use status of each parcel based on the land use codes contained in the 2002 tax roll data. We assumed that the parcel was undeveloped prior to the year built date. To estimate the supply of vacant residential land at the start of the pre-plan and post-plan time periods, we identified land as vacant if it was categorized as vacant residential or unimproved agricultural land by the county property appraiser in the base year of the time period, or if it was categorized as improved residential land use subsequent to the base year.

To What Extent Did Land Use Changes Differ Within and Outside of Hurricane Hazard Zones?

This question required us to determine whether or not a given property parcel lay within or outside of the CHHA or the HVZ. We did so by overlaying the CHHA and HVZ boundaries with the property parcel polygons.

Evacuation zones are typically defined for each county by the county emergency management staff, based on models that predict the areas likely to be inundated by storm surge from hurricanes of different intensities. Unlike surge zones, however, which model storm surge flooding based on topography, evacuation zones are more generalized boundaries that utilize recognizable features such as major roads to define boundaries that are more readily understood by the public at large when an evacuation order is given. In some counties, separate evacuation zones are not mapped in the hurricane evacuation study for each hurricane intensity level. Thus, for example, in Volusia County, only two hurricane evacuation zones are mapped: Zone A and Zone B. Zone A is evacuated for a Category 1 or 2 hurricane, while Zone B is evacuated for a Category 3, 4, or 5 hurricane.

The extent to which individual communities are encompassed by these hurricane hazard zones varies substantially from one part of Florida to another. At one extreme

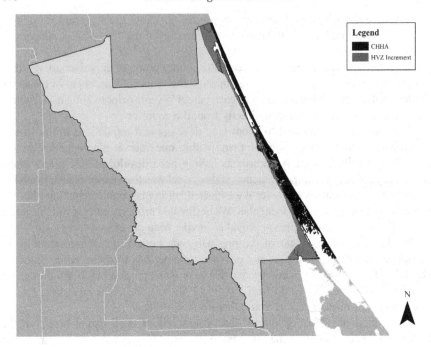

Figure 11.2   Volusia County hazard zones

are counties along the northeast Atlantic coast of Florida, such as Volusia, that have high dune ridges along narrow barrier islands and narrow hurricane hazard zones. In these counties, the area encompassed by the CHHA is largely limited to a strip along the ocean, with a fringe of HVZ along the sounds behind the barrier islands (see Figure 11.2). It is for this reason that the hurricane evacuation zones are aggregated in Volusia. In some cases, however, there are communities located on barrier islands that have no land outside the HVZ. In the eastern Florida Panhandle, in counties such as Bay, relatively high bluffs along the Gulf of Mexico limit the extent of the CHHA along the open Gulf to a fairly narrow strip, but the CHHA and HVZ are more extensive along the margins of the interior bays (see Figure 11.3). Due to steeper coastal margins along some portions of these bays, there are some communities that do not have a CHHA, but do have an HVZ, i.e., they are not flooded by a Category 1 storm but would be flooded by a Category 3. At the other extreme are counties such as Lee, along the southwest coast of Florida, where, because of low topographic relief, the CHHA and HVZ cover extensive areas (see Figure 11.4). In these instances, entire communities may lie within the CHHA.

How Did Changes Within Hurricane Hazard Zones Affect Exposure to Hurricane Flooding?

We defined five principal measures of residential exposure to hurricane flooding for each jurisdiction: 1) number of residential units, 2) total 2002 just value of residential

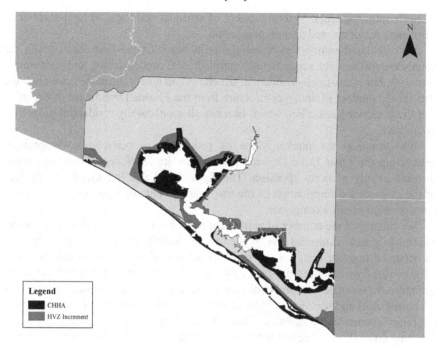

Figure 11.3 Bay County hazard zones

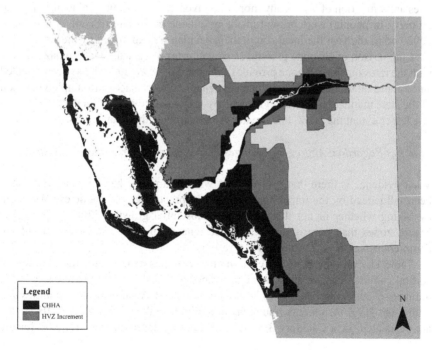

Figure 11.4 Lee County hazard zones

property improvements, 3) total number of seasonal residents, 4) total number of permanent residents, and 5) total population.

We calculated numbers of residential units based on the land use codes in the county tax roll data for single family, mobile home, condominium, and cooperative land uses. For multi-family residential structures and hotels and motels, we obtained data on the number of units per structure from the Florida Department of Business and Professional Regulation, which licenses all multi-family residential properties in the state.

We estimated the market value of residential property improvements by subtracting the *Land Value* for each parcel from its *Total Just Value* as assessed by each county property appraiser. The *Total Just Value* is typically 85 percent of the assessor's determination of the true market value of a property prior to the consideration of any exemptions.

We calculated the number of permanent residents for each residential parcel by multiplying each residential unit by the average household size and the proportion of occupied housing units for the census tract in which it was located. We used 1990 census figures for our pre-plan estimates and 2000 census figures for our 2002 estimates. To estimate the number of seasonal residents, we used separate multipliers for tourist units and seasonal residential units. The first multiplier was obtained from the regional hurricane evacuation studies prepared by or for each county's regional planning council. The second multiplier was obtained from the US Census.

Once we had determined the above parameters for each jurisdiction, we calculated growth rates within and outside of the two hurricane hazard zones. These growth rates are a function of how many more of a given measure (units, value of improved property, or people) were present inside or outside of the hazard zone of interest in 2002 versus the year the local comprehensive plan was approved by the state.

Comparison of absolute growth, as well as rates of growth, within a hazard zone with the rates outside the zone provides a measure of where growth has been directed since approval of the local comprehensive plan. *A priori*, successful implementation of the state mandate would be evidenced by very low growth within the CHHA, and to a lesser extent the HVZ, and greater growth outside these hazard zones.

Findings Regarding Aggregate Changes in Exposure to Hurricane Flood Risks

What evidence is there that growth rates within hurricane hazard zones may have been influenced by the implementation of comprehensive plan policies? We begin by asking whether or not there has been substantial growth within the hurricane hazard zones that were required to be designated in the coastal elements of the comprehensive plans. The answer is a resounding, yes!

Table 11.2 presents sums for each of our five exposure measures for all of the 89 coastal communities we analyzed. Our estimates reveal substantial increases in the numbers of residential structures both within the CHHAs of these communities and within the HVZ increments (the areas in addition to those within the CHHAs that are evacuated for a Category 3 hurricane), totaling approximately 153,000 for the

Table 11.2   Aggregate changes in residential hurricane exposure between plan approval year and 2002

Exposure Parameter	CHHA (n = 77)	HVZ Increment (n = 77)	Total HVZ
Number of residential units	76,084	76,613	152,697
Seasonal population	46,876	19,098	65,974
Permanent population	123,946	155,846	279,792
Total population	170,822	174,944	345,766
2002 just value of residential structures	$18.9 billion	$10.5 billion	$29.4 billion

two hurricane hazard zones combined.[7] Given the comparable average 1990–2000 population growth rates for our sample (26 percent) and for all 35 coastal counties in the state (27 percent), a rough approximation for the state is 425,000 new residential units within the total HVZ.[8] The associated increase in total population (permanent plus seasonal) for the two hazard zones within our sample is in excess of 345,000 people, while the sum of the increase in 2002 just value of residential property improvements is nearly $30 billion. Applying a similar extrapolation approach yields rough estimates of total increases within the hurricane hazard zones of the state's 35 coastal counties of approximately 958,000 new residents and $80 billion in new residential structures subsequent to state approval of local comprehensive plans between 1988 and 1991.

These are sobering figures against which we take a closer look at the aggregate changes in exposure patterns within and outside of the CHHAs of our sample. The CHHA is the most hazardous area and is the area within which the most restrictive policies are supposed to be implemented pursuant to the state's growth management planning mandate.

An examination of growth rates offers a means of comparing the extent to which development has occurred within CHHAs relative to those areas entirely outside the CHHAs. Table 11.3 reports the aggregate growth rates following comprehensive plan approval for each of our five exposure parameters for our full sample. The aggregate growth rate in numbers of residential units within the CHHAs of our 77 communities was 23 percent. This is a substantial growth rate. However, when compared to the aggregate growth rate in areas outside of CHHAs (39 percent), it appears there may have been some moderation in growth within CHHAs. These figures offer a potentially positive spin on the absolute numbers reported in Table 11.2.

Looking more closely at Table 11.3, the growth rates are virtually identical for total population and numbers of residential structures because population is a multiple of numbers of residential units. The lower growth rate in seasonal population relative to permanent population in areas outside the CHHA may reflect

---

7   The smaller sample sizes for the separate CHHA and HVZ analyses reflect the fact that some communities lie only within one hazard zone or the other.

8   This estimate is based on the proportion of total 2000 population in the 35 coastal counties that is accounted for by the 15 counties in our sample (36 percent). This extrapolation implicitly assumes that our sample is also proportionately representative of the spatial extent of CHHAs and HVZs throughout the state.

Table 11.3   Aggregate growth rates for residential hurricane exposure between plan approval year and 2002

Exposure Parameter	Inside CHHA (n = 77)	Outside CHHA (n = 73)
Number of residential units	23%	39%
Seasonal population	23%	33%
Permanent population	24%	41%
Total population	24%	40%
2002 just value of residential structures	46%	71%

Table 11.4   Aggregate growth rates for numbers of residential units before and after comprehensive plan approval within and outside of CHHAs

	Prior to Plan Approval	After Plan Approval	After/Prior Ratio
Within CHHA (n = 74)*	77%	23%	0.30
Outside CHHA (n = 68)*	115%	39%	0.34

* The sample sizes are lower in this comparison than in Table 11.3 because complete pre-plan land use data were not available for several jurisdictions.

some change over time in the proportion of non-tourist residential units that were vacant and seasonal. The rates for 2002 just value are considerably higher than those for numbers of residential units, which is a reflection of rapid increases in the market value of property in Florida. It is no surprise that the ratio of the just value increase relative to the increase in number of residential structures is somewhat higher in the CHHAs (2.00) than in areas further landward (1.82).

Table 11.4 presents a spatial-temporal comparison of aggregate growth rates for residential units within and outside of CHHAs both before and after comprehensive plan approval, as well as ratios for the growth rates prior to and after plan approval. All else being equal, we would expect lower growth rates within CHHAs after plan approval than before, if coastal communities were effectively implementing policies to limit growth within CHHAs. Table 11.4 shows this to be the case, but the table also shows a similar decline in the aggregate numbers of residential units built outside of CHHAs. This parallel trend both within and outside of CHHAs may be evidence, therefore, that the post-plan-approval decline within CHHAs was due to broader phenomena such as the overall impacts of comprehensive plan implementation, independent of policies directed specifically at CHHAs, or other phenomena such as reduced rates of population growth, slower economic growth, or reduced vacant land supply. In fact, reference to Table 11.1 shows that population growth in these 15 counties was significantly greater between 1980 and 1990 (45 percent increase) than between 1990 and 2000 (25 percent). Thus the observed declines in residential growth rates after plan approval may not reflect any direct impacts of plan implementation.

It is worth noting, however, that the ratio of new residential units built after plan approval to the number built over a comparable time period prior to plan approval is somewhat lower for the aggregate area within the CHHAs of the communities

Table 11.5   **Aggregate changes in residential growth density before and after comprehensive plan approval within and outside of CHHAs**

	Growth Density Prior to Plan Approval (units/acre)	Growth Density After Plan Approval (units/acre)	After/Prior Ratio
Within CHHA	0.99	0.74	0.75
Outside CHHA	0.31	0.30	0.97

in our sample (0.30) than for the areas outside of CHHAs (0.34). This difference offers another signal that differential comprehensive plan policies might have had an effect.

To assess the possible effects of land supply, we calculated residential growth densities. These represent the total numbers of new residential units built within a given time period, normalized for the amount of vacant residential land present at the start of that time period.[9] In Table 11.5 we display aggregate residential growth densities before and after comprehensive plan approval within and outside of CHHAs in our sample jurisdictions. Not surprisingly, growth densities were higher within the CHHAs, immediately adjacent to the coast, than in more inland areas, both before and after plan approval. Beyond this distinction, however, we again find stronger evidence of growth patterns consistent with the state's mandate to direct populations away from CHHAs: There was a substantial decrease in aggregate growth density within the CHHA following plan approval, whereas there was no substantial change in residential growth density outside that hazard zone.

Impacts of Land Use Changes on Shelter Demand and Evacuation Times

Changes in land use within coastal communities alter the numbers of people living within hurricane hazard zones, as well as the types of people and their associated evacuation and sheltering behavior. Thus land use changes can affect the numbers of cars that attempt to evacuate, the numbers of residents who are likely to respond to an evacuation order and when they do so, and the numbers who are likely to seek emergency shelter in public facilities. Evacuation clearance time is a function of how many cars travel along a specified route in response to an evacuation order, the physical capacity of the road network that constitutes the evacuation route, and the manner in which the road network is managed. Both the adequacy of the public shelter capacity and the evacuation clearance time depend in part on the intensity of the hurricane which, in turn, influences both the numbers of people who evacuate and the supply of shelters that can safely be used.

Hurricane evacuation and shelter studies are conducted at the county and regional levels on a periodic basis under the auspices of eight of Florida's ten regional

---

9   We coded the land use of a property parcel as vacant residential if it was designated as vacant residential land or unimproved agricultural land with no primary residence in the property appraiser tax roll. We also assumed that parcels coded as residential in 2002 were vacant residential land prior to the year in which the current residential structure was built.

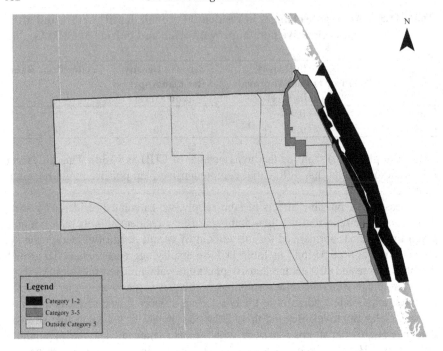

Figure 11.5　Indian River County traffic evacuation zones

planning councils. Traffic evacuation zones (TEZs), which are defined within each county, serve as the spatial unit of analysis (see, for example, Figure 11.5). TEZ boundaries are coterminous with the boundaries of the county's evacuation zones for hurricanes of different intensities. However, they are not coterminous with municipal boundaries within counties.

Within each TEZ, numbers of evacuating residents and vehicles are estimated for three land use types: (1) permanent residential, (2) mobile homes, and (3) seasonal residential. Different assumptions are made about evacuation participation rates for each of these land use types for different storm intensities, including what is called "shadow evacuation," i.e., people who evacuate from areas that are not ordered to evacuate. Evacuation orders for permanent and seasonal residences are assumed to be directed only at residents of the TEZs within the evacuation zone defined for a given hurricane intensity. As a general rule, however, it is assumed that all residents of mobile homes within a county are ordered to evacuate for hurricanes of any intensity. In areas where there is a distinct tourist season, different assumptions are made about occupancy rates of seasonal residential units for the "high" and "low" seasons. Evacuating vehicles from each TEZ are allocated to the evacuation network and, therefore, may flow through several TEZs before exiting the county. Evacuation clearance times typically are calculated for critical roadway segments that present bottlenecks due to capacity constraints.

Relatively complex spreadsheet models are used to estimate shelter demand and evacuation clearance times for the regional hurricane evacuation studies. For some regional planning councils, contractors have developed "abbreviated transportation

Table 11.6 Contributions to 2004 emergency shelter deficits from residential growth between plan approval year and 2002

County	Increments in Shelter Demand by Category (persons)			2004 Deficit Category 4/5	Deficit Due to Growth
	Category 1	Category 3	Category 4/5		
Bay	1,480	2,212	3,394	7,445	46%
Pasco	3,199	6,643	10,269	40,454	25%
Indian River	465	751	751	721	104%
Volusia	432	909	3,085	21,368	14%
St. Johns	1,017	2,421	2,598	2,509	104%
Sum:	6,593	12,936	20,097	72,497	28%
Median:	1,017	2,212	3,085	7,445	46%

models" (ATMs) that can be used to make quick estimates of the effects of changes in residential populations or behavioral assumptions for a given evacuation network. We employed several of these ATMs from the most recent hurricane evacuation studies to assess the impacts of post-plan-approval residential growth on both shelter demand and evacuation clearance times on critical roadway segments. We analyzed five coastal counties, Bay, Pasco, Indian River, St. Johns, and Volusia, which encompass 22 of the 89 coastal communities in our full sample.

We entered our land use data into the ATM spreadsheet model to define a baseline scenario for each TEZ for the year in which the comprehensive plan was approved for the principal jurisdiction in which the TEZ is situated. Using the ATM to aggregate the evacuating residents and vehicles for all TEZs, we estimated pre-plan-approval shelter demand and evacuation clearance times at the county level. We then compared those values to estimates based on the 2002 residential land uses in each of the TEZs. Because we used the most recent ATMs to conduct these analyses, the net values we calculated represent the increments in shelter demand and evacuation clearance times, based on the roadway capacities incorporated in the ATM, that are attributable to the residential growth that occurred between the comprehensive plan approval dates and 2002.

Table 11.6 shows that for the five counties analyzed, the median increases in shelter demand due to post-plan-approval growth ranged from approximately 1,000 people for a Category 1 hurricane, to 2,000 for a Category 3 storm, and 3,000 for a Category 4/5. Shelter deficits are defined in terms of the worst-case, Category 4/5 storm (Florida Department of Community Affairs, Division of Emergency Management, 2004). As shown in the table, post-plan-approval growth has resulted in an increased demand of over 20,000 persons in these five counties combined for a Category 4/5 storm. Based on a standard of 20 square feet of shelter space per person, these deficits translate into nearly 1.5 million square feet of needed shelter capacity. Comparison to 2004 deficits for a Category 4/5 hurricane reveals that growth that has occurred between approval of local comprehensive plans and 2002 has been responsible for between 14 and 104 percent of the individual county shelter deficits and for 28 percent of the aggregate deficit for these five counties.

Table 11.7   Contributions to 2002 evacuation clearance times from
             residential growth within hurricane vulnerability zones between
             plan approval year and 2002

County	Category 1			Category 3			Category 4/5		
	2002 Time	Post-Plan Increment	Post-Plan Share	2002 Time	Post-Plan Increment	Post-Plan Share	2002 Time	Post-Plan Increment	Post-Plan Share
Bay	14.3	1.6	12%	22.0	2.8	13%	24.5	2.2	9%
Pasco	4.0	0.1	3%	8.7	0.3	3%	14.3	0.4	3%
Indian River	7.3	1.8	25%	8.0	1.9	14%	8.0	1.9	14%
Volusia	7.7	0.3	4%	10.0	0.4	4%	13.8	0.5	4%
St. Johns	11.0	2.4	22%	17.0	3.6	21%	16.9	3.6	21%
Mean:	8.9	1.2	13%	13.1	1.8	11%	15.5	1.7	10%

Shaded values equal or exceed the state guideline of 16 hours for maximum clearance time.

Table 11.7 displays the aggregate effects of post-plan-approval growth within the
HVZs of individual coastal jurisdictions on the maximum 2002 evacuation clearance
times of the five counties we analyzed. The increments reported are for the most
constrained critical roadway segment in each county for a given hurricane intensity.
Mean increments due to growth within HVZs range from 1.5 hours for a Category
1 hurricane to 2.2 hours for a Category 3 and 3.0 hours for a Category 4/5. Such
increases are clearly not in concert with the intent of the state's planning mandate to
maintain or reduce hurricane evacuation clearance times within HVZs in particular
or within counties as a whole.

Increases in permanent and seasonal populations within HVZs between plan
approval and 2002 account for an average of between 15 and 18 percent of estimated
2002 evacuation clearance times for all three hurricane categories. Estimated 2002
evacuation clearance times for Bay and St. Johns counties equal or exceed the
state's recommended maximum of 16 hours for both Category 3 and Category 4/5
hurricanes. In St. Johns County, the post-plan increments within the HVZ account
for all of the margins that exceed the state guideline. In Bay County, the post-plan
increments within the HVZ account for about half of the margins that exceed the
state guidelines. For both the Category 3 and the Category 4/5 storms, the pre-plan
base evacuation clearance times that we calculated for Bay County already exceeded
the state guideline of 16 hours.

Discussion

Our findings present a mixed picture of the possible impacts of Florida's hazard
mitigation planning mandates on development that has occurred within the areas
most prone to hurricane flood damage, "coastal high hazard areas" and "hurricane
vulnerability zones". The general pattern that emerges from an examination of
aggregate development trends for all 89 communities in our sample is one of slower
residential growth rates within CHHAs than in areas outside of CHHAs, both prior
to and after comprehensive plan approval and implementation. When we control
for vacant land supply, we find higher growth densities within CHHAs than outside

both before and after plan implementation, but we also find a much greater decline in those densities after plan approval within CHHAs.

The raw numbers are cause for concern, however. While the aggregate growth densities within CHHAs have slowed, the total increases in residential exposure that have occurred since comprehensive plan approval are very substantial. Rough extrapolations based on the estimates from our sample indicate that approximately 425,000 new housing units were constructed within the combined area of the CHHAs and HVZs in the state between the years in which individual comprehensive plans were approved and 2002. These new residential units added approximately $80 billion in 2002 just value of property improvements within these hazard zones and exposed an additional 958,000 residents to the hazards of hurricane flooding.

Analysis also reveals considerable impacts of this residential growth on public shelter demand and evacuation clearance times in a subsample of five counties. Post-plan growth through the year 2002 accounts for between 14 and 104 percent of the reported 2004 shelter deficits for a category 4/5 hurricane in these five counties. Post-plan growth within the HVZs added 1.6 and 2.4 hours, respectively, to the evacuation clearance times for a Category 1 hurricane in two counties—Bay and St. Johns—and 2.2 and 3.6 hours, respectively, for a Category 4/5 hurricane. This growth accounted for most or all of the evacuation clearance time increments for a Category 3 or a Category 4/5 hurricane that pushed these two counties beyond the state's recommended maximum of 16 hours.

It is apparent that there may be some significant constraints operating to limit the extent to which local governments are able or willing to reduce growth within CHHAs and HVZs. A number of the coastal cities in Florida face a serious dilemma in implementing policies to restrict development within these areas. Almost a quarter (21 percent) of the 77 communities in our sample that have land within a CHHA have no land outside the CHHA. Letter-of-the-law compliance by these communities with the state's requirement to limit development within CHHAs and direct populations away from those areas would essentially require a moratorium on future growth. A full 25 percent of the 77 communities that have an HVZ have no land *outside* the HVZ. These communities face an analogous dilemma regarding the state's mandate to reduce or maintain hurricane evacuation clearance times, although they have the option of increasing infrastructure capacity as an alternative to limiting growth.

Even where communities may have attempted to follow both the letter and the spirit of the law, at least three other constraints may have limited their ability to do so:

1. conversion of seasonal residences to permanent residences;
2. vesting of developments that had already received various forms of development approval prior to adoption and approval of the local comprehensive plans between 1988 and 1991; and
3. perceived political or legal constraints to down-zoning property within the CHHA or HVZ so as to reduce the allowable densities of residential development.

Vesting arises under stipulations of Florida's growth management statute (*Fla. Stat.* §163.3233, 2006) that require that a local government's laws and policies governing

Growth Management in Florida

Lynn Haven
Number of Units Built by Year and Location

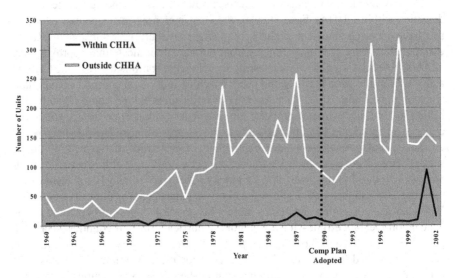

Figure 11.6 Post-plan-approval development trend that may be indicative of vested development within CHHA

the development of land at the time of the execution of a development agreement "shall govern the development of the land for the duration of the development agreement." Local governments may only apply subsequently adopted laws and policies to a development that is subject to a development agreement where "[t]hey are not in conflict with the laws and policies governing the development agreement and do not prevent development of the land uses, intensities, or densities in the development agreement."

Most communities in the state had zoning ordinances in effect prior to adoption and approval of comprehensive plans pursuant to the state's 1985 mandate. In some coastal communities, those ordinances permitted residential densities greater than the intensities of residential development that existed at the time the new comprehensive plans were approved. There is no central record of the number of developments that were vested under those zoning provisions. However, in the absence of specific numbers, it may be possible to ascertain from observations of development trends whether or not vesting may have been a contributing factor to post-plan-approval development patterns.

If a few residential developments were vested in a community prior to plan approval, one might expect to see scattered peaks in residential development after plan approval that represent execution of vested development authority. Figure 11.6 provides an example of such a pattern in the City of Lynn Haven. Based on similar visual time trend analyses, we find evidence of possible vested developments having been permitted after plan approval in 16 of the 61 communities in our sample that have land both within and outside of a CHHA.

Niceville
Number of Units Built by Year and Location

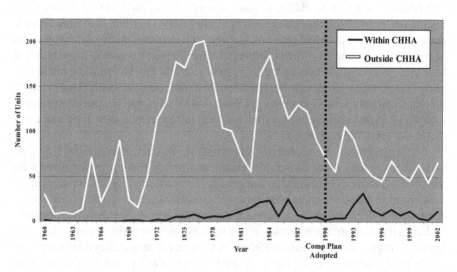

Figure 11.7 **Post-plan-approval "business as usual" development trend within CHHA**

If a substantial portion of the vacant residential land in a community was vested at the time the comprehensive plan was approved, that condition might be reflected in no significant change in residential development trends after plan approval. Figure 11.7 provides an example of such a "business as usual" pattern for the City of Niceville. We see evidence of such a development pattern in more than half of the 61 communities which have land both within and outside of a CHHA. Other factors, however, may explain a "business as usual" development trend, including failure to adopt or implement policies to limit growth within the CHHA.

Regardless of the existence of vested developments, many communities are now facing pressure to allow for more intense redevelopment where their zoning ordinances still allow higher densities in coastal areas than the as-built densities at the time their comprehensive plans were approved (V. Hubbard, personal communication, February 2006). In other communities, where there was still substantial vacant land in the early 1990s, local governments face political pressure to permit new development at or close to the maximum densities permitted in the zoning ordinances. Some of this pressure is market driven. However, there also are provisions in the state's comprehensive plan regulations governing the future land use element that promote higher density redevelopment and infill. These include requirements that local governments include objectives and policies in their future land use elements that discourage the proliferation of urban sprawl (*Fla. Admin. Code* §§ 9J-5.006(3)(b)(8), 2006) and that they review proposed plan amendments to determine whether or not they discourage sprawl (*Fla. Admin. Code* §§ 9J-5.006(5), 2006).

In both circumstances, most local governments are reluctant to down-zone property because of the threat of litigation under the state's 1995 Bert Harris Act (*Fla. Stat.* §70.001, 2006), which requires compensation to landowners for regulations noticed for adoption or adopted after May 11, 1995, that "inordinately burden" property. Published case law has not yet interpreted many of the key terms in the Act, so it remains difficult to predict what facts or economic impacts might lead to a government action losing in a lawsuit. While few claims have made it to circuit court, many have been filed and settled before going to court (University of Florida College of Law Conservation Clinic, 2004). Thus, the Act presents possible costs in legal and settlement expenses for local governments, even for those cases that never reach the courtroom.

Further research is needed to determine if, in fact, the policies in local comprehensive plans that address development within CHHAs are correlated with the observed changes in residential exposure to hurricane flooding that have occurred subsequent to comprehensive plan approval. Such an analysis would provide more compelling evidence that the apparent trends towards lower intensity development within CHHAs versus areas outside are, in fact, attributable to the implementation of the state's hurricane hazard mitigation mandate. Detailed case studies that reconstruct the implementation stories that lie behind the observed patterns of changes in residential land use within hurricane hazard zones would provide additional insight into the factors that have operated to dictate how growth has occurred in Florida's coastal communities.

Acknowledgements

This research was conducted under the auspices of the Florida Sea Grant College Program with support from the National Oceanic and Atmospheric Administration, Office of Sea Grant, US Department of Commerce, Grant No. R/C-P-26. Additional funding was provided by the Florida Department of Community Affairs and the DeVoe Moore Center at Florida State University. We are deeply indebted to four cohorts of graduate research assistants whose enthusiasm and diligence made this work possible: Michelle Freeman, Daniel Harris, Alex Joyce-Peickert, Ramona Madhosingh, Brandie Miklus, John Richardson, Audrey Smith, Preeti Solanki, Lara Mae Webster, and Chris Whittaker.

References

Berke, P. R., and French, S. P. (1994). The influence of state planning mandates on local plan quality. *Journal of Planning Education and Research 13*(4): 237–250.

Blake, E. S., Jarrell, J. D., Rappaport, E. N., and Landsea, C. W. (2005). *The deadliest, costliest, and most intense United States tropical cyclones from 1851 to 2004 (and other frequently requested hurricane facts)*. (NOAA Technical Memorandum NWS TPC-4). Miami, FL: National Hurricane Center, National Oceanic and Atmospheric Administration.

Burby, R. J., and Dalton, L. C. (1994). Plans can matter! The role of land use plans and state planning mandates in limiting the development of hazardous areas. *Public Administration Review 54*(3): 229–238.

Burby, R. J., and May, P. J. (1997). *Making governments plan.* Baltimore: Johns Hopkins University Press.

Chapin, T. S., Deyle, R. E., and Baker, E. J. (2006). *A parcel-based GIS method for evaluating conformance of local land use planning with a state mandate to reduce exposure to hurricane flooding.* Manuscript submitted for publication.

DeGrove, J. M., and Miness, D.A. (1992). *The new frontier for land policy: Planning and growth management in the United States.* Cambridge, MA: Lincoln Institute of Land Policy.

Deyle, R. E., and Smith, R.A. (1998). Local government compliance with state planning mandates: The effects of state implementation in Florida. *Journal of the American Planning Association 64*(4): 457–469.

Florida Department of Community Affairs, Division of Community Planning. (2005). *Protecting Florida's communities: Land use planning strategies and best development practices for minimizing vulnerability to flooding and coastal storms.* Tallahassee, FL: The Department.

Florida Department of Community Affairs, Division of Emergency Management. (2003). *Emergency reimbursement system executive financial report.* Tallahassee, FL: The Department.

Florida Department of Community Affairs, Division of Emergency Management. (2004). *Statewide emergency shelter plan.* Tallahassee, FL: The Department.

Florida Department of Community Affairs, Florida Coastal Management Program. (2000). *Florida assessment of coastal trends.* Tallahassee, FL: The Department.

May, P. J., Burby, R. J., Ericksen, N. J., Handmer, J. W., Dixon, J. E., Michaels, S., et al. (1996). *Environmental management and governance: Intergovernmental approaches to hazards and sustainability.* New York: Routledge.

University of Florida College of Law Conservation Clinic. (2004). Implementation of the model land development code for Florida springs protection. Gainesville, FL: The Law Conservation Clinic.

Chapter 12

Urban Containment and Neighborhood Quality in Florida

Arthur C. Nelson, Casey J. Dawkins, Thomas W. Sanchez,
and Karen A. Danielsen

Rapid suburbanization since World War II in America has created many of the challenges we face today. Roads intended to relieve congestion have become congested. Cookie-cutter subdivisions have replaced scenic landscapes. Once-vital downtown stores have been abandoned as shoppers transferred their allegiance to convenient suburban malls. The spread of low-density residential development made public transit impractical, making the automobile virtually the only choice for transportation. Automobile dependence has degraded the air in some places to alarming levels. Once-tranquil communities with their own unique character have been overwhelmed by more people, automobiles, and shopping centers. But the problem is not growth per se; the problem is how to manage growth in ways that minimize costs and maximize benefits to both individuals and the public at large.

Urban containment is an attempt to confront the reasonable development needs of the community, region, or state, and accommodate them in a manner that preserves public goods, minimizes fiscal burdens, minimizes adverse interactions between land uses while maximizing positive ones, improves the equitable distribution of the benefits of growth, and enhances quality of life. At its heart, urban containment aims to achieve these goals by choreographing public infrastructure investment, land use and development regulation, and deployment of incentives and disincentives to influence the rate, timing, intensity, mix, and location of growth. Broadly speaking, urban containment programs can be distinguished from traditional approaches to land use regulation by policies that are explicitly designed to limit the development of land outside a defined urban area, while encouraging infill development and redevelopment inside it.

In response to concerns about contemporary development patterns, some American states and metropolitan areas have attempted to contain the outward expansion of urban development. Although the idea of urban containment is not new in America—some New England townships in the 17th Century forbade homes from being built in the nearby farmland—its modern form arose only as recently as the late 1950s. Lexington and Fayette County, Kentucky, are credited with implementing the nation's first effort to contain urban sprawl, chiefly by limiting development to an area within an urban service line and preventing urban-scale residential development in the Bluegrass area around.

During the 1970s, urban containment emerged in a few more metropolitan areas, including Miami-Dade County, Minneapolis-St. Paul, Boulder, Sarasota, and Sacramento, and in one state—Oregon. Florida's growth management legislation in the middle 1980s enabled local governments to adopt various forms of urban containment strategies. Washington State adopted Oregon-style containment laws in the early 1990s and applied them to the most urbanized counties.

Beginning in the 1970s and continuing through the rest of the 20th Century, numerous metropolitan areas saw individual local governments pursue containment on their own, chiefly throughout coastal metropolitan California. Half a century ago there was but one clear example of urban containment in metropolitan America. Our research has revealed there are now more than one hundred metropolitan areas with at least one example of metropolitan-wide or local government containment. Examples are not limited to Sunbelt or West Coast regions where growth pressures are legion; Sioux Falls, South Dakota has one of the nation's oldest programs.

Urban containment seems to promise many things (see Nelson, 2000). Here, we focus our attention on neighborhood quality, an area of urban containment research that has not received much attention. Our particular focus is whether urban containment in Florida can be associated with neighborhood quality improvement since the advent of statewide growth management implemented in earnest in the late 1980s. We describe Florida's urban containment approach generally based on a review of Florida statutes, review issues of understanding neighborhood quality and how containment should relate to improving conditions over time, present our model and data, report results of data analysis, and offer implications for policy and research.

Urban Containment in Florida

Nelson and Dawkins (2004) created a typology of urban containment programs based on a content analysis of more than 100 containment plans, many from Florida. Their typology classifies most containment efforts in Florida as "strong-accommodating." That is, while plans must accommodate projected development needs, they must do so while also protecting natural systems and natural-resource based industries such as farming and fishing.

The genesis of Florida's approach to contain the outward expansion of urban development is found in the State Comprehensive Plan, adopted in 1985 and codified as Chapter 187. This Act requires all of Florida's 67 counties and 410 municipalities to adopt Local Government Comprehensive Plans that direct future growth and development based on 26 goals, with several hundred policies that guide state, regional, and local government planning and decision-making. While many goals address such areas as education, health, and the elderly, many others relate directly to land-use and facility planning in the context of urban containment. In our view, Florida planning statutes are clearly restrictive in attempting to prevent the outward expansion of urban development but also accommodating in meeting development needs. A sample of Chapter 187 goals and policies read like textbook principles of urban containment in that they encourage and sometimes require local plans to achieve compact development patterns, preserve natural systems, maintain agriculture, create land uses that support transit options, achieve mixed-use development, and deliver public facilities and services efficiently. Florida Administrative Code Rule 9J-5 fleshes out how these and other State Comprehensive Plan goals and policies must be addressed by local governments.

Despite these provisions, however, urban containment—perhaps the embodiment of the planning principles espoused in the state planning acts—is not applied throughout the state. Nelson and Dawkins (2004) found that about a third of Florida counties had urban containment policies either countywide or in some cities. Does growth management generally or urban containment specifically achieve desired outcomes? Our review of many plans approved by the state allowed acreage homesites and planned communities in rural areas, created incentives for rural development where urban areas were at infrastructure capacity, and did not truly achieve the efficient land-use patterns desired. Several commentators have made similar observations (Chapin and Connerly, 2003; Porter, 1999; Sakowicz, 2004). We conclude that Florida-style growth management may not be much different from traditional land-use planning, except for those jurisdictions pursuing urban containment.

Urban Containment Influences on Neighborhood Quality

In many ways the debate surrounding urban containment is related to attitudes toward growth, even if in some instances containment policy may be more favorable to housing production (Nelson, Dawkins, and Sanchez, 2005; Nelson, Sanchez, and Dawkins, 2004). Growth advocates consider growth the route to economic well being that brings positive attributes of job opportunities, an increased tax base, and cultural and other amenities (See also Downs, 1998). But the anti-growth crowd declares growth harmful to the quality of life and to the environment, with the unintended consequence of straining local governments to provide additional services.

Smutny (1998) remarks that, "growth is a complicated phenomenon." The public demonstrates a great deal of subtlety about growth. When the National Association of Home Builders polled households on their opinion of growth and growth management, 75 percent of the respondents indicated that they wanted to plan and manage growth but only 12 percent wanted to pass laws to restrict growth and only two percent favored laws to stop it (Ahluwalia, 1999).

Many detractors of urban containment do not necessarily endorse status quo development but they believe that urban containment policies make existing problems, such as crime and congestion, worse (Burnett and Villarreal, 2004; Campbell, 1998, O'Toole, 2001). To their way of thinking, density is the enemy.

Pro-growth advocates argue that if the public wanted more urban containment, the market would soon meet the demand (Gordon and Richardson, 1998; Staley, 1999). Some have even suggested that sprawl-free growth can be achieved if plans guiding development are based on urban containment principles (Heid, 2004). Also pro-managed growth scholars have noted that sprawl does provide many benefits for a large number of people despite its inefficiencies and high costs (Burchell and Mukherji, 2003; Downs, 1998).

But what are the major quality of life problems that sprawl generates? These quality of life killers are traffic congestion, open space consumption, excessive energy use, lack of infrastructure provision and improvement, NIMBYism, crime, concentrated poverty, poor schools, inadequate public services, and scarce fiscal resources (Downs, 1998; Yu, Johnson, and Zhang, 2004).

The urban containment movement seeks to reform state growth management legislation and is a reaction against the perceived ills of urban sprawl. There are two interest groups associated with urban containment. Pro-growth developers want to ensure a supply of land, reduce building costs, and be allowed to build a wide variety of housing types. Planners and environmentalists want to create more compact development, spur revitalization and infill, and encourage less dependence on cars. Both groups see livability as a goal.

Planners have begun to embrace many of the ideals of the New Urbanists as one means to achieve livability (Godschalk, 2004). Offering their prescriptions to developers for increasing livability in communities, New Urbanists advocate a specific set of desirable neighborhood characteristics that urban containment supporters have adopted as part of their principles to improve neighborhood quality of life.

There is some evidence that these characteristics contribute to a higher quality of life in neighborhoods. In studies of urban and suburban Portland neighborhoods, Lund (2003) and Song and Knaap (2004) both found that perceptions of neighborhood quality were influenced by having many of New Urbanist design characteristics in the neighborhoods.

Managed Growth, Urban Containment, and Quality of Life

Despite its unpopularity in some quarters, growth management, as well as its sister movement, urban containment, is seeking to balance the interests of both planners and developers with the needs of the community, state, and region. Hirschhorn (2000) concludes that urban containment means growth that is supportive of quality of life and quality-of-place, not slow- or no-growth.

As Nelson and Dawkins (2004) argue, growth management should meet development needs in a manner that preserves public goods, minimizes fiscal burdens, minimizes adverse interactions between land uses while maximizing positive ones, improves the equitable distribution of the benefits of growth, and enhances the quality of life. Yet there exists no comprehensive research looking at whether or not growth management programs have made a difference in the quality of life in those areas that have adopted them compared to areas that have not. Much of the advocacy literature on both sides of the sprawl and urban containment debate discusses declines in quality of life as a fact, but there is no empirical basis to state what exactly decline means. Hirschhorn (2000), for instance, claims that it has taken years of unmanaged, rapid growth to see the consequences of suburbanization, with many of them difficult to quantify individually but nonetheless profound when considered collectively.

We turn our attention now to issues of measuring quality of life and more particularly neighborhood quality.

An Overview of Quality of Life as a Concept

Quality of life is a complicated phenomenon with many definitions. For instance, Hirschhorn and Souza (2001) indicate traffic congestion, loss of open space, environmental impacts, and threats to economic growth and government budgets as the major contributors to a loss of quality of life. Myers (1988) measures quality of life in terms of the different levels of satisfaction of citizens in different residential locations based on traffic, crime, job opportunities, and parks. He goes on to note that a high quality of life results typically from growth because it creates jobs and often leads to higher wages. Further complicating the picture is that rapidly growing communities see older quality of life assets (such as open spaces and sense of place) transition to newer ones such as more shopping and restaurant options, and more economically and socially vibrant places. Longtime residents often do not appreciate the new quality of life assets yet the new assets can have the effect of making a community more attractive.

A recent study indicates that growth can fuel quality of life decline. Gabriel, Mattey, and Wascher (2003) found that there was substantial deterioration of the quality of life in some states that experienced rapid population growth during the 1981–1990 period. The analysis reveals that quality of life declined due to reduced investments in infrastructure, increased traffic congestion, and air pollution. In areas that did not experience fast growth, quality of life either remained relatively high or improved.

How should planning improve quality of life through managed growth? Myers (1988) believes that the role of planning is to improve quality of life by mitigating the negative impacts of growth. Planners can do this by slowing the rate at which these negative effects are felt or simply slowing the pace of physical change. If quality of life is maintained, he argues, the higher cost of locating there will be worth paying, and improving urban amenities could offset the loss of small town or environmental assets. For instance, Kahn (2001) found that while fast-growing metropolitan areas have falling rents (his measure of quality of life decline), real estate prices have risen in such slow-growing areas as San Francisco, where growth controls dominate the planning systems.

Myers (1988) remarks that the planning profession has been slow to adopt the quality of life concept partially because of the ambiguity of the idea. (Ambiguity is a characteristic the urban containment debate seems to encourage.) This is borne out in the literature: Most of the recent quality of life studies are not found in planning journals but rather in those devoted to geography, sociology, marketing, real estate, or economics. Among planners, there is some recent interest in general neighborhood quality (Greenberg, 1998), but most of their work assesses a particular planning technique (such as whether New Urbanism works) or is done for a policy evaluation purpose (such as building neighborhood indicators) (Lund, 2003; Sawicki and Flynn, 1996; Song and Knaap, 2004). Two studies on general neighborhood quality appeared in the planning literature around the late 1980s (Landis and Sawicki, 1988; Myers, 1988). Some recent empirical studies (Gabriel et al., 2003; Kahn, 2001) have investigated the decline of regional quality of life due to growth, but it is not clear whether their findings are generalizable to smaller geographical units.

Despite all the attempts to improve quality of life through growth management, quality of life may not be improving partly because of the way that these programs were enacted. Smutny (1998) reveals that state growth management legislation was enacted in two waves. The first wave in the 1970s was influenced by the environmental movement. The second wave, enacted in the 1980s and later, focused on the more global sprawl issues of economic development, housing infrastructure, quality of life and environmental concerns.

Do all the growth controls actually create better quality of life? There is anecdotal evidence that the second wave growth controls enacted in the 1990s may be hurting some communities' quality of life. For example, 2004 marks the third straight year that Coloradoans have repealed growth plans or reworked existing programs (Siebert, 2004). Loudoun County, Virginia has found itself on the verge of repealing its 1990s-style growth controls as well (Laris, 2001; Laris, 2004). Perhaps the real question is which type of growth control actually produces better quality of life?

Measuring Quality of Life

Quality of life is usually characterized as multidimensional, making measurement somewhat complicated. Measurements require consideration of both empirical or objective measures and qualitative or subjective measures (Greenberg, 1999; Sirgy, Rahtz, Cicic, and Underwood, 2000). Different types of data are often gathered and pooled in the analysis because there is often assumed to be a potentially significant relationship between objective and subjective measures. Subjective measures are the stronger correlates of neighborhood quality in most studies (Sirgy et al., 2000). Lund (2003) states that she uses subjective, personal level attributes in order to remove confounding variables and to reflect the causal priority of those that remain.

She also discovered, to her surprise, that attitudinal variables were the most powerful variables for predicting the empirical factors. Empirical measures, or "hard measures," include such variables as housing prices, employment rates, and numbers of facilities. According to Riecken and Yavas (2001), empirical measures are relatively easily defined and measured such as through easily counted units and can be reasonable proxies for measures of achievement and well-being. Empirical measures in quality of life studies are usually classified into five, somewhat overlapping, dimensions: Economic, Physical, Social, Satisfaction, and Personal.

Qualitative measures are also divided into the same five dimensions as above but use less objective measurements. They often involve surveying respondents regarding their individual subjective and self-reported perceptions, evaluations, attitudes, feelings, and levels of satisfaction in terms of their quality of life (Riecken and Yavas, 2001; Sirgy et al., 2000). Most qualitative measures involve surveys with scaled questions to measure the attitude strength.

Measuring Neighborhood Quality

Sawicki and Flynn (1996) warn of the methodological pitfalls of trying to construct neighborhood indicators. As with other quality of life measures, the unit of analysis must be carefully considered. Geographical concepts such as neighborhood are part of a nested hierarchy of data, and care must be taken to ensure that the data gathered are for the correct unit of analysis and capture what residents believe to be their neighborhood. This is not an easy task; unfortunately what constitutes a neighborhood is difficult to establish without dispute (Song and Knaap, 2004). Widgery (1995), for example, defines a neighborhood as the geographic area considered by most individuals as their nearest social and psychic space outside the home. This definition does not lend itself very well to empirical definition. Song and Knaap observed that while their neighborhood construct relied on data-driven boundaries defined by the census tracts—block groups and sub-blockgroups—quantitatively derived definitions based on standardized empirical units do not always capture what individuals actually perceive as their neighborhood. The ambiguity of the neighborhood's definition is why it is important to use both subjective and objective indicators in measuring quality of life.

Other methodological concerns include the need to measure neighborhood change and the predictive capacity of neighborhood variables. Sawicki and Flynn argue that a well constructed neighborhood indicator should generate data that can measure meaningful change in neighborhoods over time. Greenberg, Schneider, and Choi (1994) argue that while it may seem inconsistent that neighborhood pleasantness is not included among the determinants of quality of life, such a variable would likely be unpredictable.

All this begs the question, though, about whether the neighborhood is even the appropriate locus for measuring the effect of urban containment on quality of life. For example, it may be possible that if urban containment stimulates infill and redevelopment, neighborhoods become denser; and to some, this could mean an erosion of quality of life. The process of infill and redevelopment may also take years or decades, during which time residents are impacted adversely such as through construction, rerouting of traffic, reduced sales among local businesses, and uncertainty as to how the changes will affect their lives, thus lowering quality of life. Despite temporal erosion in neighborhood quality of life, the quality of life in the larger community and perhaps the state could improve. Indeed, this is one of the tensions created through Florida's Growth Management Act: To elevate overall quality of life, the status quo may need to be disrupted and with it the local sense of quality of life.

More problematic for measurement is that even if quality of life erodes, it could possibly erode more without urban containment policies in place. Florida is one of the nation's fastest growing states. During the period 2000 to 2030, new development of all kinds could equal in volume the amount of development that existed in 2000. If this projected development is left unmanaged, quality of life could be threatened at all levels, from the neighborhood to the state. Nevertheless, growth management approaches such as urban containment may not be able to elevate future quality of life above a past level even if the alternative would have made people worse off.

If one assumes the neighborhood is the appropriate level of quality of life analysis, the question remains how to measure it. For instance, there does not appear to be much, if any, research using the American Housing Survey (AHS) data to assess neighborhood quality—despite the fact that this survey is the nation's most robust attempt to track housing, neighborhood, household, and socioeconomic changes over time. Some researchers (Greenberg et al., 1994; Song and Knaap, 2004) use the AHS questions as inspiration for their own surveys, but they have not compared their results to corresponding data collected in the AHS.

In their study of environmentally degraded neighborhoods, Greenberg and Schneider (1996) used AHS data as baseline measures of some demographic and neighborhood quality variables. It is difficult to determine the comparability of their work to the present study, however, because they included additional data in their analysis. They also limited their focus to a few neighborhoods in New Jersey and specifically looked for poor neighborhood quality.

Model and Data

As we are interested in the association between urban containment and neighborhood quality, we tested for the following:

Neighborhood Quality = *f* [*Housing Attributes, Neighborhood Attributes, Market and Location Controls, Urban Containment Period*]

Before specifying each vector, let us review the source of data used in our assessment: the American Housing Survey. For nearly 30 years, the Census Bureau and the Department of Housing and Urban Development have supported the AHS. It is conducted biennially across the nation and periodically among more than 40 metropolitan areas. Since 1983 (and most recently in 2003), the AHS interviews 50,000 to 80,000 households nationally for information about their socioeconomic characteristics, housing attributes, and neighborhood quality, among other things. Our analysis includes all AHS national survey years from 1985 through 2003, ten in all. The year 1983 is excluded for lack of immediately available electronic data. The survey also includes metropolitan area identifiers. We used these data for two assessments, as follows.

Regional and Florida Comparisons

In order to determine if urban containment in Florida is effective in improving neighborhood quality over time, we compared quality indicators among metropolitan areas in states within the same region over time. Florida is grouped in the Census Bureau's Southeast Division, along with District of Columbia, Georgia, Maryland, North Carolina, South Carolina, and Virginia. With 25 million residents added between 1970 and 2000 and another 30 million projected for the period 2000 to 2035, it is the nation's fastest growing region in population. The region enjoys a mostly sprawling landscape in terms of development (aside from the Appalachian Mountain Range in the extreme western portions of some states) and is largely interconnected through transportation routes and economic linkages.

It is also a region with some important variations in planning regimes. Two states, Florida and Maryland, are among the nation's leaders in state-level planning, but Maryland's efforts are recent and not as comprehensive in scale as Florida's. Georgia has state-mandated planning but none of the plan-making rigor seen in Florida or Maryland. North Carolina has one of the more rigorous coastal zone management programs; but as most of that state's growth occurs between the Appalachians and the Coastal Plain, state-level involvement in local planning and development decision-making is limited. South Carolina and Virginia have little state-level involvement in local planning or development decision-making, although urban containment exists in a handful of Virginia counties. These variations allow for interesting comparisons among the states' planning regimes in terms of neighborhood quality.

As AHS data for cases are coded for metropolitan areas, we could readily sort them by state, except where the metropolitan areas span state boundaries. In such instances, we assigned the case to the state with the greatest share of the metropolitan population. Washington, DC, which is the largest metropolitan area in the region (and fourth largest in the nation), spans parts of three states and the District. We used it as the "referent" to which the other states were compared. We also pooled all Florida metropolitan AHS respondents. While data did not allow us to compare individual metropolitan areas they did allow us to compare cases over time.

We had a certain expectation about the outcome. In recent work for the Fannie Mae Foundation, we (Nelson, Dawkins, Sanchez, and Danielsen, 2005) found that urban containment improves neighborhood quality over time. That is, while improvements could not be detected in early years of urban containment over a period of time extending about a generation (20–30 years) there is a positive association between urban containment and improving neighborhood quality. We surmise that it takes about a generation to redirect investment from sprawling landscapes to ones that are more central. As this happens neighborhood stability is enhanced and a sense of overall improvement in neighborhood quality is seen.

Specification

Since we were attempting to predict only the association between neighborhood quality and state planning regime (in the Southeast cross-section analysis) and between neighborhood quality and presence of statewide containment planning over time (in the Florida analysis), we used ordinary least squares regression. Based on our prior work for the Fannie Mae Foundation (Nelson, Dawkins, Sanchez and Danielsen 2005) we found that the variables reported in the following tables were appropriate for the analysis. For reasons we explain in our earlier work, the dependent variable is "neighborhood quality" as scored by respondents on a 1 to 10 scale, with 10 being of highest quality. This is a self-assessment exercise for which respondents are not given descriptions of differences between levels of quality. This is one of the limitations of panel data, but as our earlier research noted, this data is nonetheless a reasonable way in which to compare neighborhood quality across space and time. A brief description of the variables and how they are measured is included in the regression tables themselves. All data come from the American Housing Survey.

The experimental variables include Florida for the Southeast analysis and the urban containment periods for both analyses. The period 1991–1997, which includes AHS national surveys for 1991, 1993, 1995, and 1997, corresponds to the time when (nearly) all local governments had prepared plans deemed acceptable to the state

Table 12.1   Neighborhood quality regression results: Owner occupants in the Southeast Region

Variables and Measurement Scale	Overall Neighborhood Quality		
	Coefficient	t-Ratio	One-Tailed Sig.
Constant	7.547	0.452	0.326
Housing Attributes			
Unit adequate (1,0)	-0.101	-1.619	0.053
Housing unit quality (increasing from 1 to 10)	0.578	68.676	0.000
Age of unit (2004 minus year built)	0.001	0.763	0.223
Attached unit (1,0)	-0.140	-3.363	0.001
Number of units in bldg. (continuous)	0.001	2.332	0.010
Number of bedrooms (continuous)	-0.043	-1.999	0.023
Number of bathrooms (continuous)	0.083	2.786	0.003
Room air conditioning (1,0)	-0.004	-0.081	0.468
Central air (1,0)	0.161	2.592	0.005
No heat source (1,0)	0.150	1.446	0.074
Unit square feet (continuous)	0.000	2.342	0.010
Lot square feet (continuous)	0.000	0.530	0.298
Neighborhood Attributes			
Percent rental units in zone (continuous)	-0.370	-0.985	0.162
Percent African-American in zone (continuous)	-1.011	-3.037	0.001
Average family income in zone (continuous)	0.000	0.287	0.387
Average years of residence in zone (continuous)	-0.013	-0.895	0.186
Percent in zone reporting crime problems (continuous)	0.202	0.479	0.316
Percent in zone reporting noise problems (continuous)	0.132	0.151	0.440
Percent in zone reporting litter problems (continuous)	-2.333	-2.431	0.008
Percent in zone reporting undesirable land uses (continuous)	0.992	0.895	0.186
Market and Location Controls (DC is referent)			
Percent new units in zone (continuous)	-0.028	-0.041	0.484
Maryland (1,0)	-0.061	-0.587	0.279
Virginia (1,0)	0.126	0.875	0.191
North Carolina (1,0)	0.035	0.208	0.418
South Carolina (1,0)	-0.150	-1.282	0.100
Georgia (1,0)	-0.001	-0.013	0.495
Florida (1,0)	-0.219	-1.625	0.052
Central city (1,0)	0.066	0.790	0.215
Survey year (continuous)	-0.002	-0.215	0.415
Urban Containment Period			
1991–1997 survey year (1,0)	-0.261	-3.426	0.001
1999–2003 survey year (1,0)	-0.233	-1.968	0.025
Regression Statistics			
F			179.485
Adjusted R-square			0.298
N			13020

Dependent variable: Overall neighborhood quality (increasing from 1 to 10).

planning agency, the Florida Department of Community Affairs. We call this the "early implementation" period. The period 1999 through 2003 includes surveys for 1999, 2001 and 2003, and we call this the "middle implementation" period. Future analyses may include a "mature implementation" period but we suspect that will not be until at least three surveys after 2005. These periods also apply to Georgia, which adopted its statewide planning program in 1989, and marginally to Maryland, which adopted its in 1995.

Results

Table 12.1 reports results of the Southeast analysis while Table 12.2 reports results for Florida. We found no problematic collinearities. The coefficients of determination are modest but not unreasonable for studies of this sort.

According to the data presented in Table 12.1, there was no statistically significant association ($p < 0.01$) between state planning regime and neighborhood quality using metropolitan Washington, DC, as the referent. (This level of significance was selected because of the very large sample size.) We found a negative relationship between urban containment and neighborhood quality, but it was significant only in the first (1991–1997) period of the implementation of urban containment policies. We also found no general relationship with respect to survey year—an overall longitudinal measure. Similar results were obtained in the Florida analysis, as shown in Table 12.2. In effect, neighborhood quality eroded during the first third of a generation of urban containment (20–30 years after implementation), and urban containment policies appeared to have had a neutral or ambiguous effect during the second third. If our previous work (Nelson, Dawkins, Sanchez, and Danielson, 2005) proves out, the mature third period (and presumably thereafter) may see a positive association.

Interpretations and Implications

The first analysis compared neighborhood quality as perceived by AHS respondents over time for all metropolitan areas in the Southeast, controlling for the fixed effects of each state. The fixed-effects approach allowed us to determine differences in neighborhood quality between states with and without statewide growth management efforts. The model included variables for the presence of urban containment as well so that we could assess its impact even in states without statewide growth management

The first analysis indicated no significant difference in neighborhood quality change between the states with or without growth management. Indeed, in all states, overall neighborhood quality declined during the study period and there seems to be no discernable difference in the trend between them.

On the other hand, we did find a negative association when considering urban containment in early years but not later. We surmised from a recent study (Nelson, Dawkins, Sanchez, and Danielsen, 2005) that it takes about a generation (20 to 30 years) for positive effects to emerge. Most urban containment schemes in the Southeast and all but one in Florida (Miami-Dade) are younger than this. That

Table 12.2   Neighborhood quality regression results: Owner occupants in Florida MSAs

Variables and Measurement Scale	Overall Neighborhood Quality		
	Coefficient	t-Ratio	One-Tailed Sig.
Constant	-1.230	-0.048	0.481
Housing Attributes			
Unit adequate (1,0)	-0.152	-1.573	0.058
Housing unit quality (increasing from 1 to 10)	0.575	46.47	0.000
Age of unit (2004 minus year built)	0.002	1.272	0.102
Attached unit (1,0)	0.023	0.349	0.364
Number of units in bldg. (continuous)	0.001	1.025	0.153
Number of bedrooms (continuous)	-0.065	-1.857	0.032
Number of bathrooms (continuous)	0.244	4.962	0.000
Room air conditioning (1,0)	-0.089	-1.223	0.111
Central air (1,0)	0.138	1.417	0.078
No heat source (1,0)	0.044	0.397	0.346
Unit square feet (continuous)	0.000	0.912	0.181
Lot square feet (continuous)	0.000	-0.267	0.395
Neighborhood Attributes			
Percent rental units in zone (continuous)	-1.743	-2.568	0.005
Percent African-American in zone (continuous)	-1.984	-2.993	0.002
Average family income in zone (continuous)	0.000	0.291	0.386
Average years of residence in zone (continuous)	0.018	0.394	0.347
Percent in zone reporting crime problems (continuous)	0.495	0.817	0.207
Percent in zone reporting noise problems (continuous)	4.115	2.120	0.017
Percent in zone reporting litter problems (continuous)	-6.182	-3.559	0.000
Percent in zone reporting undesirable land uses (continuous)	2.583	0.751	0.227
Market and Location Controls			
Percent new units in zone (continuous)	1.570	1.266	0.103
Survey year (continuous)	0.340	2.270	0.012
Central city (1,0)	0.002	0.186	0.427
Urban Containment Period			
1991–1997 survey year (1,0)	-0.327	-2.765	0.003
1999–2003 survey year (1,0)	-0.348	-1.924	0.027
Regression Statistics			
F			101.534
Adjusted R-square			0.303
N			5788

Dependent variable: Overall neighborhood quality (increasing from 1 to 10).

neighborhood quality eroded in early years of containment but not later suggests that in future years containment may be associated with positive quality changes— similar to our national findings.

The second analysis looked at Florida specifically. Because by definition all jurisdictions in Florida pursue growth management, the major differences between them would be in the use of urban containment. As in the study of the Southeast, neighborhood quality was found to have eroded during the study period. In areas with urban containment policies, neighborhood quality eroded more during the early years. In later years, however, there was no significance difference between containment and neighborhood quality change. We suspect this is for the same reasons we surmised for the Southeast as a whole.

This preliminary analysis does not show definitively that urban containment in Florida improves neighborhood quality. This does not mean that planning or urban containment per se reduces overall quality of life, however, because it is not possible to know the counter-factual: whether Florida neighborhoods would be better or worse off without the intervention. For instance, the "growth accommodation" orientation of Florida's Growth Management Act requires that growth projections be met—stopping or slowing growth to preserve perceived quality of life is not really an option for most Florida jurisdictions. Many neighborhoods are likely to be impacted as a result.

The analysis raises numerous methodological and policy implications. Methodologically, the literature (Lund, 2003; Sirgy et al. 2000; Song and Knaap, 2004; Widgery, 1995) suggests that quality of life studies are better when there is a mix of objective and subjective measures. Widgery (1995) goes on to speculate that there may be "various intervening variables interacting between the 'real world' and human perception. Such intervening elements may be found in differing aspiration levels or expectations by the observer of the objective world." He also discusses the fact that a person's "standards of comparison" and "accommodation" to surroundings may be intervening factors. On this score we note that the AHS does not include the number and variety of personal attitudinal variables that would be needed to allow for inferences about the strength of respondents' perceptions and attitude. Similarly, a study by the National Association of Home Builders (Ahluwalia 1999) found that 89 percent of the households in their survey were somewhat to very satisfied with the quality of life in their neighborhood—but compared to what?

Examining the effects of urban containment policies on quality of life makes explicit the struggles between existing and newly developing areas. To some, neighborhood quality may increase if development pressures are diverted from established neighborhoods to suburban fringe and rural areas. Yet, displacing development in this manner may exacerbate the very externalities (congestion, pollution, fiscal stress, destruction of sensitive landscapes) that erode state, community, and in many cases neighborhood-scale quality of life.

While we cannot conclude from this study that urban containment in Florida may eventually improve neighborhood quality, evidence from recent years indicate that it may not compromise it. It appears from analysis that in the intermediate term (after a decade of implementing adopted urban containment policies) growth management seems likely to improve existing neighborhood quality. This is consistent with Oregon's experience. Oregon's 1973 statewide planning legislation called for all jurisdictions to prepare and implement plans meeting state goals and guidelines based on a Year 2000 planning horizon. It took the state until the mid–1980s to have all the planning finished (Knaap and Nelson, 1992). Two books assessing Oregon's progress in achieving desired outcomes published in the 1990s were only mildly supportive of the regime given the modest indications of its of effectiveness (see Knaap and Nelson 1992; Abbott, Adler and Howe 1993). It was not until about ten years after full implementation of Oregon's plans (and 25 years after initial adoption) that results were measurable, in large part because objective data (such as census, transportation, pollution, etc.) were not available until then (see Nelson, 2000). The association of Florida's Growth Management Act with perceptions of increased neighborhood quality is remarkable given that Florida is the fastest growing of the states. Equally impressive is that this association appears to have occurred earlier in Florida's period of implementation than in Oregon's.

One axiom in medicine when treating a patient is "do no harm." Yet the purpose of medicine is to intervene when possible to elevate quality of life. Evidence seems to be emerging that in Florida the growth management "treatment" has begun to make neighborhoods better on the whole than they may have been without it.

References

Abbott, C., Adler, S., and Howe, D.. (1993). *Planning the Oregon way: A twenty year assessment*. Corvallis: Oregon State University Press.

Ahluwalia, G. (1999). Public attitudes towards growth. *Housing Economics 37*(5), 7–12.

Burchell, R. W., and Mukherji, S. (2003). Conventional development versus managed growth: The costs of sprawl. *American Journal of Public Health 93*(9), 1534–1540.

Burnett, H. S., and Villarreal, P. (2004). *Urban containment=Crime, congestion and poverty* (Brief Analysis No. 473). Washington, DC: National Center for Policy Analysis.

Campbell, F. (1998). Urban containment, stupid policy. *Regulation 21*(2), 10–12.

Chapin, T., and Connerly, C. (2003). *Attitudes towards growth management in Florida: Comparing citizen support in 1985 and 2001*. Tallahassee: Florida State University, DeVoe L. Moore Center.

Downs, A. (1998). The big picture. *The Brookings Review 16*(4), 8–11.

Gabriel, S. A., Mattey, J. P., and Wascher, W. L. (2003). Compensating differentials and evolution in the quality of life in the United States. *Regional Science and Urban Economics 33*(5), 619–649.

Godschalk, D. R. (2004). Land use planning challenges: Coping with conflicts in visions of sustainable development and livable communities. *Journal of the American Planning Association 70*(1), 5–13.

Gordon, P., and Richardson, H. W. (1998). Prove it: The costs and benefits of sprawl. *The Brookings Review 16*(4), 23–25.

Greenberg, M. (1998). Age, perceptions, and neighborhood quality: An empirical test. *Human Ecology Review 5*(2), 10–18.

Greenberg M., Schneider, D., and Choi, D. (1994). Neighborhood quality. *Geographical Review 84*(1), 1–7.

Greenberg, M. R. (1999). Improving neighborhood quality: A hierarchy of needs. *Housing Policy Debate 10*(3), 601–624.

Greenberg, M. R., and Schneider, D. (1996). *Environmentally devastated neighborhoods: Perceptions, policies, and realities.* New Brunswick, NJ: Rutgers University Press.

Heid, J. (2004). *Greenfield development without sprawl: The role of planned communities* (ULI Working Paper on Land Use Policy and Practice).Washington, DC: Urban Land Institute.

Hirschhorn, J. S. (2000). *Growing pains: Quality of life in the new economy.* Washington, DC: National Governors Association.

Hirschhorn, J. S., and Souza, P. (2001). *New community design to the rescue: Fulfilling another American dream.* Washington, DC: National Governors Association.

Kahn, M. E. (2001). City quality of life dynamics: Measuring the costs of growth. *Journal of Real Estate Finance and Economics 22*(2/3), 339–352.

Knaap, G. J., and Nelson, A. C. (1992). *The regulated landscape: Lessons on state land use planning from Oregon.* Cambridge, MA: Lincoln Institute of Land Policy.

Landis, J. D., and Sawicki, D. (1988). A planner's guide to the *Places rated almanac. Journal of the American Planning Association 54*(3), 336–346.

Laris, M. (2001, July 24). Loudoun adopts strict controls on development: Deputies rein in crowd gathered for 7–2 vote. *The Washington Post*, B01.

Laris, M. (2004, April 22). Two minds on growth. *The Washington Post*, LZ01.

Lund, H. (2003). Testing the claims of the new urbanism: Local access, pedestrian travel and neighboring behaviors. *Journal of the American Planning Association 69*(4), 414–429.

Myers, D. (1988). Building knowledge about quality of life for urban planning. *Journal of the American Planning Association 54*(3), 347–358.

Nelson, A. C. (2000). Smart growth or business as usual? Which is better at improving quality of life and central city vitality? In S. M. Wachter, R. L. Penne, and A. C. Nelson (Eds.), *Bridging the divide: Making regions work for everyone* (pp. 83–106). Washington, DC: US Department of Housing and Urban Development.

Nelson, A. C., and Dawkins, C. J. (2004). *Urban containment in the United States: History, models and techniques for regional and metropolitan growth management.* Chicago: American Planning Association.

Nelson, A. C., Dawkins, C. J., and Sanchez, T. W. (2005). The effect of urban containment and mandatory housing elements on racial segregation in US metropolitan areas. *Journal of Urban Affairs 26*(3), 339–350.

Nelson, A. C., Dawkins, C. J., Sanchez, T. W., and Danielsen, K. (2005). *Does urban containment influence neighborhood and housing quality?* Unpublished final report for the Fannie Mae Foundation.

Nelson, A. C., Sanchez, T. W., and Dawkins, C. J. (2004). Urban containment and racial segregation. *Urban Studies 41*(2), 423–440.

O'Toole, R. (2001). The folly of urban containment. *Regulation 24*(3), 20–25.

Porter, D. R. (1999). Rethinking Florida's growth management system. Retrieved March 2, 2006, from National Association of Realtors Web site: http://www.realtor.org/SG3.nsf/Pages/retgrowth?OpenDocument

Riecken, G., and Yavas, U. (2001). Improving quality of life in a region: A survey of area residents and public sector implications. *International Journal of Public Sector Management 14*(6/7), 556–568.

Sakowicz, J. C. (2004). Urban sprawl: Florida's and Maryland's approaches. *Journal of Land Use 19*(2), 377–422.

Sawicki, D. S., and Flynn, P. (1996). Neighborhood indicators: A review of the literature and an assessment of conceptual and methodological issues. *Journal of the American Planning Association 62*(2), 165–183.

Siebert, T. (2004, April 25). Town revisits growth curbs: Lagging tax base forces second look at rules born amid '90s sprawl. *The Denver Post.*

Sirgy, M. J., Rahtz, D. R., Cicic, M., and Underwood, R. (2000). A method for assessing residents' satisfaction with community based services: A quality-of-life perspective. *Social Indicators Research 49*(3), 279–316.

Smutny, G. (1998). Legislative support for growth management in the Rocky Mountains: An exploration of attitudes in Idaho. *Journal of the American Planning Association 64*(3), 311–323.

Song, Y., and Knaap, G. J. (2004). Measuring urban form: Is Portland winning the war on sprawl? *Journal of the American Planning Association 70*(2), 210–225.

Staley, S. R. (1999). *The sprawling of America: In defense of the dynamic city* (Policy Study No. 251). Los Angeles: Reason Public Policy Institute.

Widgery, R. (1995). Neighborhood quality of life: A subjective matter? In M. J. Sirgy and A. C. Samli (Eds.), *New dimensions in marketing/quality-of-life studies* (pp. 267–278). Westport, CT: Quorum Books.

Yu, T., Johnson, V., and Zhang, M. (2004). Urban sprawl: Myth or reality. *Journal of American Academy of Business 4*(1/2), 1–8.

PART III
Innovations and Limitations of the Florida Growth Management Experiment

Chapter 13

Transportation Concurrency: An Idea Before its Time?

Ruth L. Steiner

When concurrency was incorporated into the 1985 Growth Management Act (GMA), it was seen as a critical experiment in planning that would tie capital improvements to land development and planning to implementation. John DeGrove (1989), first secretary of the Florida Department of Community Affairs (DCA), characterized it as "the most powerful policy requirement built into the growth management system" (p. 5). Patricia McKay (1989), the Planning Director of 1000 Friends of Florida since 1988, considered concurrency "the cornerstone of an effective growth management program" (p. 51).

The requirement for concurrency appears to be quite simple and straightforward. Local governments are required to write comprehensive plans that include a capital improvements element with the "estimated public facility cost, including a delineation of when facilities would be needed, the general location of the facilities and projected revenue sources to fund the facilities [FSA §166.3177 (3) (a) 2]." Further, each local government issues land development regulations to ensure that public facilities and services[1] satisfy the comprehensive plan requirement that they be "available when needed for the development orders, and permits are conditioned on the availability of public facilities and services necessary to serve proposed development [FSA §166.202 (2) (g)]." This requirement is called concurrency because in 1986 the statutes were further amended to state: "It is the intent of the Legislature that public facilities and services needed to support development shall be available concurrent with the impact of development" (Powell, 1993, p. 292).

Transportation concurrency has remained controversial since the legislation was passed. Downs (2003), for example, argues that concurrency cannot be successful in Florida because it requires increases in capacity and those increases in capacity will only lead to more traffic congestion. Calthorpe and Fulton (2001) contend that concurrency cannot be successful because it requires regional cooperation, which they say is not likely to occur.

---

1 The term "public facilities" was included in the 1985 legislation. However, the types of facilities intended were not clarified in legislation until the 1993 amendments to the GMA. These facilities include: roads, sanitary sewer, solid waste, storm water, potable water, parks and recreation, and mass transit [FSA § 163.3180 (1) (a)]. This chapter will address only transportation concurrency.

This chapter argues that concurrency is not a complete failure; instead, it may simply be a policy that was legislated before local governments, the DCA, and the Florida Department of Transportation (FDOT) had the institutional capacity to implement it in a context-sensitive manner. Transportation concurrency is still evolving and challenges to its implementation remain. Many of the tools necessary to implement a transportation concurrency management system that addresses the diversity of situations faced by local governments in Florida were simply not available when the legislation passed in 1985. In the last 20 years, transportation planning practice has changed significantly and new tools have been developed to address some problems with the implementation of transportation concurrency.

Florida's Transportation Concurrency Approach as Originally Designed

When Florida placed the concurrency requirement in the GMA in 1985, it began an ambitious experiment in growth management (Pelham, 1992). Concurrency is based on the adequate public facilities ordinances (APFO) that many local governments throughout the country had passed as a part of growth controls (Kelly, 1993; White, 1996). The policy reasons for APFOs and concurrency are similar—they require infrastructure to be in place concurrent with the impact of development. However, no other state required concurrency for all local governments as a part of statewide growth management, and even today only the state of Washington has a similar requirement for local governments participating in its growth management program.

Transportation concurrency is to be implemented by local governments under the guidance and review of two state agencies—the DCA and the FDOT. The local government and each agency have a defined role in the implementation of the concurrency management system within the comprehensive plan and land development regulations.

Each local government develops a comprehensive plan, which includes goals, objectives, and policies for managing growth, and land development regulations that implement the comprehensive plan. Development orders should not be issued unless transportation infrastructure is available at the time of their issuance. The comprehensive plan includes a transportation concurrency management system, which requires consistency among three mandatory elements: the future land use plan element (including the land use map), the transportation/traffic circulation, element,[2] and the capital improvements plan element. Among other functions, the future land use plan identifies areas for future development and redevelopment and, by extension, areas where transportation and other roadway improvements might be needed. The transportation element defines the location of existing and future roadways to support the existing and future land uses and the level of service standards for each major roadway segment. Level of service represents a ratio of the volume of traffic

---

2   The transportation element is required for local governments located in urbanized areas with populations over 50,000 and under the jurisdiction of metropolitan planning organizations (see FSA Section 339.175), while a traffic circulation element is required of other local governments. The transportation element is more comprehensive.

on a roadway compared to the total capacity available on that roadway, as defined in a level of service handbook (Florida Department of Transportation, Systems Planning Office [FDOT], 1992, 1998, 2002). The local government establishes the level of service standards for each of its major roadways on a scale from "A" to "F," with "A" representing a free flow of traffic and "F" representing gridlock, and then monitors these standards on an ongoing basis.

The third mandatory element, the capital improvements plan, includes a list of proposed public facilities, their locations, estimated costs, dates needed, and anticipated revenue sources for, at minimum, a five-year period. The CIE is required to be updated annually. The local government then issues land development regulations to implement the comprehensive plan. Thus, a well-defined transportation concurrency management system involves the coordination of land development with transportation investment and the establishment of a plan for funding the transportation investments through the Capital Improvements Plan.

When a developer proposes a project, a planner reviews it to determine the number of trips generated using formulas in a trip generation handbook (Institute of Traffic Engineers [ITE], 2001a, 2001b). The trip generation rate is based upon studies of the number of trips expected from a specific type of land use and is measured in the number of automobile trips for a specific unit of development. This total is compared to the number of trips available on the adjacent arterials to determine if the capacity exists to accommodate the new development. If adequate capacity exists, the project meets the concurrency requirement and is reviewed against other land development regulations. If the roadway does not have capacity, a strict interpretation of concurrency requires the local government to turn down the development and impose a moratorium on all new development until additional capacity is available.

At the state level, the DCA is responsible for developing the guidelines for the implementation of the GMA and for the review and approval of local comprehensive plans. The FDOT is responsible for developing guidelines for establishing the level of service standards, providing technical assistance to local governments, and reviewing local and regional developments that access the state highway system.

Problems with the State's Approach to Transportation Concurrency

From the very beginning, state and local government officials struggled with the implementation of transportation concurrency. In particular, level of service standards on the state highway system, the standards used for roadway concurrency, the perception that transportation concurrency was causing sprawl, the long lead time for building roads, and the backlog of transportation projects were of concern. In addition, the meaning of the requirement that facilities be "available concurrent with development" complicated concurrency implementation. (See Boggs and Apgar, 1991; Pelham, 1992; Powell, 1993, 1994; Rhodes, 1991.)

A key challenge in the implementation of concurrency has been the lack of funding for a backlog of needed transportation improvements and the ongoing costs of building new facilities. While the failure of the Legislature and the Governor to develop a stable source of funding for local infrastructure is extremely important, this aspect of concurrency is not the focus of this chapter. The funding issue has been addressed elsewhere (for example, DeGrove, 1989; Nicholas and Steiner, 2000; Pelham, 1992; Siemon, 1986, as well as Chapter 5 of this book). Rather, transportation planning practice was initially not adequate to address the diversity of development situations in Florida counties and the tools were inadequate to accommodate the diversity of needs in the transportation system.

The first evidence of problems with transportation concurrency surfaced in 1989 when the first local comprehensive plans were being completed. Pinellas, Dade, Hillsborough, and Broward counties all had major deficiencies in transportation infrastructure that put some development at risk of moratoria (Weaver, Hanna, Smolker, Carraway, and Craine, 1989; Weaver, Hanna, Carraway, Craine, and Smith, 1990). Local governments that were attempting to plan infill, redevelopment, or downtowns struggled to match their local goals with the concurrency system. For example, in response to the desire for continued development, Broward County proposed the use of compact deferral areas and Pinellas County developed a service protection area and deferral areas (Weaver et al., 1989, 1990). These deficiencies existed even though these local governments had prepared or updated their comprehensive plans and land development regulations using accepted transportation planning methods, such as those in the *Highway Capacity Manual* and the ITE *Trip Generation* report (Hall, 1990; Lincks, 1990); yet the goals of their plans were incompatible with the state's transportation concurrency mandate.

In 1991, Governor Lawton Chiles created the third Environmental Land Management Study Committee. In its final report the Committee dedicated an entire chapter to "concurrency and public facilities," and twenty of its 174 recommendations concerned this topic (State of Florida, Environmental Land Management Study Committee, 1992). Concurrently, the Bureau of Economic and Business Research at the University of Florida conducted a survey of local governments to understand the status of the implementation of growth management, generally, and the concurrency management systems, in particular. This study found that local governments took a variety of approaches to implementing concurrency and that many faced deficits in transportation facilities. Among responding local governments, 47 percent indicated a deficit in state roads, 32 percent in county roads, and 14 percent in city roads. About two-thirds of all jurisdictions exempted projects from concurrency review and fewer than half monitored the impact of these projects on available capacity. Approximately one-quarter of jurisdictions did not have adequate staff to manage and administer concurrency and approximately two-thirds relied on intergovernmental agreements to provide required facilities (Audirac, O'Dell, and Shermyen, 1992, pp. 11–13).

In 1998, the Legislature appointed a Transportation and Land Use Study Committee (TLUSC) to evaluate transportation and land use planning coordination in Florida. Several key issues were identified by the Committee, including:

- Florida must have true multi-modal planning and transportation systems
- Regional mobility should not adversely affect community livability
- Transportation is essential to economic vitality
- Better land use planning will lead to better transportation systems

- Reward development in the right place at the right time with the right form
- Focus on performance outcomes, not micro-managing local processes.

Two years later, a Florida Growth Management Study Commission (2001), appointed by the governor, concluded:

> The fundamental flaw of concurrency is the fundamental assumption that population growth can somehow be controlled by the regulation of land development activities on a project-by-project basis. At best, it keeps a problem with overburdened infrastructure from becoming exacerbated in one location while encouraging growth in other areas with excess capacity as measured by adopted level of service indexes.... While traffic concurrency has some appropriate flexibility, the exemptions are cumbersome to implement, thus discouraging development in urbanized areas where communities might realize economies of scale for other infrastructure and other benefits of thriving urban areas (pp. 8–9).

Refinements to the State's Approach to Transportation Concurrency

The transportation concurrency requirements in the GMA have been amended four times to address the problems in its implementation. In 1992, transportation concurrency management areas were created (Powell, 1994). Their purpose is to promote infill development in portions of an urban area that will support more efficient mobility alternatives, including public transit. A transportation concurrency management area may be established in "a compact geographic area with an existing network of roads where multiple, viable alternative travel paths or modes are available for common trips [FSA Sec. 163.3180 (7)]." Furthermore, in these areas, an areawide level of service may be established for facilities with similar functions serving common origins and destinations.

In 1993, the Legislature created several exceptions to the transportation concurrency regulations in order to address concerns about sprawl, and to provide incentives for redevelopment, revitalization, and infill development. These exceptions can be characterized as project-specific or area-specific (Durden, Layman, and Ansbacher, 1996). The new amendments included the following project-specific exceptions: (1) urban redevelopment projects; (2) *de minimis* projects; (3) projects that promote public transportation; (4) part-time projects; and (5) projects for which private contributions are made. Urban redevelopment projects are not subject to the concurrency requirement for up to 110 percent of the roadway impacts generated by prior development. *De minimis* projects are those with impacts that do not significantly degrade the existing level of service or require concurrency review. A local government may exempt projects promoting public transportation, such as office buildings that incorporate transit terminals and fixed rail stations, by setting standards for granting this exception in the local comprehensive plan. Projects such as stadiums, performing arts centers, racetracks, and fairgrounds that are located within urban infill, urban redevelopment, existing urban service areas, or downtown revitalization areas and that pose only special part-time demands on the roads may be exempt from concurrency. A local governments may also allow a developer to proceed with the development of land, notwithstanding failure to meet concurrency, and avoid a claim of a temporary taking, if the developer is willing to pay a "fair share" of the cost of providing the transportation facility necessary to serve the proposed development.

Two area-specific exceptions were added in 1993: long-term transportation concurrency management systems (LTCMS) and transportation concurrency exception areas (TCEA). The purpose of a long-term concurrency management system is to allow a community to gradually address its backlog of projects so that additional development can proceed. A comprehensive plan with a ten to 15 year time horizon is established that: (1) designates specific areas where a deficiency exists, (2) provides a financially feasible means to ensure that these deficiencies will be corrected within the time horizon, and (3) demonstrates how development can be accommodated through investments (including roads and public transit) that mitigate these deficiencies.

The purpose of a transportation concurrency exception area is to reduce the impact of transportation concurrency on urban infill and redevelopment efforts. Exception areas allow for urban infill, urban redevelopment, and downtown revitalization in areas where road facilities are congested and levels of service are low. In areas designed to promote urban infill, no more than ten percent of the land can be developable vacant land. Specific development density and intensity thresholds must also be met.

During the 1999 session of the Legislature, several recommendations of the Transportation and Land Use Study Committee were incorporated in amendments to the Florida Statutes. Of greatest relevance to this study, legislation (1) allowed urban infill and redevelopment areas as justification for transportation concurrency exception areas; (2) authorized the establishment of multimodal transportation districts; and (3) provided that the concurrency requirement does not apply to public transit facilities.

Multimodal transportation districts are another areawide exception, with the goals of facilitating the use of multiple modes of transportation and reducing automobile use and vehicle miles traveled. These districts may be established for : (1) development in existing areas, such as a central core of a municipality, where the focus is on guiding redevelopment and infill opportunities; and (2) proposed development located outside of the traditional municipal area (Guttenplan, Davis, Steiner, and Miller, 2003). A multimodal transportation district should include community design features that provide adequate multimodal mobility and accessibility as well as have a wide variety of land uses, including a solid residential base. In addition to the appropriate scale and mix of land use, the district should promote transit, bicycle and pedestrian travel and good intermodal connection and take the form of an urban center, regional center, or traditional town or village (FDOT, 2003).

In the Growth Management Reform Act of 2005, several more changes were made to the transportation concurrency provisions. Three aspects of the amendments directly impact the transportation concurrency system. First, local governments were now required to incorporate proportionate fair-share funding into their transportation concurrency management systems. This funding method, which allows developers to proceed with their projects after they pay their share of the required infrastructure (even if it is not available), had previously been optional in local concurrency regulations. Second, the requirement that all local governments update their capital improvements element on an annual basis will be more carefully monitored. Local governments that fail to update their capital improvements element will also be prevented from updating their comprehensive plan for any other purposes.

Third, local governments with transportation concurrency exception areas are required to update their exception areas to ensure that they are planning for and implementing transportation strategies to enhance mobility within the designated area. Specifically, local governments are required to establish guidelines within their comprehensive plans that implement strategies promoting the purposes of the exception. The local governments are also required to show that they support and fund mobility, and address the urban design, appropriate land use mixes, and network connectivity needed to promote urban infill, redevelopment, and downtown revitalization. They were also required to coordinate with FDOT to mitigate the impacts of these areas on the Strategic Intermodal System (SIS).

Concurrency within the Broader Context of Transportation Planning

Concurrency implementation is best understood within the broader context of transportation planning as it has evolved over the last twenty years. The late 1980s and early 1990s was a period of major change in transportation planning. With the completion of the interstate highway system, public interest groups lobbied to refocus the federal regulations and transportation practice on livability rather than mobility. The Transportation Land Use Study Committee noted that "mobility refers to the ability to travel between and through communities. Livability is defined by a set of characteristics that make better communities including variety, safety, convenience, commerce, recreation, aesthetics, ... a sense of place, and a sense of community" (TLUSC, 1999, p. 7). Focusing too much on mobility and traffic movement can result in six-lane highways that cut through neighborhoods, reducing accessibility to goods and services that are within walking distance and putting pedestrians and bicyclists at risk.

The change in emphasis began with the passage of the 1991 Intermodal Surface Transportation Efficiency Act (ISTEA) and its successors: the 1998 Transportation Efficiency Act for the Twenty-First Century (TEA-21) and the 2005 Safe, Accountable, Flexible, Efficient Transportation Equity Act—A Legacy to Users (SAFETEA-LU). This shift in emphasis can be seen when one considers how concurrency has been implemented in Florida. As the Transportation Land Use Study Commission concluded, the concurrency system "has been implemented almost exclusively as 'motor vehicle concurrency' because most professionally accepted transportation methodologies and level of service measurements focus on motor vehicle mobility" (TLUSC, 1999, p. 18).

It is helpful to consider how a planner accounts for capacity in a typical concurrency management system that does not use any areawide exceptions. Recall that the level of service for roadways is calculated as a ratio of the volume to capacity, where capacity comprises three components: (1) available capacity, (2) reserved capacity,

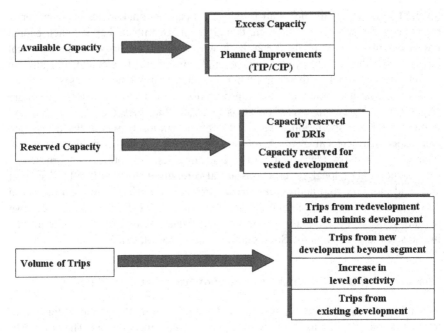

Figure 13.1 Components of roadway volume and capacity
Source: Steiner, Waterman, Randall, Metler, et al., 1999

and (3) the volume of traffic (see Figure 13.1). Under concurrency, planners need to measure both the current and future capacity of a given segment of a roadway. In the case of current level of service, measurement is a relatively simple matter of comparing current trips to the capacity for a given road segment. Measuring future level of service is more complex, however, requiring consideration of new development and changes to roadway segments To estimate the future volume of traffic, a planner must consider the natural growth in the number of trips by current users as well as the growth due to new users who drive through the corridor. A planner must also reserve capacity for development that has been permitted but not yet built, including small scale projects and developments of regional impact that have vested rights to build. Road capacity may also need to account for various planned improvements.

While at first glance the measurement of the level of service appears to be a simple calculation of volume to capacity, in the concurrency management system it is complicated by the need to consider the timing of new development and new roadway capacity. This relationship becomes even more complex as one considers the impact of traffic signals and the multiple users of the transportation system: The system may need to track not only vehicles but the people in those vehicles. The standard practice in the transportation profession at the time the concurrency requirement was passed was to measure roadway capacity using the methodology in the *Highway Capacity Manual* (Transportation Research Board [TRB], 1985) and the *Level of Service Handbook* (FDOT, 1992) and to estimate the number of vehicle trips from new development using the *Trip Generation* report (ITE, 1991). Florida practice differed from national practice in that the FDOT began to develop simplified tables representing "typical" conditions and methods to reflect the impact of signalized intersections on arterials (G. Sokolow, personal communication, December 21, 2004). Thus, roadway capacity in Florida was initially measured without consideration of multiple modes of transportation or differences in demand placed upon major urban arterials by different land use-transportation configurations.

The potential for reduced traffic impact is based upon a developing body of research on the relationships between land use and transportation. Beginning in the late 1980s, Cervero (1989) found that in suburban office parks, a mix of uses was associated with less use of the automobile. In 1992, Downs argued that regions could not build their way out of traffic congestion by simply expanding highway capacity. A wide range of studies have since been conducted that explore the role of density, diversity, design (i.e., connectivity), and regional destination in reducing the number of trips by automobile and providing additional transportation capacity through other modes of transportation. Ewing and Cervero (2001) summarize over 50 of these studies, concluding that reduced numbers of automobile trips and higher levels of walking, bicycling, and transit usage occurred in some contexts. Many of these analyses considered only a small number of neighborhoods in a single city, though, so the reductions are not universally accepted.

By the early 1990s, then, transportation researchers were beginning to understand the impact of alternative land use-transportation configurations on the transportation system. Specifically, they began to understand how different development types can have different transportation impacts. For example, conventional suburban development is characterized by separation of land uses, a less connected system of roadways, low density, and lack of facilities for transit, bicycles, and pedestrians. In contrast, traditional neighborhood development[3] has a mix of land uses, an interconnected street network, high density of development, and pedestrian and transit-oriented features. This design, which is a major component of multimodal planning, has the potential to reduce the impact of development on the transportation system generally, and on the adjacent arterials more particularly, through: (1) reduced automobile trip generation, (2) higher rates of internal capture (i.e., more trips on the local street network and *not* on adjacent arterials), (3) more trips by alternative modes of travel, (4) more trip chaining (an activity pattern that chains a series of trips together), and (5) reduced trip distance.

Reducing the impact on adjacent arterials, many of which are state or federal highways in Florida, is important. These roadways are the workhorses of the transportation system in that they support both regional traffic flow and local accessibility. Developments that reduce their impact on adjacent arterials by providing local accessibility allow these arterial to better meet regional mobility needs.

---

3 Traditional neighborhood development is also called neo-traditional development and New Urbanism. The differences between these three terms are beyond the scope of this chapter. For convenience, traditional neighborhood design will be used to refer to all of these concepts and the associated land use and transportation configuration.

Over the last 15 years, the measurement of highway capacity has shifted in emphasis from freeways, rural and suburban highways, and major urban arterials (TRB, 1994) to a wider range of transportation facilities (e.g., two-lane highways, ramps, transit, bicycle and pedestrian routes) and it now includes more areawide analysis (TRB, 2000). FDOT has taken a leadership role nationally in this transition. The difference in practice is most noticeable in the Florida's LOS Handbook. FDOT has changed the name from the 1998 LOS Handbook (FDOT, 1998) to the 2002 Quality/Level of Service Handbook (QLOS) (FDOT, 2002). The 2002 edition provides "the first successful multimodal approach unifying the nation's leading automobile, bicycle, pedestrian and bus Q/LOS evaluation techniques into common transportation analysis" (FDOT, 2002, p. i). This approach is a recognition that various users value different aspects of travel: While speed, throughput, and travel time may be important to automobile users, pedestrians may prefer a leisurely walk and bicyclists may prefer facilities with fewer trucks and lower speeds. The handbook is the culmination of a series of research projects that developed tools for multimodal analysis and that form the basis for the analysis of multimodal transportation districts (FDOT, 2003; see, e.g., Guttenplan, Landis, Crider, and McLeod, 2001; Landis, Vattikutti, and Brannick, 1997; Landis, Vattikutti, Ottensberg, McLeod, and Guttenplan, 2001; TRB, 1999).

The measurement of trip generation is also changing. For instance, internal capture rates, or the trips that stay within a development, are now calculated for projects with multiple complementary uses. Internal capture rates are used to reduce both the number of trips attributed to new development and the associated development fees. The first discussion of internal capture rates for multiuse projects appeared in the Fifth Edition of the ITE Trip Generation Report (ITE, 1991). By 1997 when the first Trip Generation Handbook was published, internal capture for multi-use developments had been qualified to include only projects that are "planned as a single real estate project, typically between 100,000 and 2 million square feet in size, contain two or more land uses, some trips are between on-site land uses, and trips between land uses do not travel on major street system" (ITE, 1997, p. 1). Multi-use developments that were not to qualify for internal capture rates included central business districts, suburban power centers, shopping centers, and large projects with some on-site mixed use, such as office building with some retail on the lower floors. Although this definition of internal capture appears to apply to a narrow list of projects, in practice, reductions for internal capture are routinely used; and there is concern that "local and state transportation planners lack a comprehensive, credible data set that can be used to confirm or deny the soundness of proposed internal capture estimates (TRB, 2005)." The National Cooperative Highway Research Program is therefore currently funding additional research to clarify the situations under which internal capture rates may be used and to develop guidelines for setting the rates (TRB, 2005).

Transportation models, which are generally used for regional planning and sometimes for large-scale projects, can be important in the coordination of land use and transportation. Early in the 1990s, national studies concluded that most regional transportation models incorporated only automobiles and transit; few regions were modeling bicycle and pedestrian activity (Harvey and Deakin, 1993; Parsons Brinckerhoff Quade and Douglas, Inc., 1993). In 1999, a guidebook documented a variety of techniques to estimate demand, relative demand potential, and supply quality analysis, and proposed supporting tools and techniques for bicycle and pedestrian planning (United States Department of Transportation, Federal Highway Administration, 1999a, 1999b). Yet even the newer basic transportation models are not sensitive to alternative transportation-land use configurations, because the traffic zones aggregate land uses in a large geographic area and the highway network includes only major roadways. To analyze alternative land use configurations and multimodal transportation alternatives would require a more detailed network that includes facilities for bicycles and pedestrians or a conversion to a geographic information system (GIS) framework (Steiner, Bond, Miller, and Shad, 2004).

In summary, transportation planning research suggests that different land use-transportation configurations have different impacts on the transportation network. Transportation practice has slowly adapted its measurements to reflect the shift in emphasis from vehicles and roadway capacity to the movement of people by multiple modes and the associated multimodal capacity, from the simple movement of goods and people to the balancing of the multiple goals of the transportation system. Consistent with this, the basic concurrency system has emphasized roadway capacity with two exceptions: where local governments choose to use the full range of tools provided in the level of service handbooks and in exception areas designed to accommodate the movement of traffic while addressing other community needs.

In Table 13.1, the four areawide exception tools—the long-term concurrency management system (LTCMS), transportation concurrency management areas (TCMA), transportation concurrency exception areas (TCEA), and multimodal transportation districts (MMTD)—are compared. The table provides an assessment of each tool, using the general objectives of transportation concurrency identified by White and Paster (2003) and modified to include other characteristics of coordinated land use-transportation planning.. The first two groups of objectives—controlling the density, rate, timing, amount, location, types of land use and the discouragement of sprawl—are considered pertinent to any transportation concurrency management system. The second two groups of objectives are generally seen as contributing to the multimodal characteristics of an area.

This table suggests that each of the four types of areawide exceptions has a defined niche in concurrency planning, and collectively they address all of the objectives. (Accommodating development at the established level of service standards is a logical omission, because an exception is used only when an area is not meeting its standards.) The long-term concurrency management system most directly addresses the need to coordinate land development by funding additional roadway capacity; the statutes and administrative rules contain no discussion of efficient land use development patterns.

The transportation concurrency management area (TCMA) is the least well defined of all of the areawide exceptions. TCMAs may be established in a compact geographic area with an existing network of roads where multiple, viable alternative travel paths or modes are available for common trips. TCMAs are intended to promote infill development in portions of an urban area that support more efficient mobility alternatives, including public transit.

Table 13.1　Comparison of multimodal characteristics of concurrency areawide tools

Objective or Characteristic	LTCMS	TCMA	TCEA	MMTD
Controls density of development			X	X
Controls rate of development	X			
Controls timing of new development	X			
Controls amount of development	X			(1)
Controls location of development	X			(1)
Controls type of development	X		X(2)	X
Accommodates development at established roadway level of service standards				
Coordinates facility and service capacity	X			
Discourages sprawl development		X		X(3)
Encourages efficient development pattern		X	X(2)	X
Does not cause reduction in automobile LOS	X			
Addresses connectivity		X(4)	X(2)	X
Considers land use mix		(1)	X(2)	X
Promotes redevelopment		X	X	X
Promotes infill development		X	X	(5)
Promotes downtown revitalization			X	
Promotes public transit		X	X	X
Promotes bicycling and walking			X(2)	X
Uses Areawide level of service		X		X
Uses Multimodal level of service				X
Designates minimum size of area		X(4)		(1)

(1) These elements are discussed in the statutes, but no measure is provided.
(2) These elements were added in Growth Management Reform Act of 2005.
(3) These elements are not discussed in statutes, but are in implementation guidelines.
(4) The TCMA may be established in "a compact geographic area with an existing network of roads where multiple, viable alternative travel paths or modes are available for common trips." [163.3180 (7) FSA]
(5) MMTDs can be used in redevelopment and infill areas and for proposed development outside of the traditional municipal area.
Source: Adopted from White and Paster (2003) and Steiner, Li, Shad, and Brown (2003)

　　Transportation concurrency exception areas (TCEAs) were established to promote urban infill and redevelopment in primarily downtown urban cores. The 2005 amendments added requirements that made TCEAs more similar to multimodal transportation districts. The major difference between transportation concurrency exception areas and multimodal transportation districts is that they do not require local governments to track trips and monitor levels of service on local roadways. As such, the multimodal transportation districts can be seen as a second generation of the transportation concurrency management areas and the transportation concurrency managements systems. They can be used in a variety of development contexts from infill and redevelopment to new development outside the municipal boundary and they incorporate areawide and multimodal level of service measures, and a mix of land uses that are likely to lead to internal capture of trips onto an interconnected multimodal network.

The use of areawide exceptions by local governments has been relatively modest, given that all 67 counties and approximately 460 cities are required to have a transportation concurrency management system. TCEAs have been the most widely used of the areawide approaches, with approximately 28 applications statewide. The scale of these exception areas varies from 128,000 acres in Miami-Dade County to 98 acres in Pompano Beach (Florida Department of Community Affairs, 2006). Four of the exception areas (Hallendale, Miami Springs, Ocala, and Pensacola) are of the scale that was initially proposed for multimodal transportation districts—namely, between two and four square miles) incorporate large areas of their respective cities into exception areas. However, many communities that have had transportation concurrency exception areas approved in their comprehensive plans have not taken the steps to implement them (Florida Department of Community Affairs, 2007). In addition, most of the 25 communities in Miami-Dade County that are a part of that exception area do not even mention it in their plans. Only one long-term concurrency management system is currently in use, in Pinellas County. No one has systematically tracked transportation concurrency management areas, but at least five have been identified: in Hialeah, Miami Beach, DeLand, and Miami-Dade and Collier counties. Only one community (Destin) is currently implementing a multimodal transportation district, though others are under review.

Policy Lessons from Florida Experience with Transportation Concurrency

Concurrency is a simple concept: Public facilities and services to support new development should be planned and built concurrent with the impact of the development. The state of Florida has been implementing concurrency for twenty years with mixed success. Most notably, concurrency has been criticized for contributing to sprawl, emphasizing motor vehicle mobility over multimodal accessibility, emphasizing regional mobility at the expense of community livability, taking a one size fits all approach, hindering creativity and micro-managing the local process, and providing disincentives to redevelopment, downtown revitalization, and infill development. Despite these criticisms, the system has changed over time, evolving from a system that emphasizes highway capacity to one that supports multimodal planning in certain contexts. At its core, transportation concurrency attempts to coordinate land use with transportation and provide transportation funding to support land development. Any assessment of transportation concurrency needs to consider the diversity of situations throughout the state, as well as the conflicts between state and local governments over the use and management of the transportation system.

Of Florida's approximately 460 cities, 75 percent are under 10,000 in population and many are also located outside of urban areas (TLUSC, 1999, p. 5). These cities often do not have traffic congestion when measured by the traditional highway capacity standards. Some communities have had slow rates of growth, while others have grown very rapidly. Still others may have a desire to redevelop their Main Streets, even if located on the state highway system. At the other extreme, cities and counties with large populations have high levels of congestion and a desire to redevelop existing urban areas. From the beginning, these communities could not easily implement transportation concurrency because they already exceeded the level of service standards used in traditional transportation planning practice; their situation required the use of transportation concurrency exception areas or transportation concurrency management areas. Some communities would like to continue to develop at a slow rate and have lowered their level of service standards to allow for this. In other places, the concurrency management system could be used to ensure that additional development does not take place at all.

With such diverse situations, finding a single method to implement a concurrency system has been extremely difficult. Over the last 20 years, FDOT has developed an array of tools to address the diversity of situations from the basic transportation concurrency management system to the areawide exceptions. FDOT began by developing simplified methodologies that addressed the "typical" situation throughout Florida, then developed more complex methodologies for local governments with greater institutional capacity. Communities that are interested in redevelopment, infill, and downtown revitalization can use transportation concurrency exception areas. Others with well-developed street networks can adopt transportation concurrency management areas. Those needing time to overcome project backlogs can develop long-term concurrency management systems. Finally, there is the option of using multimodal transportation districts in other locations.

At its core, the evolution of transportation concurrency is about conflicts over how best to meet the needs of people in the community in the comprehensive planning process. The state and local governments often have conflicting goals in the transportation system. The FDOT has attempted to protect the capacity on the state highway system, which includes many of the urban arterials in the state. The stated purpose of the Florida Intrastate Highway System is "to provide efficient, high-speed networks for intercity movement of people and freight" (TLUSC, 1999, p. 24); yet this goal is compromised in urban areas, where these roads serve as "the Main Street" and local governments have zoned the land along these corridors for commercial development. To protect the capacity on major arterials for regional and intercity travel, local governments need to build much better grid street networks for local traffic, better mixes of land uses to allow greater internal capture, and higher density developments to increase the number of trips that can be taken by other modes of travel. Although the multimodal transportation districts encourage—and the transportation concurrency exception areas now require—roadway networks that support local accessibility, the state has generally had a difficult time getting local governments to build them.

Transportation concurrency has proven to be more complex than other forms of concurrency. Like roadways, schools and parks are congestible facilities (Nicholas and Steiner, 2000), but unlike roadways these facilities can provide for the needs of new development through substitutes or reduced levels of service. For example, a resident of a new development can be asked to pay for membership in a recreation club to substitutes for publicly-provided parks and recreational land. No simple substitutes exist for roadway access and capacity. While some might argue that walking, bicycling, and living in a walkable neighborhood can substitute for the use of the automobile, most would concede that only a small percentage of the population does not use the automobile for most trips. Thus, the concurrency management system needs to continue to accommodate the automobile, as well other modes of transportation for people who do not drive.

Support for both regional mobility and local accessibility is reflected in the evolution of the concurrency management system from a relatively simple system that considered only roadway capacity, to a more complex system still focused on roadways but with exceptions to concurrency, to the current even more complex system that measures multimodal capacity and alternative land use-transportation configurations. To address these varied, complex, and sometimes competing concerns, transportation planning (as well as the research supporting it) has become more sophisticated over the years. However, neither the planning tools and strategies, nor the level of funding was adequate to successfully implement concurrency at its inception. Thus, transportation concurrency may indeed have been an idea before its time.

References

Audirac, I., O'Dell, W., and Shermyen, A. (1992). Concurrency management systems in Florida: A catalog and analysis. *Bureau of Economic and Business Research Monographs* (No. 7). Gainesville: University of Florida.

Boggs, H. G., and R. C. Apgar, R. C. (1991). Concurrency and growth management: A lawyer's primer. *Journal of Land Use and Environmental Law*, 7, 1: 1–27.

Calthorpe, P., and Fulton, W. (2001). *The regional city: Planning for the end of sprawl.* Washington, DC: Island Press.

Cervero, R. (1989). *America's suburban centers: The land use-transportation link.* Boston: Unwin Hyman.

DeGrove, J. M. (1989). Florida's greatest challenge: Managing massive growth. In B. Brumback and M. J. Marvin (Eds.), *Implementation of the 1985 Growth Management Act: From planning to land development regulations* (pp. 1–8). Fort Lauderdale: Florida Atlantic University/Florida International University Joint Center for Environmental and Urban Problems.

Downs, A. (1992). *Stuck in traffic: Coping with peak-hour traffic congestion*, Washington, DC: Brookings Institution.

Downs, A. (2003). Why Florida's concurrency principles (for controlling new development by regulating road construction) do not—and cannot—work effectively. *Transportation Quarterly*, *57*(1), 13–18.

Durden, B., Layman, D., and Ansbacher, S. (1996). Waiting for the go: Concurrency, takings and Property Rights Act. *Nova Law Review, 20*, 661–682.

Ewing, R., and Cervero, R. (2001). Travel and the built environment: A synthesis. *Transportation Research Record, 1780*, 87–113.

Florida Department of Community Affairs. (2006). Statewide transportation concurrency exception area list June 9, 2006. Retrieved July 10, 2006, from http://www.dca.state.fl.us/fdcp/dcp/transportation/tcea.pdf.

Florida Department of Transportation, Systems Planning Office. (1992). *1992 level of service handbook*. Tallahassee, FL: Author.

Florida Department of Transportation, Systems Planning Office. (1998). *1998 level of service handbook*. Tallahassee, FL: Florida Department of Transportation Systems Planning Office.

Florida Department of Transportation, Systems Planning Office. (2002). *2002 quality/level of service handbook*. Retrieved December 1, 2006, from http://www. dot.state.fl.us/planning/systems/sm/los/pdfs/QLOS2002.pdf.

Florida Department of Transportation, Systems Planning Office. (2003, November). *Multimodal transportation districts and areawide level of service handbook*. Retrieved December 1, 2006, from http://www.dot.state.fl.us/planning/systems/ sm/los/pdfs/MMTDQOS.pdf.

Florida Growth Management Study Commission. (2001). *A livable Florida for today and tomorrow*. Retrieved December 20, 2004, from www.floridagrowth. org/gmsc.pdf

Guttenplan, M., Davis, B., Steiner, R., and Miller, D. (2003). Planning level areawide multi-modal level of service (LOS) analysis: Performance measures for congestion management. *Transportation Research Record 1858*, 61–68.

Guttenplan, M., Landis, B. W., Crider, L., and McLeod, D. S. (2001). *Multimodal level of service (LOS) analysis at a planning level* (TRB Paper No. 01-3084). Tallahassee, FL: Florida Department of Transportation.

Hall, R. (1990). Everything you wanted to know about transportation level of service. In *Proceedings of the Florida Bar Continuing Education Committee, Environmental and Land Use and Local Government Law Section: Course No. 6806. Land use conflicts: Remedies and enforcement* (pp. 7.1–7.12). Tallahassee, FL: The Florida Bar.

Harvey, G., and Deakin, E. (1993). *A manual of regional transportation modeling practice for air quality analysis* (Version. 1.0). Prepared for the National Association of Regional Councils. Washington, DC: National Association of Regional Councils.

Institute of Transportation Engineers. (1991). *Trip generation: An informational report, 5th Edition*. Washington, DC: Author.

Institute of Transportation Engineers. (1997). *Trip generation handbook: An ITE recommended practice*. Washington, DC: Author.

Institute of Transportation Engineers. (2001a). *Trip generation: An informational report, 7th Edition*. Washington, DC: Author.

Institute of Transportation Engineers. (2001b). *Trip generation handbook: An ITE recommended practice*. Washington, DC: Author.

Kelly, E. D. (1993). *Managing community growth: Policies, techniques, and impacts*. Westport, CT: Praeger.

Landis, B., Vattikutti, V. R., and Brannick, M. (1997). Real-time human perceptions: Towards a bicycle level of service. *Transportation Research Record*, 1578, 119–126.

Landis, B., Vattikutti, V. R., Ottensberg, R., McLeod, D., and Guttenplan, M. (2001). *Modeling the roadside pedestrian level of service*. *Transportation Research Record*, 1773, 82–88.

Lincks, T. F. (1990). Transportation considerations. In *Proceedings of the Florida Bar Continuing Education Committee, Environmental and Land Use and Local Government Law Section: Course No. 6906. Mechanics of zoning and land use law* (pp. 1.1–1.20). Tallahassee, FL: The Florida Bar.

McKay, P. S. (1989). Capital improvements planning in Florida. In B. Brumback and M. J. Marvin (Eds.), *Implementation of the 1985 Growth Management Act: From planning to land development regulations* (pp. 39–53). Fort Lauderdale: Florida Atlantic University/Florida International University Joint Center for Environmental and Urban Problems.

Nicholas, J. C., and Steiner, R. L. (2000). Smart growth and sustainable development in Florida. *Wake Forest Law Review, 35,* 645–670.

Parsons Brinckerhoff Quade Douglas, Inc. (1993). *The pedestrian environment.* Portland, OR: 1000 Friends of Oregon.

Pelham, T. (1992). Adequate public facilities requirements: Reflections on Florida's concurrency system for managing growth. *Florida State University Law Review, 19,* 974–1053.

Powell, D. L. (1993). Managing Florida's growth: The next generation. *Florida State University Law Review, 21,* 223–339.

Powell, D. L. (1994, November). Recent changes in concurrency. *The Florida Bar Journal, 68,* 67–70.

Rhodes, R. M. (1991). Concurrency: Problems, practicalities, and prospects. *Journal of Land Use and Environmental Law, 6,* 241–254.

Siemon, C. L. (1986). What goes around, comes around. In J. M. DeGrove and J. C. Juergensmeyer (Eds.), *Perspectives on Florida's Growth Management Act of 1985.* (pp. 115–131). Cambridge, MA: Lincoln Institute.

State of Florida, Environmental Land Management Study Committee. (1992). *Building successful communities: Final report.* Tallahassee, FL: Author.

Steiner, R., Waterman, J., Randall, K., Mettler, C., Wright, S., and Hatfield, J.(1999). The impact of concurrency management and the Florida Growth Management Act on transportation investments. Tallahassee, FL: Florida Department of Transportation, Office of Policy Planning.

Steiner, R. L., Li, I., Shad, P., and Brown, M. B. (2003). Multimodal trade-off analysis in traffic impact studies. Tallahassee, FL: Florida Department of Transportation, Office of Systems Planning.

Steiner, R., Bond, A., Miller, D., and Shad, P. (2004). Future directions for multimodal areawide level of service handbook research and development. Tallahassee, FL: Florida Department of Transportation, Office of Systems Planning.

Transportation Land Use Study Committee. (1999). *Final report of the Transportation Land Use Study Committee.* Tallahassee, FL: Florida Department of Transportation.

Transportation Research Board. (1985). *Highway capacity manual.* Washington, DC: Author, National Research Council.

Transportation Research Board. (1994). *Highway capacity manual, Special report 209* (3rd ed., rev.). Washington, DC: Author, National Research Council.

Transportation Research Board. (1999). *Transit capacity and quality of service manual.* Washington, DC: Author, National Research Council.

Transportation Research Board. (2000). *Highway capacity manual 2000.* Washington, DC: Author, National Research Council.

Transportation Research Board. (2005). National cooperative highway research program (projects since 1989). Retrieved December 10, 2005, from http://www4. trb.org/trb/crp.nsf/NCHRP+projects.

United States Department of Transportation, Federal Highway Administration. (1999a). *Guidebook on methods to estimate non-motorized travel: Overview of methods.* (Publication No. FHWA-RD-98-165). Retrieved December 1, 2006 , from http://www.fhwa.dot.gov/tfhrc/safety/pubs/vol1/title.htm

United States Department of Transportation, Federal Highway Administration. (1999b). *Guidebook on methods to estimate non-motorized travel: Supporting documentation* (Publication No. FHWA-RD-98-166). Retrieved December 1, 2006, from. http://www.fhwa.dot.gov/tfhrc/safety/pubs/vol2/title.htm

Florida Department of Community Affairs (2007). A guide for the creation and evaluation of transportation concurrency exception areas (Prepared by University of Florida, Department of Urban and Regional Planning under Contract No. 06-DR-73-13-00-22-137). Tallahassee, FL: Florida Department of Community Affairs.

Weaver, R., Hanna, J., Smolker, D., Carraway, C., and Craine, B. 1989. It's a mad, mad, mad, mad, mad world. In *Proceedings of the Florida Bar Continuing Education Committee, Environmental and Land Use and Local Government Law Section: Course No. 6639. Every which way but loose: Concurrency revisited* (pp. 2.1–2.67). Tallahassee, FL: The Florida Bar.

Weaver, R. L., Hanna, J. E., Carraway, C. B., Craine, B., and Smith, D. M. (1990). Everything you wanted to know about transportation level of service. In *Proceedings of the Florida Bar Continuing Education Committee, Environmental and Land Use and Local Government Law Section: Course No. 6806. Land use conflicts: Remedies and enforcement* (pp. 7.1–7.12). Tallahassee, FL: The Florida Bar.

White, M. S. (1996). *Adequate public facilities ordinances and transportation management* (Planning Advisory Services Report No. 465). Chicago: American Planning Association.

White, M. S., and Paster, E. L. (2003). Creating effective land use regulations through concurrency. *Natural Resources Journal, 43,* 753–779.

Chapter 14

Why Do Florida Counties Adopt Urban Growth Boundaries?

Randall G. Holcombe

Florida's 1985 Growth Management Act (GMA) was passed by the state's legislature to deal with problems that many Floridians perceived were the result of the state's rapid population growth. As detailed in Chapters 2 and 3, the GMA requires all local governments to submit local comprehensive plans that must meet the approval of Florida's Department of Community Affairs. The local comprehensive plans must include a future land use map (FLUM) stating allowable uses for all land in the government's jurisdiction; and when the Department of Community Affairs began reviewing plans in the late 1980s, many plans were rejected because they did not adequately control urban sprawl. In effect, as laid out in a local government's comprehensive plan, the FLUM can effectively establish an urban growth boundary for that jurisdiction.

Florida's land use planning process does not require explicitly-drawn urban growth boundaries, and because the land use map already serves this purpose, such boundaries might be viewed as redundant or as unnecessary. Nevertheless, 28 of Florida's 67 counties have explicitly included some type of urban growth boundary in their planning process. This study examines characteristics of those counties that have urban growth boundaries to try to explain factors that prompt counties to adopt them. While the land use planning process can be seen as a method of rationally directing growth and development, it is governed by a political process and is created with public input. The principal goal of this chapter is to gain a better understanding of the political factors that motivate local jurisdictions in Florida to institute urban growth boundaries.

The Economics of Florida's Growth Management Act

Florida had passed growth management legislation in 1975, a decade earlier than the 1985 Act, but the general perception was that it was ineffective at controlling the negative effects of growth. The GMA attempted to address a wide range of growth-related problems, including inadequate infrastructure to support growth, environmental degradation, affordable housing, and decaying downtown areas resulting from people fleeing the inner cities for the suburbs. Despite the wide range of problems blamed on growth, the issues that seemed most important to the typical Florida resident were increasing traffic congestion and dramatic changes in the character of many communities, as small towns grew into larger and more homogeneous cities with strip malls and suburban development. Florida's GMA was intended to mitigate a wide range of growth-related problems, and although there was some opposition, the 1985 Act was largely supported by diverse groups of Floridians ranging from environmentalists to developers (See Chapter 6).[1]

Despite the complexities in the details of Florida's growth management process, the main tool of growth management in Florida is the land use map, and it works by restricting what landowners are able to do with their property. If landowners want to develop their land in a way that is in conflict with the way it is designated on the land use map, the development is not allowed. Florida's growth management policies do not enable anybody to do anything that could not be done without those policies; they only restrict the ability of people to develop and use their property in certain ways.

Because of their design, Florida's growth management policies act as a restriction on the supply of developable land. A restriction in supply raises the price, which creates winners and losers. Higher land prices benefit landowners with developable and developed land, and benefit homeowners, because the market value of their property increases. Losers in the process are renters, people who are moving into an area and want to either rent or buy a residence, and people whose property is outside the developable area according to the land use map. As such, growth management policies can be viewed as creating a transfer of income from poorer Floridians to richer Floridians. Those who own their own homes and who own other real estate tend to be better off financially than those who rent and who do not own real estate.

While a land use map acts as a restriction on the supply of developable land, local comprehensive plans can be amended as often as twice a year, so the growth boundaries created by the land use map are subject to change. To the degree that the amendment process allows landowners the flexibility to develop their land as they choose, the restrictive effects of the land use map are weakened.[2] One way to strengthen those effects is to include an explicit urban growth boundary in the local comprehensive plan. Then not only does the land use map have to be modified, the urban growth boundary must also be enlarged.

Nelson, Pendall, Dawkins, and Knaap (2002) note that the primary factors that determine the price of housing are supply and demand. Factors that restrict supply or increase demand raise prices, while factors that increase supply or reduce demand lower prices. Nelson et al. also note that, broadly considered, growth management policies do not uniformly raise housing prices; and Pendall, Martin, and Fulton (2002,

---

1   Audirac, Shermyen, and Smith (1990) note that policies implied in Florida's 1985 Growth Management Act run counter to people's preferences for low-density lifestyles and find insufficient evidence to argue that the more compact urban form pushed by Florida's growth management policies will provide any economic benefits to Floridians. See Holcombe and Staley (2001, chap. 8) for more background on growth management in Florida.

2   Despite the ability to modify the plans, Nelson (1999) concludes that Florida's growth management policies have been effective at increasing population densities, indicating that growth management does have an effect similar to the establishment of an urban growth boundary.

p. 35) rightly point out that urban containment policies should be viewed as a wider array of tools than just growth boundaries. Policies that increase housing density, for example, can increase supply and lower housing prices. However, as Pendall et al. (p. 30) observe, one of the policy purposes behind urban growth boundaries is to reduce the supply of land and raise its price for the purpose of increasing density; the more expensive the land, the more densely it will be developed. Dawkins and Nelson (2002), however, conclude that urban containment policies do not increase density sufficiently to offset the reduction in the supply of developable land, so urban growth boundaries cause higher housing prices.

Urban growth boundaries analyzed in isolation unambiguously reduce the supply of developable land and so have the effect of raising housing prices. This benefits homeowners and owners of other developed and developable property through an increase in the market value of their property. Logan and Zhou (1989) find that growth controls have only a minimal impact on growth, and then only in "higher-status" areas. Their findings do not apply to Florida, where local comprehensive plans under the 1985 GMA did not come into effect until the 1990s, but the results are suggestive of restrictions on growth being implemented to reduce growth in high income areas.

Politics and Urban Growth Boundaries

In an ideal world run by benevolent government officials, public policies would always be designed to be in the public interest. In the real world, public policies are the product of the political process, and individuals tend to support policies that benefit them—policies that sometimes are not in the public interest. A number of studies, including Donovan and Neiman (1995), Knaap (1987), and Bollens (1990), conclude that self-interest is a determining factor in political support for growth management policies. If this is true, then building upon the previous section, one would expect that higher income areas, which have more people that gain from urban growth boundaries, would be more likely to impose urban growth boundaries. If policies are created to further the economic interests of constituents, then higher-income areas would be more likely to have urban growth boundaries.

Of course, urban growth boundaries might be demanded for other reasons. The 1985 GMA was passed because of the pressures imposed on the state because of growth, so it stands to reason that areas that are growing faster or that already have higher population densities would be more likely to be suffering the effects of growth and therefore would favor the imposition of explicit urban growth boundaries. Baldassare (1990) concludes that rapidly-growing areas in California are more likely to support no-growth policies, for example. Because urban growth boundaries are created through the political process, the political structure of county governments might also have an effect on whether urban growth boundaries are imposed. Clingermayer and Feiock (1990) note that institutional factors, including form of government, can influence the adoption of growth policies.

One can debate what effects might be expected to be produced by urban growth boundaries and whether urban growth boundaries would be desirable. In the real world, urban growth boundaries and other growth management policies are a product of the political process, and whether such policies get sufficient political support to be implemented depends on whether citizens accept and embrace them. Citizens might support policies that impose costs on them to further some greater public good; but as Downs (1957) argued, most citizens will be rationally ignorant of much of government policy, and special interests tend to drive the process based on what benefits them. The empirical work that follows suggests that private interests dominate public interests in determining where urban growth boundaries are established.

The literature indicates three broad hypotheses as to why urban growth boundaries might be supported and implemented by local governments in Florida. First, people might support urban growth boundaries because they receive private benefits from them. The literature suggests that these private benefits tend to be correlated with income, and that higher-income individuals benefit at the expense of lower-income individuals in the form of higher property values, the exclusionary properties of growth boundaries, and the creation of more homogeneous communities that enhance their lifestyles. This self-interest hypothesis suggests that higher-income jurisdictions would be more likely to implement urban growth boundaries. Second, people might support urban growth boundaries because of the negative effects of growth, which suggests that places with greater population growth and higher population densities would tend to support these boundaries. Third, political and institutional factors may play a role in whether urban growth boundaries are implemented. While one can speculate on why urban growth boundaries might be imposed, this chapter examines these ideas empirically, after first discussing the nature of these boundaries.

Urban Growth Boundaries in Florida's Counties

Because urban growth boundaries are not required by Florida's growth management laws, individual counties can implement them if they want and in a manner of their choosing. As a result, there is some variety in the types of urban growth boundaries imposed in Florida, and even some ambiguity as to what constitutes an urban growth boundary. Nelson and Dawkins (2003, p. 13) define an urban growth boundary as a line on a map based on an explicit policy to prevent the extension of key public facilities, especially water and sewer line and urban development, without plan amendments. Urban growth boundaries in Florida predate Florida's 1985 Growth Management Act. In 1975 both Dade and Sarasota counties created urban growth boundaries: Dade enacted a growth boundary and Sarasota created boundaries with three levels for development; urban, semi-rural, and rural (Nelson and Dawkins, 2003, p. 9). Florida counties can also effectively create urban growth boundaries by establishing urban service areas beyond which water, sewer, and other municipal services will not be extended, as in Leon County, or by setting aside land in rural preservation areas, as in Palm Beach County. Because each county draws up its own local comprehensive plan, no two urban growth boundary policies will be the same.

For the purpose of this study, Florida counties with urban growth boundaries were identified from three sources. Nelson and Dawkins (2003) list Florida counties with urban growth boundaries, which provides one source. Another list was provided

Table 14.1   Florida counties by urban growth boundary status

County	Boundary	County	Boundary
Alachua	Yes	Lee	No
Baker	Yes	Leon	Yes
Bay	Yes	Levy	No
Brevard	Yes	Liberty	No
Broward	Yes	Madison	Yes
Calhoun	No	Manatee	Yes
Charlotte	Yes	Marion	Yes
Citrus	No	Martin	Yes
Clay	Yes	Miami-Dade	Yes
Collier	Yes	Monroe	No
Columbia	Yes	Nassau	Yes
DeSoto	No	Okaloosa	Yes
Dixie	No	Okeechobee	No
Duval	Yes	Orange	Yes
Escambia	No	Osceola	Yes
Flagler	Yes	Palm Beach	Yes
Gadsden	No	Pasco	No
Gilchrist	No	Pinellas	No
Glades	No	Polk	Yes
Gulf	No	Putnam	No
Hamilton	No	Santa Rosa	No
Hardee	No	Sarasota	Yes
Hendry	No	Seminole	Yes
Hernando	No	St. Johns	No
Highlands	No	St. Lucie	Yes
Hillsborough	Yes	Sumter	No
Holmes	No	Suwannee	No
Indian River	Yes	Taylor	No
Jackson	Yes	Union	No
Jefferson	No	Volusia	No
Lafayette	No	Wakulla	Yes
Lake	No	Walton	No
		Washington	No

by Florida's Department of Community Affairs, and another was created by a survey done by the DeVoe Moore Center at Florida State University. The lists were similar, but not identical, perhaps because of definitional differences. In cases where two or all three of the sources identified a county as having an urban growth boundary, that county was counted as having one. In the eight cases where only one source identified an urban growth boundary for the county, a phone call was made to the county's growth management department to ask directly. Five of the counties confirmed they did have an urban growth boundary and three said they did not. Table 14.1 identifies those counties in Florida that have an urban growth boundary as of 2006.

Testing Hypothesis #1: Income and Urban Growth Boundaries

Several OLS and probit regressions were run to see if income is related to the adoption of urban growth boundaries by Florida counties. The existence of an urban growth boundary, treated as a binary dependent variable, was regressed separately on per capita income in 1990, on per capita income growth 1990–2000, and on both independent variables together. Per capita income is positive and statistically significant in all cases, while per capita income growth is found to be statistically insignificant. Considered in isolation, the strong positive relationship between per capita income and the existence of an urban growth boundary suggests that higher-income counties tend to adopt urban growth boundaries. This lends support to the self-interest hypothesis.

Testing Hypothesis #2: Growth Pressure and Urban Growth Boundaries

Florida's growth management measures were passed in response to the pressures of population growth, so one would expect that counties with more population growth or more population density would be more likely to use a strong containment policy like an urban growth boundary. Table 14.2 shows the association between various measures of population and the existence of urban growth boundaries. Regression results in this and the following tables are from probit regressions, but all regressions were also run using OLS, with essentially identical results.[3] One might expect more populous and faster-growing counties to be more likely to adopt urban growth boundaries, and Table 14.2 shows that to be the case for some measures of population.

The first regression in Table 14.2 simply shows the relationship between urban growth boundaries and a county's total population in 1990. Run by itself, there is a strong positive relationship showing that counties with larger populations are more likely to have urban growth boundaries. There is, however, a tenuous theoretical relationship between total population and growth management, as total population might be a proxy for political variable. Within a county, a larger population would mean that each individual would have a smaller influence on that county's policies. It is possible that citizens in more populous counties might favor urban growth boundaries as a method of placing an additional check on their elected officials.

The next two regressions in Table 14.2 are more relevant to growth management. The second regression details the relationship between population growth from 1990 to 2000 and the existence of an urban growth boundary. This relationship could not be weaker, with a z-statistic of zero and Pseudo R^2 of zero. While one might think that counties with greater rates of population growth would be more likely to implement urban growth boundaries, that turns out not to be the case. However, the third regression used population growth from 1980 to 1990 as an independent variable

---

3   The z-statistics in the probits were always close to the t-statistics in the OLS regressions, the signs were always the same on significant coefficients, every coefficient that was statistically significant at the .05 level in the probits was also significant in the OLS regressions, and no coefficients were significant in one type of regression but not in the other.

Table 14.2   **Population and urban growth boundaries**

Independent Variables	(1)	(2)	(3)	(4)
Constant	-.6271	-.1697	-.6699	-.4811
	(-3.03)	(-.51)	(-2.46)	(-2.50)
Population, 1990	2.78e-06			
	(3.10)			
Pop. Growth, 1990–2000		.000		
		(0.00)		
Pop. Growth, 1980–1990			.0132	
			(2.23)	
Pop. Density, 1990				.0017
				(2.81)
Pseudo R^2	.161	.000	.060	.090

Dependent Variable: Urban Growth Boundary.
(z-statistics in parentheses.)

and found a strong positive association. This suggests that counties with higher population growth rates prior to the introduction of growth management were more inclined to implement urban growth boundaries. The combined results of regressions 2 and 3 suggests that counties with urban growth boundaries had higher population growth in the decade prior to the implementation of Florida's GMA but not in the decade after the Act took effect. This suggests that urban growth boundaries may have been effective at slowing growth in the fastest-growing counties. An additional regression (not shown here), which included population growth during both 1980–1990 and 1990–2000, found that growth in the earlier period remains statistically significant while growth in the latter period remains insignificant, reinforcing the results shown in regressions (2) and (3).

The fourth regression investigates the relationship between urban growth boundaries and population density in 1990. This relationship is strongly positive, as might be expected. Counties where the population density is higher are more likely to implement urban growth boundaries as a restriction on further growth. In another regression not shown in the table, all four population variables were included, and only population and population growth from 1980 to 1990 are statistically significant at the .05 level. Because there is a greater variation in county population than county size, the counties with the largest populations also have the highest population densities, hence the population variable minimizes the impact of the population density variable. The hypothesis that urban growth boundaries are implemented in response to high population density is questionable on statistical grounds, and it appears that total population and population growth are the primary determining population factors.

Demographics and Urban Growth Boundaries

It is common to account for demographic differences when looking at determinants of public policies, and regressions were run using the median age, the percentage of the population aged 65 and older, and the percentage of the population enrolled in public schools as independent variables to try to explain the existence of urban growth boundaries. None of these variables were statistically significant, run by themselves or with other variables. There is no evidence that these demographic variables are correlated with the existence of an urban growth boundary.

Two other variables were examined to see if they had any relationship with urban growth boundaries: whether a county was on the coast, and the percentage of the county's presidential votes going to Gore in the 2000 presidential election. The hypothesis on coastal counties was that they tend to have higher population densities and also have a natural boundary adjacent to the (typically) most desirable locations in the county, which could give rise to a demand for urban growth boundaries. The percentage of votes to Gore was used as a proxy for political orientation, with the thought that more liberal constituents might favor more government activism and therefore urban growth boundaries. Neither variable was significant and the results are not reported for that reason.

Characteristics of Government and Urban Growth Boundaries

Two characteristics of county government were examined as possible determinants of the existence of urban growth boundaries. The first characteristic is form of government, which specifies whether the county is run by an administrator appointed by the county commission or whether it has an elected chief executive. The form of government is identified in Jeong (2004, p. 52). Forty-four of Florida's 67 counties have a commission-chief administrator form of government, with the remainder having an elected chief executive. The second government variable is home rule. Under Florida law, counties are limited in the services they can provide to citizens, but counties with home rule authority can expand their scope of services beyond those listed in the Florida Statutes. Such services might otherwise be provided by cities, but home rule authority recognizes that county-provided services might be reasonable in increasingly-urbanized unincorporated areas. Information about the existence of home rule authority comes from Jeong (2004, p. 53) and reveals that 18 counties had home rule authority as of 2004. However, four counties that obtained home rule authority in 1996, 1998, and 2002 (for two counties) were not counted as home rule counties.

In univariate regressions (not shown here), both the form of county government and home rule were found to be associated with the existence of urban growth boundaries. When run together, the home rule variable retained its statistical significance but form of government did not, suggesting that home rule is more closely correlated with the existence of urban growth boundaries than whether the county is run by an appointed administrator or elected chief executive. Political institutions do appear to make a difference as to whether urban growth boundaries are adopted, as suggested by Clingermayer and Feiock (1990).

The Determinants of Urban Growth Boundaries

Having looked at a number of variables that might influence the existence of an urban growth boundary, the single variable that is most closely correlated with the

Table 14.3   Determinants of urban growth boundaries

Independent Variables	(1)	(2)	(3)	(4)
Constant	-2.970	-1.915	-3.064	-2.016
	(-3.53)	(-1.69)	(-3.54)	(-1.73)
Per Capita Income, 1990	.0002	-.0002	.0002	.0002
	(2.89)	(2.90)	(2.87)	(2.90)
PCI Growth, 1990–2000		-.0234		-.0231
		(-1.35)		(-1.30)
Population, 1990	9.01e-07	9.16e-07	3.97e-06	.4.03e-06
	(.93)	(.88)	(1.74)	(1.74)
Pop. Density, 1990			-.0028	-.0028
			(-1.70)	(-1.67)
Home Rule	.5539	.4824	.8098	.7253
	(.97)	(.84)	(1.33)	(1.18)
Pseudo R²	.283	.304	.316	.335

Dependent Variable: Urban Growth Boundary.
(z-statistics in parentheses.)

existence of urban growth boundaries is per capita income. This relationship is examined further in the regressions in Table 14.3, which combined the other variables that may impact urban growth boundaries with per capita income to see which retain their explanatory power. Regression (1) used per capita income, population, and home rule as independent variables. Per capita income remains highly significant, but population and home rule are no longer statistically significant. The Pseudo R^2 of .283 is only slightly higher in this regression than the Pseudo R^2 of .237 when population is used as the only independent variable. The low z-statistics on population and home rule indicate that those variables are correlated with per capita income, and that their significance is eliminated when per capita income is taken into account.

The next regression added in per capita income growth to the other independent variables to see whether its inclusion removes the statistical significance of per capita income. Per capita income growth is not statistically significant, and the z-statistic on per capita income is about the same with the addition of this variable. Regression (3) removed the income growth variable and added population density, which was also significant by itself. Holding these other factors constant, population density has a negative coefficient (although not significant at the .05 level), suggesting that if anything, urban growth boundaries are associated with lower population densities once other factors are controlled for. Per capita income remains statistically significant. The fourth regression included all of the independent variables in the table. The z-statistic on per capita income is the highest in this specification, while none of the other independent variables are statistically significant at the .05 level.

Table 14.4 reports estimates of the impact of population growth on the existence of urban growth boundaries when per capita income is taken into account. The results in Table 14.2 showed that greater population growth in the decade preceding the implementation of the 1985 GMA was positively correlated with the existence

Table 14.4   Population growth and per capita income as determinants of urban growth boundaries

Independent Variables	(1)	(2)	(3)
Constant	-3.486	-2.869	-2.967
	(-4.26)	(-3.38)	(-3.41)
Per Capita Income, 1990	.0003	.0002	.0002
	(3.73)	(2.23)	(2.28)
Pop. Growth, 1980–1990	.0041	.0066	.0058
	(.65)	(.99)	(.85)
Population, 1990		1.20e-06	4.14e-06
		(1.13)	(1.80)
Pop. Density, 1990			-.0027
			(-1.62)
Home Rule		.4497	.7113
		(.77)	(1.14)
Pseudo R²	.242	.294	.324

Dependent Variable: Urban Growth Boundary.
(z-statistics in parentheses.)

of an urban growth boundary. Regression (1) in Table 14.4 shows that when both per capita income and population growth from 1980 to 1990 are used together as independent variables, the statistical significance of population growth disappears and only per capita income remains significant. Higher-growth counties tend to be those with higher per capita incomes, and it appears that the growth variable was acting as a proxy for per capita income.

When looked at individually, several independent variables appear to be correlated with the existence of an urban growth boundary. However, regression equations (2) and (3) reveal that whenever per capita income was included as an explanatory variable, it became the only variable to retain its significance in the regression. In regressions not shown in the tables, various combinations of independent variables were run, and in every case per capita income was found to be statistically significant at the .05 level. The regressions in Tables 14.3 and 14.4 are illustrative of this general result. While per capita income is correlated with the other independent variables that are statistically significant when run by themselves, including per capita income in the regression pushes their statistical significance below the .05 level. None of these other variables is statistically significant at the .05 level when per capita income is included.

These regressions were intended to examine three hypotheses regarding the existence of urban growth boundaries. The per capita income variable was a measure of the degree to which the self-interest of higher-income individuals was responsible for the existence of urban growth boundaries, and it is the only variable that is significant when run in combination with all others. Urban growth boundaries are a product of political decision-making, and those jurisdictions that implement them are the ones with the highest per capita incomes, suggesting that urban growth boundaries are established to further the interests of upper-income Floridians.

Because Florida's urban growth policies were driven by the perceived negative effects of population growth, one might have expected that urban growth boundaries would be associated with greater population growth and/or higher population densities, but this turns out not to be the case. While population growth in the decade prior to the implementation of growth management is correlated with the existence of urban growth boundaries, as is population density, when run in a regression with per capita income, both are statistically insignificant. Counties with higher growth rates and population densities tend to have higher per capita incomes; and when per capita income is controlled for, population density has no effect on whether a county adopts an urban growth boundary.

The same is true of institutional and political factors. While some of these variables appear to have an impact when examined by themselves, when combined in a regression with per capita income, only per capita income is statistically significant. Demographic characteristics also are not statistically significant. The empirical evidence shows that when all factors are taken into account, per capita income is the only robust factor in the determination of adoption of urban growth boundaries by Florida counties.

Conclusion

While many factors are related to the existence of urban growth boundaries in Florida counties, when all variables are included, the only factor that appears to have any significant robust impact is the county's level of income. The empirical results in this chapter suggest that urban growth boundaries are created through the political support of higher-income groups for their benefit, and not in response to pressures associated with higher growth rates. If urban growth boundaries were imposed in response to the pressures of growth, then the growth-related variables would have remained statistically significant even after taking into account income differences.

This conclusion is consistent with much of the literature on the adoption of growth controls. Donovan and Neiman (1995) found that growth controls in California had a minimal impact on population growth, but did affect community characteristics and make communities more racially homogeneous. Baldassare and Protash (1982) found that in California growth controls are more likely to be passed in jurisdictions with a greater percentage of white collar population. Knaap (1987) found that Oregon voters tended to support growth controls based on their economic interests. The literature is not unanimous on this point. For example, Gottdiener and Neiman (1981) found broad support for growth controls in Riverside, California, and directly called into question the thesis that they are passed by the affluent who want to limit access to their community. However, as Feiock (1994) notes, growth controls tend to have a negative impact on economic development, and Bollens (1990) argues that people with a more direct connection to the local economy tend to show the least support for growth controls. The findings in this chapter buttress a conclusion that has been supported in the previous literature: People support growth controls based on their economic interests.

When these conclusions are applied directly to Florida, urban growth boundaries appear to be created in higher-income counties to further the interests of higher-income Floridians. These urban growth boundaries may have minimal effects because the land use map that is a required component of each county's local comprehensive plan already delineates how land can be used within the jurisdiction. However, as Nelson and Moore (1996, p. 242) note, the actual effects of growth management policies are more closely related to local political demands and how those plans are implemented than to the actual policies or plans themselves. If so, the existence of an explicit urban growth boundary to supplement the county's land use map may indicate the demand for more limiting growth policies in those higher-income jurisdictions. However, the data show that Florida's urban growth boundaries are not created in response to growth, but rather are created in response to the demands of higher-income Floridians.

The relationship between income and urban growth boundaries surely is not just a case of high-income people using the political process for their financial gain. As income increases, so does support for environmental protection and other goals of growth management. While selfish motives may be at play, altruistic motives may also play an important role. But while upper-income people may have altruistic motives for favoring more restrictive growth policies, it is easier to favor them when they benefit financially and when the costs of those policies are pushed onto others. It is telling that, when all factors are accounted for, only income remains a statistically significant predictor of the existence of urban growth boundaries, and that measures related to growth (such as population density or population growth) do not explain the adoption of boundaries by counties in the state.

With regard to the growth management process more generally, these results suggest that growth management policies may be the result of a political process driven by the interests of upper-income individuals. Even if their interests are partly based on altruism—as upper-income people are more likely to favor stronger environmental protection—their interests still are at odds with those of lower-income individuals who weigh environmental protection less heavily and weigh affordable housing more heavily. Equally noteworthy, from the perspective of narrow self-interest, more restrictive growth policies tend to produce private benefits for upper-income people and tend to impose private costs on lower-income people.

In political decision-making, people tend to weigh their own interests heavily; and it appears that in the case of urban growth boundaries in Florida, economic self-interest trumps concerns about growth.[4] When self-interest and the general public interest coincide, as will often be the case, the results can be desirable, but there will be a minimal role for government policy because individuals pursuing their own self-interests will be led by an invisible hand to do what is best for everyone. When self-interest and the general public interest do not coincide, the results of this study complement findings from other growth management research that suggests that policies are more likely to be driven by the self-interest of politically powerful constituents than to reflect the general public interest.

---

4 Of course, growth management is not unique in this regard. There is a substantial literature showing that special interests have an undue influence in all areas of public policy.

Acknowledgements

The author gratefully acknowledges the research assistance of DeEdgra Williams and helpful comments from Tim Chapin and Harrison Higgins. Financial support for this research was provided by the Federal Home Loan Bank of Atlanta.

References

Audirac, I., Shermyen, A. H., and Smith, M. T. (1990). Ideal urban form and visions of the good life: Florida's growth management dilemma. *Journal of the American Planning Association 56*(4), 470–482.

Baldassare, M. (1990). Suburban support for no-growth policies: Implications for the growth revolt. *Journal of Urban Affairs 12*(2), 197–206.

Baldassare, M., and Protash, W. (1982). Growth controls, population growth, and community satisfaction. *American Sociological Review 47*(3), 339–346.

Bollens, S. A. (1990). Constituencies for limitation and regionalism: Approaches to growth management. *Urban Affairs Quarterly 26*(1), 46–67.

Clingermayer, J. C., and Feiock, R. C. (1990). The adoption of economic development policies by large cities: A test of economic, interest group, and institutional explanations. *Policy Studies Journal 18*(3), 539–552.

Dawkins, C. J., and Nelson, A. C. (2002). Urban containment policies and housing prices: An international comparison with implications for future research. *Land Use Policy 19*(1), 1–12.

Donovan, T., and Neiman, M. (1995). "Local growth control policy and changes in community characteristics. *Social Science Quarterly 76*(4), 780–793.

Downs, A. (1957). *An Economic Theory of Democracy.* New York: Harper and Row.

Feiock, R. C. (1994). The political economy of growth management. *American Politics Quarterly 22*(2), 208–220.

Gottdiener, M., and Neiman, M. (1981). Characteristics of support for local growth control. *Urban Affairs Quarterly 17*(1), 55–73.

Holcombe, R. G., and Staley, S. R. (Eds.). (2001). *Smarter growth: Market-based strategies for land-use planning in the 21st century.* Westport, CT: Greenwood Press.

Jeong, M.-G. (2004). *Local land use choices: An empirical investigation of development impact fees in Florida.* Unpublished doctoral dissertation, Florida State University, Tallahassee.

Knaap, G. J. (1987). Self-interest and voter support for Oregon's land use controls. *Journal of the American Planning Association 53*(1), 92–97.

Logan, J. R., and Zhou, M. (1989). Do suburban growth controls control growth? *American Sociological Review 54*(3), 461–471.

Nelson, A. C. (1999). Comparing states with and without growth management: Analysis based on indicators with policy implications. *Land Use Policy 16*(2), 121–127.

Nelson, A. C., and Dawkins, C. J. (2003). Urban containment—American style(s). Unpublished manuscript, Virginia Polytechnic Institute and State University, Alexandria, VA..

Nelson, A. C., and Moore, T. (1996). Assessing growth management implementation. *Land Use Policy 13*(4), 241–259.

Nelson, A. C., Pendall, R., Dawkins, C. J., and Knaap, G. J. (2002). *The link between growth management and housing affordability: The academic evidence.* Washington, DC: Brookings Institution Center on Urban and Metropolitan Policy.

Pendall, R., Martin, J., and Fulton, W. (2002). *Holding the line: Urban containment in the United States.* Washington, DC: Brookings Institution Center on Urban and Metropolitan Policy.

Chapter 15

Paying for the "Priceless": Florida Forever, Managing Growth, and Public Land Acquisition

Harrison T. Higgins and Neil B. Paradise

Since 1950, Florida has experienced a population increase averaging four percent annually. In that period, eight million acres of forest and wetlands, just less than a 25 percent of the state's entire land mass, have been urbanized to accommodate these population increases. Resulting from this growth, researchers have documented disturbances to natural systems, ranging from alterations of the flow regimes of the state's rivers, to the frequency of fires in its woodlands, to changed drainage patterns, disturbed dune systems, and the introduction of exotic species.

In response, Florida has pursued two different sets of policies to stem the tide of natural resource degradation and depletion: 1) growth management and 2) various public lands acquisition programs, the latest incarnation dubbed Florida Forever. Unlike those of other growth management states (notably Maryland's) Florida's land acquisition program, while the best funded, has minimal connections to state and local land use planning goals and planning processes. This chapter argues that, despite extensive and consistent state funding for Florida Forever and a growth management approach committed to conservation, the effects of both the acquisition program and the growth management regime on resource protection have been limited by the failure to establish clear linkages between them. Indeed, there is some evidence to suggest that they may be working at cross purposes despite a shared policy legacy.

From its inception Florida's 1985 Growth Management Act (GMA) has required local governments to include in their comprehensive plans both a conservation element, to set policies for the preservation of natural resources, and a recreation and open space element for the planning of recreational facilities and natural reservations. The same legislation requires local governments to assess their current and projected water needs in the context of regional water supply, thereby encouraging the preservation of water recharge areas that sustain that supply. The accompanying administrative rules that execute the GMA, Chapter 9J-5, also promulgate a series of urban form policies that seek to contain sprawling, land-consumptive, low density development patterns and to concentrate development in existing urbanized and newly urbanizing areas.

While modern strategies for mitigating the effects of human habitation on natural resources include everything from regulating building location and the material of surface parking lots to altering development density and intensity standards, inhibitions against advancing resource protection, especially species protection, through land use planning and regulation on private property are myriad (see, e.g., Dwyer, Murphy, and Ehrlich ,1995 and Press, Doak, and Steinberg, 1996). Perhaps in recognition of these constraints, Florida has not relied solely (or even largely) on growth management policy for resource protection. Instead it has developed several robust state-based public lands acquisition programs. Florida Forever, the latest program in 40 years of increasingly ambitious land purchasing efforts by the state, is now the largest of any such program at the state level and currently outpaces the land acquisition efforts of the federal government. As a result of Florida Forever and its predecessor programs, the state has conserved over 2.3 million acres acquired through fee simple acquisition and almost an additional 400,000 acres through the purchase of less-than-fee simple conservation easements (Florida Department of Environmental Protection [DEP], n.d.).

While the state's commitment to fund Florida Forever is laudable and the acreage purchased by the state is impressive, the linkage between this program and the state's growth management system has been limited. For example, the state's land planning agency, the Department of Community Affairs (DCA), receives only 22 percent of all Florida Forever funds (see Table 15.1). While DCA has used those moneys to fund the Florida Communities Trust grant program that provides funds for the implementation of a few local government comprehensive plan conservation and recreation elements, many of those funds have gone to purchase of lands for active recreation, typically in already urbanized areas where resource protection is of minimal effect.

While local governments are beginning to see opportunities to link growth management and land preservation goals through Florida Forever, and while many rely on the state land acquisition program to promote urban containment, there remain few legal or policy mechanisms for directing development to suitable places while simultaneously providing funding to conserve places where important natural resources remain intact. In reality, when connections are made between the state's growth management regime and its preservation programs, they are often transactional, trading conservation on portions of the land for increased intensities of development on other portions.

Disconnections between these two efforts have thus led to policies with possibly contradictory effects. During its 2006 session, for instance, the Florida legislature agreed to execute the largest environmental land purchase in the history of the state, the $351 million acquisition of 74,000 acres of the Babcock Ranch property in south Florida. Stretching from the shores of Lake Okeechobee west to Gulf of Mexico, the purchase was hailed by the state as another "crown jewel" acquisition in their conservation portfolio. However, this purchase was balanced by a 17,000 acre development proposal seeking to convert the remaining ranch property to 19,000 homes and retail and commercial uses in an area of Lee and Charlotte Counties that had not previously experienced urbanization.

In this chapter we expand on this argument and delve more deeply into the Babcock Ranch case. We begin by tracing the path Florida has taken in becoming a national leader in land preservation. We then provide an overview of Florida's public

Table 15.1   Distribution of Florida Forever funds, 2006

Agency	Pct of Funding	Purposes of Funds
Dept of Environmental Protection— State Lands	35.0%	To acquire and dispose of lands as directed by the Board of Trustees of the Internal Improvement Trust Fund
Dept of Environmental Protection— Recreation and Parks	1.5%	To provide resource-based recreation while preserving and restoring natural and cultural resources
Dept of Environmental Protection— Rails to Trails	1.5%	To establish a statewide system of greenways and trails for recreation, conservation, and transportation
State Water Management Districts	35.0%	To acquire and manage lands for water management purposes under the Save Our Rivers program
Dept of Community Affairs	22.0%	To assist Florida communities in meeting the challenges of growth
Florida Recreation Development Assistance Program	2.0%	To fund the acquisition or development of land and trails for public outdoor recreation purposes
FL Fish and Wildlife Conservation Commission	1.5%	To manage fish and wildlife resources for their long-term well-being and the benefit of people
Dept of Agriculture and Consumer Services	1.5%	To protect and manage the forest resources of Florida, ensuring they are available for future generations

Source: Florida Forever Status Report 2006

lands acquisition program in 2006 and describe the processes through which Florida Forever, now a part of the state's constitutional framework, operates to preserve land and resources. Following this, we detail the Babcock Ranch case, an admittedly extraordinary recent acquisition, in order to consider the possible implications of Florida Forever's implementation. In a concluding section we describe ways the state might establish stronger linkages between these land acquisition efforts and the state's growth management approach for the purposes of strengthening the effect of the state's conservation tactics and extending the purchasing power of the state's resource acquisition dollars.

Florida's Evolving Environmental Ethic

Prior to adopting its growth management legislation in 1985, Florida's development policy focused largely on making seemingly unsuitable land available for urbanization and agricultural production. This attitude was symbolized in the behavior of the state's Trustees of the Internal Improvement Fund (TIIF), which controlled the distribution of state lands. The TIIF went as far as selling submerged bay lands to developers, resulting in projects such as the dredging of much of St. Petersburg's

Boca Ciega Bay to provide fill for waterfront developments on a formerly prolific marine system (Stephenson, 1997). By 1964, the bay, formerly a pristine water body with vast meadows of turtle grass, had been transformed from a shallow coastal lagoon into a "channelized cesspool" with landfill occupying 12.5 percent of its acreage (Stephenson, 1997, 139).

Since 1970, however, the state's growth management and public land acquisition policies have attempted to address the interaction of environment and urbanization rather than focus upon overcoming environmental obstacles to development. These different efforts share roots in the political discussions of the late 1960s and early 1970s. The nature of Florida governance was radically altered by federally mandated reapportionment in 1967, shifting power from traditional rural centers in the north of Florida to the fast growing urban areas in the state's south. At the same time, Florida environmental groups, reflecting the growing power of the movement nationally, had logged successes in blocking the development of the Everglades Jetport in 1969 (Gilmour and McCauley, 1976) and in halting further development of the Cross Florida Barge Canal in 1971 (DeGrove, 2000). Additionally, in the winter of 1970–1971, Florida suffered its worst drought in 40 years. Given the lack of rainfall to recharge the Biscayne aquifer, Miami was forced to draw water supplies directly from an already diminished Lake Okeechobee, where water levels were reported to be at half of their historic levels.

In response to that crisis, Governor Reubin Askew convened a Conference on Water Management in South Florida at the Doral Hotel in Miami (DeGrove, 2000). One-hundred fifty experts attended in response to Askew's call to examine the issues of urban growth, worsening environmental conditions, and economic development. The conference yielded several recommendations, including calls for limits on South Florida's population, an end to land reclamation efforts, and the initiation of comprehensive water resource and land development planning in the state.

Following this conference, Askew appointed a 15 member Task Force on Land Use Resource Management which recommended four pieces of legislation to the 1972 Florida legislature: the Florida Environmental Land and Water Management Act (Chapter 380), the Water Resources Act (Chapter 373), the State Comprehensive Planning Act (Chapter 23) and the Land Conservation Act (Chapter 259) (DeGrove, 2000). These elements represented the first steps in Florida's continuing experiment in managing growth. However, despite the intended linkages of urbanization with resource preservation and water quality/quantity, the 1972 legislative package in many ways marks the source of Florida's separation of land and natural resource acquisition, water resource management, and growth management, a divide the state has attempted to span through several iterations of growth management reform.

History of Land Acquisition Programs in Florida

The 1972 Land Conservation Act provided $40 million for an outdoor recreation land acquisition bond program, renewing 1964 legislation that established a $20 million program for the same purposes. In the following decades, Florida legislatures would continue to renew state land acquisition programs until 1998, when the Florida

Constitution Revision Commission proposed to amend the state's constitution to provide for conservation of its natural resources through land acquisition in perpetuity. These acquisition programs evolved over time into the largest state land acquisition program in the nation.

The first major revision to the original 1972 program came in 1979 when the Florida legislature established the Conservation and Recreational Lands (CARL) Trust Fund. CARL employed funds from the document stamp tax and an excise tax to acquire lands of environmental and cultural significance, including endangered species habitats and historic sites. The legislature also allocated additional annual funding of $20 million to local governments in lieu of taxes lost. The use of documentary stamp tax funds, in essence linking open space acquisition to real estate property transfer, would become Florida's central means of funding resource property purchases. In 1981, legislation was enacted creating two additional land acquisition programs, Save Our Rivers (SOR) and Save Our Coasts (SOC). These programs focused on water management, water supply and conservation and coastal land acquisition for recreation (Southwest Florida Water Management District, n.d.).

Public land acquisition in Florida was accelerated with the adoption of Preservation 2000, signed into law by Governor Bob Martinez in 1990. A response to recommendations made by a Martinez-appointed Commission on the Future of Florida's Environment, Preservation 2000 (P2000) was a term-limited acquisition program scheduled to end in 2000. The Commission, tasked with examining the effect of population growth on Florida's natural resources, estimated that given Florida's pace of growth, approximately 3 million acres of the state's wetlands and forests would be subject to urbanization by 2020 in the absence of state intervention. The Commission concluded that the most effective means of protecting these resources would center on a strategy of land purchase: "Acquisition coupled with restoration and regulation will best serve Florida's major ecological systems" (Carricker, 2002).

Like previous efforts, Preservation 2000 was funded through an increase in the documentary stamp tax levied during property title transfers. The increased revenue supported $300 million annually in bond funds, or $3 billion across the ten-year life of the program, making it the most highly funded land acquisition program in the United States. By statute, half of all Preservation 2000 funds were targeted to purchase conservation and recreation lands, 30 percent were targeted towards water resource acquisition, 10 percent were targeted towards funding matching grants to local governments for land purchases related to the implementation of local government comprehensive plans (for the first time explicitly linking public land acquisition to growth management), and 10 percent were used to support additions to the state's wildlife management areas, state forests, state parks and greenways and trails systems. Preservation 2000 conserved almost 1.8 million acres of land throughout Florida and was credited with successfully mitigating further damage to Florida's various natural environments, including habitats for endangered species and species of special concern to the state (Florida DEP, n.d.).

Seeking to secure the legacy of Preservation 2000, Revision 5, dubbed the "Conservation Amendment," was approved by 72 percent of Florida voters in the November 1998 election. The amendment authorized the issuing of revenue bonds for the purposes of acquiring and improving land and water resource areas for conservation, restoration, and historic preservation, restricted disposition of those lands and resources, and established the Florida Fish and Wildlife Conservation Commission. During its 1999 session, the Florida legislature responded to these constitutional changes by enacting Chapter 99–247 of the Florida Statutes, creating the Florida Forever program. Governor Jeb Bush signed the legislation into law on June 7, 1999 (Carricker, 2002).

Operating Florida Forever

Florida Forever's land acquisition strategy is ultimately one of maximizing environmental benefits given available funds. Implicit in the very name "Florida Forever" is the idea that the overriding vision of the program is the preservation of pristine Florida landscapes. In practice, the program achieves this by trying to maximize environmental benefits or, to more accurately describe the current situation, minimize the loss of those benefits from the conversion of undeveloped or minimally developed land to land with urban uses and intensities of development.

As noted earlier, and in a vein similar to the state's earlier land acquisition efforts, the goals of Florida Forever are myriad and extend far beyond the environmental land acquisition objectives. According to the Florida Forever Act, the new program's goals include insulating Florida's federal military installations from various encroachments, implementing the conservation elements of local comprehensive plans, providing for state parks, and acquiring rights-of-way for the state's system of greenways and trails among others.

Table 15.1 illustrates the distribution of Florida Forever funds. Over one-third of the funds is allocated to the Florida DEP, Division of State Lands for land acquisitions to meet the state's conservation goals. An additional 35 percent of Florida Forever funds are allocated by statute to the purpose of implementing the state's water management districts' (WMDs) five-year water quality and resource management plans, including the purchase of water recharge resources. Of the remaining funds, the majority are allocated to DCA. DCA distributes these funds through their Florida Communities Trust program to Florida localities on a competitive basis for the implementation of comprehensive plan conservation and recreation and open space elements. The remaining funds are distributed to agencies within DEP, for the acquisition of areas for the development of active recreational facilities, or to the Florida Fish and Wildlife Conservation Commission (FWCC) and the Department of Agriculture and Consumer Services (DACS) for habitat and forest conservation.

Since 1990 the Division of State Lands portion of the program, established by the P2000 program and continued through Florida Forever, has preserved over 2.3 million acres of land through fee acquisition. These land acquisitions have been widely distributed throughout the state, as illustrated in Figure 15.1 (in the color plate section), which depicts the pattern of publicly owned land in Florida. Out of a total of roughly $5 billion in state funds spent on land acquisition in both programs to date, almost two-thirds ($3.2 billion) has come from Preservation 2000 program and the remainder from Florida Forever ($1.7 billion). Table 15.2 summarizes the spending

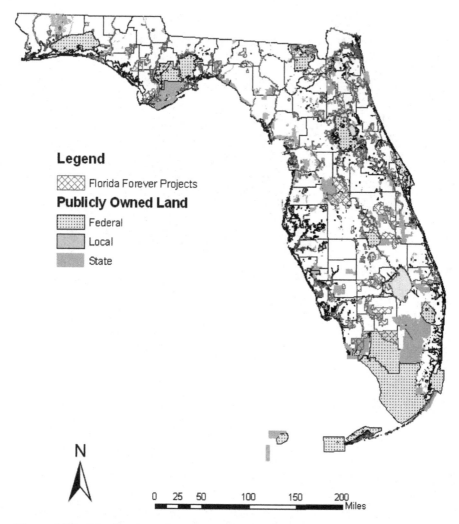

Figure 15.1 Distribution of Florida Forever land acquisition projects
Source: Florida Geographic Data Library (FGDL)

Table 15.2   Summary of P2000 and Florida Forever land acquisition efforts

	P2000		Florida Forever		Total Programs	
Fiscal Year	**Acres**	**Expenditures**	**Acres**	**Expenditures**	**Acres**	**Expenditures**
1990–1999	1,128,221	$1,931,196,210	-	-	1,128,221	$1,931,196,210
1999–2000	178,051	$367,102,902	-	-	178,051	$367,102,902
2000–2001	212,233	$406,438,299	-	-	212,233	$406,438,299
2001–2002	172,075	$215,063,001	95,202	$150,111,950	267,277	$365,174,951
2002–2003	56,211	$55,043,776	128,251	$340,309,078	184,462	$395,352,853
2003–2004	34,777	$274,097,717	68,811	$210,289,513	103,588	$484,387,230
2004–2005	-79*	$37,619,402	105,578	$303,749,575	105,499	$341,368,977
2005–2006			54,138	$329,113,999	54,138	$329,113,999
2006–2007			83,663	$496,675,858	78,231	$425,266,846
TOTAL	1,781,489	$3,286,561,308	535,643	$1,830,249,973	2,317,132	$5,116,811,281

*The negative acres amount for 2004–2005 is a result of the final reconciliation of acres.
Source: http://www.dep.state.fl.us/secretary/stats/land.htm

levels and acreage of fee simple land acquisition through these programs since 1990. During this period the state has also conserved an additional 380,000 acres through the purchase of less-than-fee conservation easements.

Florida Forever projects vary greatly and range from the Econ-St. Johns Ecosystem project that seeks to protect wetlands, extensive hydric hammocks, and riverfront along the Econlockhatchee and St. Johns Rivers, to the preservation of Florida's oldest theme park, Cypress Gardens. Table 15.3 describes the state's progress in attaining its acquisition goals for its highest priority, "Group A" projects. (The list contains only projects in-progress; already completed projects are excluded.) At over 200,000 acres, Green Swamp is the largest of the current Group A projects in terms of area. Located in central Florida and consisting of a large, water-covered raised limestone plateau, the swamp represents 870 square miles of water recharge, making it possibly only second to the Florida Everglades in terms of its hydrologic and ecologic importance to the state. The Charlotte Harbor Estuary, the second largest and most productive estuary in Florida (though threatened by phosphate mining operations), is at present the active project closest to completion, with 84 percent of the targeted land already acquired.

The Wekiva-Ocala Greenway has thus far been the most expensive of all Florida Forever projects, with over $120 million spent to acquire 66 percent of its acreage. The project seeks to create a continuous natural corridor stretching from the Wekiva Springs State Park in metropolitan Orlando northwest to the Ocala National Forest. Contributing to its expense was an agreement by state and local authorities to pay $74 million for a relatively small parcel near the Wekiva River. Seen as an essential element in satisfying legislated requirements for mitigating the growth impacts resulting from a planned expressway in the area, the owners of the 1,500-acre Neighborhood Lakes tract received over $49,000 per acre from the acquisition.

Being at its essence a "voluntary seller" program, anyone may nominate a project to the Florida Forever acquisition list. The program itself involves a sophisticated evaluation framework employed in part to respond to requirements in state law to meet various performance standards. These standards include:

Table 15.3   Progress towards acquisition of Florida's most important environmental lands

Project name	County or Counties	Year placed on list	Total acreage	Percent acquired	Cost of acquired acres
Apalachicola River	Calhoun/Gadsden/Jackson/Liberty	1991	18,940	30%	$5,717,574
Big Bend Swamp-Holopaw Ranch	Osceola	2000	59,132	6%	$3,600,000
Bombing Range Ridge	Osceola/Polk	1998	41,285	30%	$18,953,388
Brevard Coastal Scrub Ecosystem	Brevard	1993	48,387	28%	$38,504,928
Caber Coastal Connector Tract	Levy	2004	6,052	0%	$0
Camp Blanding-Osceola Greenway	Baker/Bradford/ Clay/Union	2003	153,000	0%	$0
Charlotte Harbor Estuary	Charlotte/Lee/Sarasota	1972	46,709	84%	$35,040,125
Clear Creek/Whiting Field	Santa Rosa	2004	5,026	0%	$0
Corkscrew Regional Ecosystem Watershed	Collier/Lee	1991	69,500	38%	$24,174,654
Coupon Bight/Key Deer	Monroe	1985	2,830	39%	$25,921,877
Devil's Garden	Collier/Hendry	2002	82,508	0%	$0
Escribano Point	Santa Rosa	2002	2,914	41%	$1,590,000
Estero Bay	Lee	1985	14,358	81%	$54,370,290
Etoniah-Cross Florida Greenway	Citrus/Clay/Levy/Marion/Putnam	1995	89,822	30%	$18,729,218
Fisheating Creek Ecosystem	Glades/Highlands	2000	176,876	38%	$55,628,563
Flagler County Blueway	Flagler	2003	5,015	0%	$0
Florida's First Magnitude Springs	Numerous	1991	13,980	27%	$22,847,560
Florida Keys Ecosystem	Monroe	1992	11,854	28%	$47,089,476
Florida Springs Coastal Greenway	Citrus	1995	41,108	76%	$50,569,873
Garcon Ecosystem	Santa Rosa	1995	7,735	50%	$3,353,561
Green Swamp	Lake/Polk	1992	233,598	33%	$115,321,102
Indian River Lagoon Blueway	Numerous	1998	25,843	15%	$19,783,680
Lafayette Forest	Lafayette	2004	12,800	0%	$0
Lake Santa Fe	Alachua/Bradford	2003	10,735	0%	$0
Lake Wales Ridge Ecosystem	Highlands/Lake/Osceola/Polk	1992	23,512	54%	$48,289,480

Table 15.3 Continued

Project name	County or Counties	Year placed on list	Total acreage	Percent acquired	Cost of acquired acres
Mill Creek	Marion	2003	12,285	0%	$0
Nokuse Plantation	Walton	2006	11,961	78%	$0
North Key Largo Hammocks	Monroe	1983	4,621	78%	$75,732,715
NE Florida Blueway	Duval/Flagler/St. John's	2001	25,233	53%	$30,893,950
NE Florida Timberlands and Watershed	Clay/Duval/Nassau	2001	143,499	34%	$110,119,651
Ochlocknee River Conservation Area	Leon	2004	3,105	0%	$0
Okeechobee Battlefield	Okeechobee	2001	211	0%	$0
Osceola Pine Savannas	Osceola	1995	41,121	36%	$19,183,134
Panther Glades	Hendry	2001	55,799	40%	$40,564,708
Perdido Pitcher Plant Prairie	Escambia	1995	7,661	56%	$27,299,939
Promise Ranch	Lake	2004	1,418	0%	$0
Pumpkin Hill Creek	Duval	1994	25,337	69%	$9,387,230
Spruce Creek	Volusia	1990	2,928	71%	$10,702,515
St. Joe Timberland	Numerous	2000	172,476	45%	$91,230,765
St. John's River Blueway	St. John's	2002	27,997	4%	$3,112,640
Terra Ceia	Manatee	1996	4,197	39%	$2,297,500
Tiger/Little Tiger Island	Nassau	2001	1,280	13%	$4,184,000
Upper St. Marks River Corridor	Jefferson/Leon/Wakulla	2003	15,670	0%	$0
Volusia Conservation Corridor	Flagler/Volusia	2001	79,612	54%	$35,454,145
Wacissa/Aucilla River Sinks	Jefferson/Taylor	2985	24,147	60%	$5,886,137
Wakulla Springs Protection Zone	Wakulla	1997	6,714	53%	$7,372,678
Wekiva-Ocala Greenway	Lake/Orange/Seminole/Volusia	1995	76,698	66%	$121,779,331
Yellow River Ravines	Okaloosa/Santa Rosa	2002	17,626	23%	$468,648
Totals			**1,965,115**	**34%**	**$1,185,155,035**

- the conservation of significant strategic habitat conservation areas,
- the acquisition of the highest priority conservation areas for Florida's rarest species,
- the preservation of significant landscapes, landscape linkages, and conservation corridors, giving priority to the completion of linkages,
- the acquisition of underrepresented native ecosystems, and
- the protection of large tracts of at least 50,000 acres that host intact or restorable natural communities.

These standards serve to increase the presence and viability of endangered species, threatened species, or species of special concern on publicly managed conservation areas.

A nine-member Acquisition and Restoration Council, comprised of the Secretaries of the Florida Departments of Community Affairs and Environmental Protection, representatives from the Florida Department of Agriculture and Consumer Affairs' Division of Forestry, the Florida Fish and Wildlife Conservation Commission, the Department of State's Division of Historical Resources, and four gubernatorial appointees, meets twice annually to evaluate and select projects for acquisition through the Florida Forever Trust Fund. Acquisitions may be made by purchasing either the full fee or less than the full fee (conservation easements) of a subject property. Final approval of the overall acquisition list is provided by the Florida Governor and Cabinet, acting as the Board of Trustees of the Internal Improvement Trust Fund, who may remove projects from the list but may not add them.

Acting with the Board's consent, the Division of State Lands negotiates specific purchases with land owners, paying no more than the higher of two appraised values or, where those appraisals differ appreciably and a third appraisal is demanded, no more than 120 percent of the lower of the two closest of three appraisals. While division realizes approximately $105 million annually from the sale of Florida Forever bonds, the estimated value of the properties on the acquisition list typically totals in the billions of dollars.

The targeting regime used to identify and to evaluate potential Florida Forever acquisitions has been accomplished by the Florida Natural Areas Inventory (FNAI) inaugurated for targeting CARL program purchases in the 1980s. Improvements in computer technology and the accumulation of FNAI's databases over time have made the targeting regime increasingly complex. By the late 1990s FNAI's database was sufficiently complete to produce acquisition matrices for the various projects proposed for purchase, assembling various evaluation criteria and ranking them for review by the Acquisition and Restoration Council. Today, FNAI continues to offer scientific review but has further refined its ranking methodology to provide a comprehensive natural resource analysis and has developed scoring for all potential Florida Forever projects using a Conservation Needs Assessment (Knight, Knight, and Oetting, 2000). This assessment essentially functions as a sophisticated, computerized benefit measurement metric.

FNAI has recently further automated the targeting system through the development of a Florida Forever Tool for Efficient Resource Acquisition and Conservation (FTRAC) that "provides a means for evaluating Florida Forever projects relative to a model portfolio that achieves efficient natural resource protection on the amount of land likely to be acquired during the ten year duration of the Florida Forever program" (FNAI, n.d.). Using a sophisticated set of computational tools, FNAI simulates optimal ecological land portfolios based on conservation targets. Rather than pursuing a single best scenario, FNAI works through randomly selected sets of potential properties to assemble alternative cost-feasible portfolios and obtain an optimal collection of targeted resources. The "irreplaceabilty" of any property is ascertained by the number of times it is modeled as a component of a best portfolio scenario (Oetting and Knight, 2005). Thus, Florida Forever is a land acquisition program with a level funding fully matched by its level of sophistication in targeting properties for purchase.

Buying the Irreplaceable: The Case of Babcock Ranch

Babcock Ranch represents the largest single land purchase in the history of the Florida Forever and Preservation 2000 programs. The property consists of 91,361 acres of carefully stewarded land (including a sustainable ranching operation) in southwest Florida's Charlotte and Lee counties and was, prior to its acquisition, one of the largest remaining tracts of privately owned land in Florida. From the point of view of resource preservation, the idea of acquiring the property for conservation purposes has long been attractive. The property has the potential to complete a wildlife corridor stretching from Lake Okeechobee to the Gulf of Mexico, it is habitat for threatened species (including the Florida panther and black bear), and it provides vitally needed water-recharge areas for rapidly urbanizing southwest Florida. While the scale of the Babcock Ranch acquisition renders it an extraordinary rather then representative case, this scale also offers a clearer view of some of the inherent problems implicit in the lack of connection between the state's strategies for environmental and natural resource protection and growth management, problems that have also surfaced in areas as different as the Wekiva River watershed in central Florida and Franklin County in the state's panhandle region.

The Acquisition and Restoration Council added the property to the high priority Group A Florida Forever acquisition list on December 6, 2001 (Division of State Lands [DSL], Florida Department of Environmental Protection, 2006.) At the time it had a taxable value of $52,527,237, and the property was held by a single owner, the Babcock Florida Company. Originally the Crescent B Ranch, the property was acquired as a hunting preserve by Edward Babcock, a Pittsburgh lumber magnate and politician, in 1914. The 1997 death of his son, Fred Babcock, who had managed the property since the 1930s, created tax liabilities for his heirs and paved the way for dissent among the company's remaining owners regarding the property's disposition. Despite a family history of cooperation with Florida government on preservation issues that dates from the 1940s, the Babcock heirs failed to come to terms with the state, and further discussion of state acquisition was terminated in 2005. The state's lead negotiator, Eva Armstrong, sourced the failure to the family's insistence that the state buy Babcock Florida Company stock rather than real property (an act prohibited by the state constitution) and their requirement that the state pay the family's taxes. Armstrong noted at a meeting on April 4, 2005 that the state wasn't "ready to write a check for over $700 million" for the property (Martin, 2006). With negotiations with the state government at a standstill, the owners of the Babcock Florida Company began entertaining offers from others to acquire the property, subdivide it, and develop it despite increasing activism among local and statewide environmentalists organizing in favor of preservation.

In late 2005, Kitson and Partners, a community and resort developer headquartered in West Palm Beach, intervened. Kitson sought to develop a new city of 50,000 residents with 19,500 new homes and six million square feet of retail, commercial and institutional uses. On July 31, 2006, Kitson completed its purchase of Babcock Ranch from the Babcock family and, on the same day, closed its deal with the State of Florida and Lee County for the largest single land preservation agreement in the state's history. The agreement preserves 74,000 acres of the property, allows for ranching operations to continue, and provides Kitson with 17,800 acres for development (Kitson and Partners, 2006).

The state's participation in the deal required legislative action. On June 19, 2006, Governor Jeb Bush signed into law the Babcock Range Preserve Act, committing the state to $310 million of the $350 million purchase price, the remainder coming from Lee County appropriations (FDEP, 2006). Two hundred million dollars of the $310 million state contribution to the purchase came from the allocation of Florida Forever trust funds. Despite legal challenges by the Sierra Club to the Charlotte County Comprehensive Plan amendments necessary to effect the development project on which the deal was hinged, Governor Bush called the acquisition the "high point" of his term in office.

A project of unique size and scope, Babcock Ranch is an exemplary acquisition with respect to its natural values. The property offers habitat for the Florida panther, the black bear, the red-cockaded woodpecker and many other species of concern to the state and federal governments. The land contains landscapes of high biological and ecological value, including mesic flatwoods that Florida Forever documents call "impressive" in their extent and quality (Division of State Lands, 2006). The property also serves as a water-recharge area for southwest Florida and serves to link existing conservation lands.

At the same time, however, Babcock Ranch illustrates possible inefficiencies in Florida's land resource preservation regime and tensions between the state's preservation program and the state's growth management approach. First, the state's explicit and implicit targeting strategies potentially make resource acquisition less efficient by not accounting for conversion potential or the effect of growth management policy on that potential. The state essentially discounts or outright ignores the ability of growth management controls to manage these environmentally sensitive lands. Second, the assumption that the "highest and best use" of the property will prevail in spite of growth management may lead the state to pay more for the land it seeks to acquire than it would if it relied on the policies within its growth management framework. Third, by setting the growth management framework aside in order to realize conservation acquisitions, the state has agreed to change the character of the area (by allowing urban uses on a portion of the property) and to create an important open-space amenity attractive to other development that may put nearby, currently unpreserved, environmental assets at risk.

Potential Inefficiencies in Florida Forever's Targeting Strategy

Babcock Ranch was a big, expensive purchase. Executing its acquisition in the context of limited, however generous, funds necessitates forgoing other resource preservation opportunities. Should preserving Babcock Ranch have been a state priority? Some analysts have criticized approaches like Florida Forever's for identifying and targeting lands for acquisition. Newburn, Berck, Merenlender and Reed (2005) suggest a strategy based solely on benefits, purchasing sites with the highest benefit value, using criteria that are well established in the scientific literature (e.g., contribution to aquifer recharge). Alternatively, they suggest that open space administrators could incorporate costs and benefits, targeting lands with the greatest ratio of biological benefits to land acquisition costs.

As described above, Florida Forever program administrators have employed a different strategy. Program administrators have implicitly assumed that land acquisition is the only means for protecting these assets. As a result they have largely focused upon the likelihood of biological benefit loss from the threat of future land use changes if the site were not protected. Despite its otherwise exemplary sophistication, the evaluation of potential Florida Forever acquisitions inadequately accounts for development pressure and for the effect of development threat on the opportunity costs of non-preservation. Nevertheless, there is some empirical evidence to suggest that development threat increases the state's willingness to pay for land preservation. For example, Larkin, Alavalapati, and Shrestha (2005) found significant effects associated with variables for demonstrated development potential through the application of hedonic techniques to the CARL program. Factors affecting conversion potential, including growth management polices to mitigate conversion and the presence and absence of development-enabling infrastructure, do not appear to be important considerations in the state's targeting regime.

The technology for modeling the probability of conversion has been in development for some time. Techniques for computer modeling of urban land use change, particularly in support of planning applications, have existed since the mid-1990s (see, e.g., Brail and Klosterman, 2001). Such tools generally take into account population growth, the emerging pattern of urbanization, the availability of infrastructure, land prices, and land use policy, factors largely outside the current Florida Forever targeting apparatus.

Even if conversion probability remains unmeasured, development threat appears to be a primary consideration in Florida Forever deliberations, as the Babcock Ranch case makes clear. Through a deliberative yet rough approximation of true benefit-loss-cost targeting, Florida Forever policy makers attempt to minimize the threatened loss of benefits to future land use conversion while taking acquisition costs into account. In the end, however, it is possible that the state might target different properties for acquisition if it accounted for conversion potential in a manner that included its own mitigating policies.

Substituting Threats for Facts in the Appraisal Process

How well does substitution of threat for demonstrated development potential work? According to some evaluators of Florida Forever appraisals, it does not work very well. The appraisal basis for the price at which Florida Forever negotiates land acquisitions has been heavily criticized by the State Auditor General (Davis, Walsh, and Crispo, 1997). The basis of this criticism is that the appraisal is based on "unsupported assumptions" about the potential for converting the land to urban uses and development intensities. A 1997 report by Florida TaxWatch on the transaction involving the 1993 state purchase of 2,261 acres of Keewaydin (Key) Island in Collier County also demonstrated the effect of exaggerated conversion assumptions. Taxwatch analysts found that the appraisals used as the basis for establishing the $13.2 million purchase price made several highly questionable assumptions about both the timeframe of the land use change approval process and the availability of infrastructure, both of which would have had a dampening effect on conversion (Davis et al., 1997).

The stated policies of Florida growth management should serve to restrict the urbanization of environmental lands by limiting the size and intensity of future development in resource-rich locations. Correctly implemented growth management, therefore, should reduce the cost of purchasing minimized development rights and offer a means of enhancing land acquisitions. Instead of relying on growth management and taking it into account in purchasing and price-setting decisions, however, Florida Forever policy makers and appraisers have tended to set it aside, assuming that the "highest and best use" of the land will proceed regardless of existing regulatory safeguards against resource loss.

Setting Aside Growth Management for Conservation Purposes

The Babcock Ranch case also offers an example the third potential disconnection between growth management and conservation: setting aside growth management policy on all of the land in order to effect the preservation of a large portion of it and in the process entitling development that may or may not have occurred otherwise. From the time Kitson and Partners intervened forward, the state's preservation of environmental lands became closely linked to the developers' ability to realize their development plans on portions of the property not subject to preservation. Governor Bush focused on the cost of preservation as among the most important justifications for doing whatever was necessary to entitle Kitson's development, declaring that "if this development didn't take place there's no way the state could have paid" for the purchase of the ranch, noting that he would "find it ironic beyond belief" if the state's growth management policies posed obstacles to the deal (Whitehead, 2006).

Among the considerations sought by Kitson was a finessing of the state's Rural Land Stewardship Program, which allows the transfer of development rights for agricultural land preservation, to include in the count of his allowable development units those units allocated to the portion of the property that he was about to sell the state. This in effect allowed Kitson to be compensated twice for the land, once in cash for the fee (which ostensibly included the development rights assigned to it through county planning) and a second time through the transfer of those units to the property he retained. Of equal concern was the transfer to Kitson of the water rights associated with the preserved property, allowing Kitson to use state lands for storm water retention and detention and not allotting water management resources from the lands subject to development, thereby diminishing the effects of preservation on proximate aquatic and terrestrial habitats.

The development of the Babcock Ranch project in eastern Charlotte and Lee Counties will also require the development of significant urban services and infrastructure systems, including an estimate of $517 million in new regional roads. The attraction of new services, including schools, and facilities providing greater regional mobility are almost certain to trigger the conversion of non-urbanized land to urban uses and intensities in these counties, as well as in neighboring Hendry and Glades counties. The environmental amenity created through preservation will likely attract development to this area of the state as well.

This suggests the potential for an ambiguous legacy for the Babcock Ranch acquisition. On the one hand, significant local and state funds have been assembled for the purpose of acquiring a large amount of land for preservation. On the other hand, that acquisition has enabled urbanization in a largely non-urbanized area of the state that may denigrate similar or complementary resources in the vicinity of the preserved land and largely ignore the state's policy against sprawl. Paradoxically, the Babcock Ranch deal potentially accelerates future development, and therefore resource loss, in its vicinity.

For those reasons, it is difficult to gauge whether or not the purchase represents an efficient use of state funds. The Florida Governor has argued that without the revenues provided by realizing projects like Kitson's, the state cannot afford to acquire land for preservation in perpetuity. At the same time, it could be argued that the rigorous application of the state's growth management policies, even with their incomplete treatment of the conservation of environmental resources, would afford some resource protection, while setting those policies aside appears to have potentially increased the land's development potential and the potential impact on those natural resources.

Conclusions and Policy Implications

Despite issues raised by the Babcock Ranch case, it is hard to be disappointed with Florida Forever or with the state's overall commitment to resource protection. The state's growth has been increasingly linked to the appeal of Florida's environmental assets and their contribution to the quality of life of residents and the quality of the experiences of visitors to the state. The state's fiscal commitment to environmental protection is laudable and, in the absence of that commitment, important Florida landscapes would have been diminished if not entirely lost.

Nevertheless, in the next several years, the Florida legislature will once again consider renewing the funding for the now constitutionally-mandated land acquisition program. Debates over purchases like Babcock Ranch have revealed differences of opinion about the future direction of Florida Forever. Land preservationists would like to see the program priorities remain unchanged, with more allocated funding, while others are clamoring for the purchase of park space and trail easements to make recreational opportunities more accessible to the residents of Florida's urban areas. Some policy makers have begun to wonder, after almost a half century of environmental land purchases, if the state has already acquired too much land for preservation purposes. In both 2005 and 2006, the legislature entertained entreaties from revenue-strapped Florida counties, uncompensated for the loss of property taxes related to these public acquisitions, to evaluate the state's land portfolio with a view towards declaring some holdings surplus and eventually returning them to the tax rolls. Yet another constituency views the state's growing land portfolio as a source of subsidy for badly needed housing accessible to Florida's workforce.

Between 1990 and 2010, Florida will have spent $6 billion in acquiring land in various performance-oriented programs with only tangential connections to the state's growth management program. As the Babcock Ranch case illustrates, large-scale public land purchases can potentially be at odds with other state preservation goals, environmental and resource conservation efforts, and a state mandate for urban containment. The contest over which set of values should prevail in the disposition of the Babcock Ranch case bitterly divided many of Florida's environmental and growth management advocacy groups. While some groups argued that the opportunity to preserve over 70,000 acres was too good to pass up, others believed it was a "Faustian" tradeoff, one that would bring substantial development in the form of leapfrogging sprawl to a rural area of the state. A lesson provided by the Florida experience is that land acquisition should not be viewed as naturally partnered with land use policy.

The reconsideration of Florida Forever in the coming years provides the opportunity to re-think the ordering of it objectives and the manner of its implementation. The state might consider the Maryland's Rural Legacy Program (RLP) model created in 1997. The purpose of the RLP is to provide the focus and funding necessary to protect large, contiguous tracts of land and other strategic areas from sprawl development, while enhancing natural resources. Maryland has protected over 40,000 acres with over $105 million in grant funds between the 1998 and 2003 fiscal years, and has achieved a growing ring of large greenbelt areas which have put the state well on its way towards reaching its agreement to protect 20 percent of the Chesapeake Bay watershed by 2010.

Following from our review of the program and the Babcock Ranch case, we offer the following recommendations for the improving the Florida Forever land acquisition program:

1. *Undertake an evaluation of the effects of public lands acquisition programs.* In the context of increasing land prices and the possibility of slipping support for more expensive acquisitions, it is essential that the state undertake an empirical analysis of the effects of its half century history of land resource acquisition. Such an evaluation should consider the effects of pursuing a targeting strategy for acquisitions that is focused on preserving valuable environmental benefits, while incorporating cost and the probability of land conversion within the context of existing growth management policies.

2. *Strengthen growth management's ability to deliver urban containment and conservation and provide incentives to local governments to attain it.* While urban containment achieved through "boxing in" development through tactical open space acquisition sounds attractive, it is simply unrealistic given current levels of funding. The state should consider ways to leverage growth management policy in order to stretch available preservation funds. A better-funded Florida's Community Trust program (overseen by DCA and funded by Florida Forever) should encourage local governments to identify opportunities to link conservation purchases to anti-sprawl policies in their comprehensive plans. The foci of this program should remain protecting water quality and water resources, providing habitat, reducing infrastructure costs, and protecting resource-based and tourist-based economic sectors, rather than for providing recreation facilities in urbanized areas. Where possible these funds should be leveraged for the completion of Florida Forever projects.

3. *Strengthen growth management's conservation policies.* The state's reliance on a land acquisition strategy that tends to reinforce the expectation of private owners that they have a right to develop their land to the highest possible use, regardless of its impacts on natural resources, may in fact make effecting resource preservation through acquisition more expensive (Fairfax, Gwin, King, and Raymond 2005). Rather than deferring to an overworked Florida Forever for environmental protection, state officials would be well advised to strengthen growth management while refining the lands that Florida Forever targets. The two have room to act as complements, as the state's existing growth management regime provides for land purchases at reduced prices and effective Florida Forever acquisitions ensure a high quality of life and preserved natural heritage for existing and future state residents.

4. *Reconsider the mechanism for funding Florida Forever.* Policy options should be considered that directly address undesirable economic behavior with incentives for different behavior. Some attention should be paid to the source of funding for Florida Forever, potentially replacing the current source, which depends on real estate transactions, with sources that more directly address the negative externalities associated with open space loss. The current document stamp tax is paid on all transfers of real property titles whether they occur in the inner city or at the urban fringe. A tax on open space conversions or a fee to compensate the public for the loss of environmental amenities and services would more directly affect the demand for open space conversion by making it more expensive.

5. *Use some portion of Florida Forever funds to assist Florida's revenue strapped local governments adversely impacted by public land acquisition.* State payments in lieu of taxes could be made to compensate local governments for some or all of the tax revenue lost through Florida Forever acquisitions. Again, although local governments are spared the costs of servicing preserved public lands, the loss of revenue and the loss of the opportunity for future revenue (if the land were developed) creates an incentive to work against the state's resource preservation efforts.

Ultimately, stronger growth management policy, especially in regards to urban containment, would greatly complement land acquisition by allowing key natural resources to be preserved at much lower costs. A strengthening of the tools that have had some effect on the rates of land conversion, including urban growth boundaries and open space impact fees high enough to cover the true cost of lost resources, would allow Florida Forever to succeed in concert with growth management. Land acquisition is a valuable strategy for environmental preservation and its popularity with citizens and the longstanding commitment by the state to this effort make it a very valuable tool for achieving both environmental and planning goals. However, its utility is greatest when working in an environment regulated by strong growth management. Florida Forever by itself cannot preserve enough lands to safeguard environmental resources at a reasonable cost without growth management deflecting development away from those areas until such time as they can be acquired.

References

Brail, R. and R. Klosterman. (2001). *Planning Support Systems*. Redlands, CA: ESRI Press.

Carricker, R. (2002). Florida Forever: A Program for Conservation Land Acquisition. Gainesville, FL: University of Florida Institute for Agricultural Sustainability. Accessed on November 3, 2005 from http://edis.ifas.ufl.edu/FE331.

Davis, D., M. Walsh, and N. Crispo (1997). *Florida's state lands program needs legislative and administrative overhaul*. Tallahassee, FL: Florida TaxWatch.

DeGrove, J. (2000). Florida's Growth Management Legislation: 1969 to 2000. Accessed on January 25, 2007 from http://lic.law.ufl.edu/~nicholas/GMsem/DeGrove.htm.

Division of State Lands, Florida Department of Environmental Protection. (2006). *Report of the Florida Forever program prepared for the Board of Trustees of the Internal Improvement Trust Fund of the State of Florida*. Tallahassee, FL: Department of Environmental Protection.

Dwyer, L., D. Murphy and P. Ehrlich. (1995). Property rights case law and the Endangered Species Act. *Conservation Biology 9* (4): 725–741.

Fairfax, S., L. Gwin, M. A. King, and L. Raymond. 2005. *Buying Nature: The Limits of Land Acquisition as a Conservation Strategy, 1780–2004*. Cambridge, MA: MIT Press.

Florida Department of Environmental Protection (2006). "Governor Bush Signs Bill to Purchase Babcock Ranch." Accessed on July 16, 2006 from http://www.dep.state.fl.us/secretary/news/2006/06/0619_03.htm.

Florida Department of Environmental Protection. (n.d.). *Florida Forever*. Accessed on October 31, 2005 from http://www.dep.state.fl.us/lands/acquisition/FloridaForever/default.htm.

Florida Natural Areas Inventory. (n.d.). *Florida Forever*. Accessed on November 12, 2005 from http://www.fnai.org/FLForever.cfm.

Gilmour, R and J. McCauley (1976). Environmental Preservation and Politics: The Significance of "Everglades Jetport." *Political Science Quarterly 90* (4): 719–738.

Kitson and Partners. (2006). About the Babcock Ranch. Available at http://www. babcockranchflorida.com/tabid/60/Default.aspx.

Knight, G., A. Knight, and J. Oetting (2000). *Summary Report to the Florida Forever Advisory Council.* Tallahassee, FL: Florida Natural Areas Inventory.

Larkin, S.L., Alavalapati, J.R.R., Shrestha, R.K. Estimating the Cost of preserving Private Lands in Florida: A Hedonic Analysis. *Journal of Agricultural and Applied Economics 36*: 481–490.

Martin, G. (2006, April 16). "Commission OKs Babcock City." *Charlotte Sun.* Accessed on April 16, 2006 from http://www.sun-herald.com/NewsArchive2/040506/ tp1ch6.htm?date=040506andstory=tp1ch6.htm.

Newburn, D., P. Berck, A. Merenlender and S. Reed. (2005). Economics and land-use change in prioritizing private land conservation. *Conservation Biology 19*(5): 1411–1420.

Newburn, D., P. Berck and A. Merenlender. (2006). Habitat and open space at risk: targeting strategies for land conservation. *American Journal of Agricultural Economics 88*(1): 28–42.

Oetting, J, and Knight, A. (2005). *F–TRAC Florida Forever Tool for Efficient Resource Acquisition and Conservation: Model Documentation and Project Evaluation.* Tallahassee, FL: Florida Natural Areas Inventory.

Press, D., D. Doak, and P. Steinberg. (1996). The role of local government in the conservation of rare species. *Conservation Biology 10* (6): 1538–1548.

Southwest Florida Water Management District. (n.d.) Save Our Rivers. Accessed on November 3, 2005 from http://www.swfwmd.state.fl.us.

Stephenson, R. B. (1997). *Visions of Eden: Environmentalism, Urban Planning, and City Building in St. Petersburg, FL, 1900–1995.* Columbus, OH: Ohio State University Press.

Whitehead, C. (2006, June 17). "Sierra Club Fighting to Halt Babcock Ranch Plans." *Naples Daily News.* Accessed on July 1, 2006 from http://www.naplesnews.com/ news/2006/jun/17/group_wants_halt_babcock_ranch_plans/.

Chapter 16

Affordable Housing in Florida: Why Haven't Florida's Growth Management Laws Met the Challenge of Adequately Housing All Its Citizens?

Charles E. Connerly

Although much discussion, both in this book, as well as elsewhere, has focused on Florida's approach to regulating the state's rapid growth, protecting the state's fragile environment, and meeting the state's infrastructure needs, the 1985 Growth Management Act (GMA) also directly addresses the provision of housing, including affordable, adequate housing for all its residents. Specifically, the 1985 GMA requires every jurisdiction in the state to adopt a comprehensive plan that incorporates a housing element, which includes:

1. The provision of housing for all current and anticipated future residents of the jurisdiction;
2. The provision of adequate sites for future housing, including housing for low-income, very low-income and moderate-income families, mobile homes, and group home facilities and foster care facilities, with supporting infrastructure and public facilities;
3. The creation or preservation of affordable housing to minimize the need for additional local services and avoid the concentration of affordable housing units only in specific areas of the jurisdiction (Local Government Comprehensive Planning and Land Development Regulation Act (Local Planning Act), Fla. Stat. § 163.3177.6(f)1, 1985).

These requirements mean that Florida jurisdictions are expected to ensure that there are adequate sites for all future residents regardless of their incomes and that jurisdictions should seek to avoid concentrating affordable housing in limited geographic areas. Taken together, these requirements strongly suggest that Florida communities are to ensure that adequate land is available to all income levels and that that such land—and by implication, the placement of affordable housing in newly developed areas—should be dispersed, rather than concentrated. At minimum, the Florida statutes certainly suggest that Florida communities should adopt flexible zoning approaches that ensure there is an adequate mix of lower and higher density housing, zoned for both single and multifamily occupancy. Flexible land regulations are critical because they help assure that housing is built at high enough density that housing is more likely to be affordable. Pendall's (2000) research shows that communities that zone residential areas exclusively for low density housing reduced the number of multifamily units, thereby reducing the availability of more affordable housing. In contrast, jurisdictions that seek to make housing more affordable start by making certain that sufficient quantities of land are zoned for higher density detached and attached single family housing, multifamily housing, and accessory apartments or dwellings. Portland, Oregon, for example, has adopted a flexible zoning approach that emphasizes the provision of adequate land for higher density single family and multifamily housing (Connerly and Smith, 1996).

Furthermore, Florida's growth management law encourages local jurisdictions to adopt "innovative land development regulations which include provisions such as transfer of development rights, incentive and inclusionary zoning, planned-unit development, impact fees, and performance zoning" (Local Planning Act, Fla. Stat. § 163.3202(3). 1985). This means that Florida jurisdictions are not only supposed to provide flexible zoning that will permit a mix of lower, middle, and upper income households to live in the community, but that they are encouraged to act positively and affirmatively towards achieving this goal by adopting an inclusionary zoning ordinance. Moreover, of the innovative land development regulations named in Florida's growth management law, inclusionary zoning is the only tool that pertains directly to the development of affordable housing. In a sense, therefore, the Florida growth management law privileges inclusionary zoning by specifically encouraging it as a tool for meeting the obligation of each jurisdiction to plan comprehensively for occupancy of all income groups in a dispersed, rather than a concentrated, development pattern.

The remainder of this chapter will examine how successful Florida's growth management law has been in encouraging affordable housing, specifically through inclusionary zoning. It will do this by first looking at the role that inclusionary zoning plays in not only providing affordable housing, but addressing the tendency of growth management to raise housing costs and reduce housing affordability. Because of the importance Florida assumes in the development of growth management in the US, as well as the potential for realizing inclusionary zoning in Florida's growth management laws, Florida is shown to be an important test case for the expansion of inclusionary zoning beyond a handful of states to a broader presence as a more universal approach to the provision of affordable housing. From there, the chapter will discuss why, despite its sanction by Florida law, there has been limited adoption of inclusionary zoning in Florida since passage of the 1985 Growth Management Act. The chapter will then consider whether the climate for inclusionary zoning in Florida is changing by examining a case study of inclusionary zoning adoption in the state's capital city, Tallahassee, as well as the rising tide of interest by other Florida jurisdictions in adoption.

Previous Work on Inclusionary Zoning

Inclusionary zoning takes flexible zoning regulations a step further by requiring that new residential developments set aside a fixed percentage of new dwelling units for affordable housing. Although inclusionary zoning is most commonly found in communities in California, Massachusetts, Connecticut, and New Jersey and to a lesser extent in Rhode Island, the single best-known inclusionary zoning ordinance is found in Montgomery County, Maryland, just outside of Washington, DC. Since its creation in 1973, the Montgomery County program has produced over 11,000 units of affordable housing (Montgomery County, n.d.; Porter, 2004a).

Inclusionary zoning has several key advantages. First, because inclusionary zoning is directly linked to the production of new dwelling units, it provides a straightforward mechanism for ensuring that growth includes affordable dwelling units. Although inclusionary zoning has been criticized for relying exclusively on new development, in the context of growth management this is an asset. Second, because density bonuses are frequently awarded developers of inclusionary zoning development, they enable compact development, thereby meeting a key objective of growth management. Third, by providing affordable housing in all new developments, lower income and minority households are less likely to be segregated than where new development consists exclusively of large, expensive homes. Moreover, by relying on the private or non-profit sector to actually build affordable housing, inclusionary zoning can foster an effective collaboration between government and the non-government sector. This assures that the location and type of housing will be more responsive to the price signals sent by the private market—a factor often not recognized in the more traditional government-driven public housing programs, in which the location and style of dwelling units were more reflective of politics than the private market (Burchell and Galley, 2000; Porter, 2004a).

At the same time, inclusionary zoning has several disadvantages. Most fundamental is the belief by developers that inclusionary zoning adds to the cost of the market-rate units that are developed along with the inclusionary units. Developers assert that the cost of building the affordable units is passed on to purchasers or renters of the market-rate units, a burden they believe to be unfair (Burchell and Galley, 2000). Others are concerned that a policy of inclusionary zoning that leads to the dispersal of the poor can also result in the deconcentration of minority and ethnic groups, thereby diluting their political power and culture (Burchell and Galley, 2000; Pyatok, 2004). Finally, in New Jersey, where the "builder's remedy" allows developers to build inclusionary zoning development when jurisdictions have failed to adopt affordable housing plans, such developments are accused of promoting rather than discouraging sprawl (Burchell and Galley, 2000; Carlson and Mathur, 2004; Lawrence 2001).

Overall, after over 30 years of existence, the concept of inclusionary zoning still has much promise, but has been somewhat disappointing in its impact. To some degree, this reflects the fact that in spite of the success in Montgomery County, Maryland, many of the 600 or so jurisdictions with inclusionary zoning produce relatively few units (Porter, 2004a). More distressingly, relatively few jurisdictions have even adopted inclusionary zoning laws. Most jurisdictions adopting inclusionary zoning are concentrated in a few states in the northeast and the west. Even though inclusionary zoning has been shown in Montgomery County, Maryland, to be a potentially effective producer of affordable housing, most jurisdictions prefer not to adopt it.

In Florida, as well as other states or jurisdictions which have adopted growth management, inclusionary zoning is especially important because it has the potential for mitigating the impact that growth controls have on housing costs and housing affordability. Within the Florida context, recent research has shown that adoption of the GMA has reduced housing affordability. Specifically, Anthony (2003), employing an interrupted time series analysis of Florida counties from 1980 through 1995, found that the enactment of a 1985 Growth Management Act-mandated plan had a significantly negative impact on housing affordability in the state.

More generally, Table 16.1 displays summary information on the 40 empirical studies examined by Nelson, Pendall, Dawkins, and Knapp (2002) that address the impact of growth regulation on housing prices.[1] Many of these studies were done in the 1980s and generally reflect an interest in the impact of growth regulation in communities in the west, particularly California. The overwhelming impression obtained from these studies is that growth regulation raises housing prices. Thirty-two of the forty studies, or 80 percent, report that growth regulation raises housing prices, while only 10 percent of the studies show that there is no relationship between growth regulation and housing prices. The remaining 10 percent show mixed results.

Despite this general impression, Nelson et al. urge caution in drawing conclusions that growth regulation *per se* is bad for affordable housing. Instead, they contend that if state and local governments combine growth regulation with efforts to assure the supply of affordable housing, then the supply of affordable housing is more likely to be enhanced. They cite the experience of Portland, Oregon where growth regulatory policies, including urban growth boundaries, are combined with affirmative efforts to establish minimum densities and to assure an adequate supply of multifamily housing that is generally more affordable than single family housing (Nelson et al., 2002). As indicated in Table 16.1, more recent studies of Portland have cast doubt on the claim that growth regulation in that community has had a negative impact on housing prices. If jurisdictions pay close attention to using land use regulatory techniques, as well as subsidy approaches, to assure the supply of affordable housing, then it appears they are more likely to succeed in mitigating the negative impact that growth management has on housing affordability.

Inclusionary Zoning in Florida

Given the demonstrated impact that growth management has had on housing prices and affordability and given that Florida growth management law cites inclusionary zoning as an innovation that local jurisdictions are encouraged to adopt, one would expect that Florida would be an important case for testing whether inclusionary

---

1 Anthony's research is not directly comparable to the studies shown in Table 16.1 because it measures the impact of growth management on housing affordability, not on housing prices. Nevertheless, Anthony's findings are consistent with the studies shown in Table 16.1.

Table 16.1   Empirical evidence on the impact of growth regulation on housing prices

Study	Location	Impact?
Segal and Srinivasan (1985)	National	Yes
Urban Land Institute (1977)	National	Yes
Black and Hoben (1985)	National	Yes
Guidry, Shilling, and Sirmans (1991)	National	Yes
Chambers and Diamond (1988)	National	Yes
Rose (1989)	National	Yes
Shilling (1991)	National	Yes
Dowall and Landis (1982)	San Francisco Bay Area	Yes
Dowall (1984)	Santa Rosa, Napa, CA	Yes
Landis (1986)	Sacramento, Fresno, San Jose, CA	Yes
Downs (1992)	San Diego County	Yes
Katz and Rosen (1987)	San Francisco Bay Area	Yes
Landis (1992)	California	No
Elliot (1981)	California	Yes
Schwartz et al. (1981, 1984)	Petaluma, Santa Rosa, Rohnert Park, CA	Yes
Glickfield and Levine (1992)	California	No
Mercer and Morgan (1982)	Santa Barbara County, CA	Yes
Zorn et al. (1986)	Davis, CA	Yes
Miller (1986)	Boulder, CO	Mixed
Pollakowski and Wachter (1990)	Montgomery County, MD	Yes
Porter et al. (1996)	Montgomery County, MD	Yes
Peterson (1973)	Fairfax County, VA	Yes
Beaton (1991)	New Jersey Pinelands	Yes
Real Estate Research Corp (1978)	St. Louis County, MO	Yes
Gleeson (1978)	Brooklyn Park, MN	Mixed
Nelson (1986)	Salem, OR	Yes
Correll, Lillydahl, and Singell (1978)	Boulder, CO	Yes
Nelson (1988)	Washington County, OR	Yes
Knaap (1985)	Portland, OR	Yes
Knaap and Nelson (1992)	Portland, OR	No
Phillips and Goodstein (2000)	Portland, OR	No
Downs (2002)	Portland, OR	Mixed
Frech and Lafferty (1984)	California Coast	Yes
Dale-Johnson and Kim (1990)	California Coast	Yes
Richardson (1976)	Dover Township, NJ	Yes
Parsons (1992)	Chesapeake Bay, MD	Yes
Beaton and Pollock (1992)	Chesapeake Bay, MD	Yes
Luger and Temkin (2000)	New Jersey, North Carolina	Yes
Lowry and Ferguson (1992)	Sacramento, Orlando, Nashville	Mixed
Green (1999)	Suburban Wisconsin	Yes

Source: Nelson, Pendall, Dawkins, and Knaap, 2002

zoning can be significantly expanded beyond its relatively narrow base in a few states. As Porter (2004b) noted in his study of inclusionary zoning in the US, only three states—California, Massachusetts, and New Jersey—have made significant progress in promulgating inclusionary zoning. Other key growth management states, such as Oregon, Washington, Florida, and Maryland, have paid varying degrees of lip service to affordable housing, but none have succeeded in encouraging or requiring inclusionary zoning. Porter reported that, as of 2003, no jurisdiction in Florida had adopted an inclusionary zoning program (Porter 2004b, p. 19). Finally, in 2005, 20 years after passage of Florida's Growth Management Act, Tallahassee, the state's capital city, became the first jurisdiction to adopt inclusionary zoning.[2]

Why has there been such little progress in meeting Florida's legislative commitment in the 1985 GMA to encourage inclusionary zoning? By and large, the answer appears directly attributable to the fact that the state has played a very inactive role in moving local governments towards a more aggressive approach to stimulating the development of affordable housing, including the encouragement of inclusionary housing. This has been demonstrated both in the state's approach to mandating meaningful affordable housing strategies in the various comprehensive plans adopted under the GMA, as well as its enforcement of the affordable housing provisions of the 1972 Developments of Regional Impact law.

Affordable Housing Under the 1985 Growth Management Act

In the late 1980s and early 1990s, when all jurisdictions were preparing new comprehensive plans under the 1985 Act, the state played a very limited role in making certain that these plans seriously and carefully outlined goals and strategies that would lead to decent, affordable housing for all residents. A 1993 study of ten representative jurisdictions' housing elements, each of which had been approved by the 1985 Growth Management Act's implementing agency, the Florida Department of Community Affairs (DCA), showed that none of the ten housing elements featured bold strategies for addressing housing affordability problems (Connerly and Muller, 1993). Although the jurisdictions were responsive to DCA's minimum criteria for preparation of a housing element, only two of the housing elements contained goals, objectives, and policies that were linked to their analysis of the housing market and housing needs, and none of the jurisdictions were prepared to employ creative affordable housing strategies such as inclusionary zoning.[3] Additionally, local jurisdictions failed to consider the affordable housing provisions of the state's Development of Regional Impact law, as well as the state's community redevelopment law, which permits the employment of tax increment financing as a funding source for affordable housing in areas facing shortages of affordable housing (Local Planning Act. Fla. Stat. § 163.340(10), 1985).

---

2   Actually, Key West adopted inclusionary zoning in 1986, but it has been rendered inoperative by that jurisdiction's growth limits ("Florida Jurisdictions," 2006; "Focus: Key West," 1986).

3   Dade County's housing element proposed adoption of inclusionary zoning, but the County never adopted this proposal (J. Ross, personal communication, September 13, 2004).

Follow-up investigation by the Florida Affordable Housing Study Commission in 1999 indicated that Florida's housing elements still lacked clear goals, objectives, and policies that address affordable housing problems—this in spite of the fact that since the initial round of housing elements was completed, the state had adopted the 1992 State Housing Initiatives Partnership (SHIP) housing trust fund, which provides significant new revenue for affordable housing.[4] Noting that Florida growth management law does not require jurisdictions to set specific affordable housing targets in their housing elements, the Commission found that "many local governments adopt vague goals, objectives and policies that do not provide accountability" (Florida Affordable Housing Study Commission, 1999, p. 37). As a remedy, the Study Commission recommended to the Legislature that the housing element requirements under the state's 1985 growth law be amended to specify that (a) local housing elements must have numeric goals or objectives and (b) these numeric goals or objectives must be consistent with the data and analysis reported in the housing element (pp. 39, 57). As of 2006, the Florida Legislature has not acted to adopt this recommendation. Consequently, Florida communities are effectively required only to measure affordable housing need, not to propose goals, objectives or policies that specify how these needs will actually be met.

In an investigative report for the *Miami Herald*, Peter Whoriskey (1999) considered ten South Florida cities with housing elements that had been approved by DCA. He found that two of these cities, Pinecrest and Coral Gables, defined away their affordable housing problem by claiming that apartments renting for $1,500 or less per month were affordable to low income households. Other jurisdictions (Pembroke Pines and Coral Springs) had housing elements that discussed affordable housing, but each took action after the plan was in place to make housing less affordable. Pembroke Pines officials enacted a three-year moratorium on apartment construction and afterwards adopted minimum size, parking, and landscape regulations for multifamily units. Coral Springs officials cut in half the acreage available for "dense, multifamily projects." None of the ten jurisdictions could report that apartments affordable to households with incomes under $25,000 had been built as a result of the housing elements of their comprehensive plans. Jaimie Ross, affordable housing director with 1000 Friends of Florida, a growth management advocacy group, commented, "The sad truth is that cities can come up with these policies, ignore them and no one will call them on it. The Department of Community Affairs has either been not competent or not cared enough to review and enforce the plans (Whoriskey, 1999)." In response, DCA's Chief of Local Planning claimed, "Ours is not a regulatory role. If a city is not following its plan, it's up to local citizens to

---

4 Adopted by the Florida Legislature and signed into law in 1992 as the William Sadowski Act, SHIP distributed $126 million to local governments in 2004–2005 and is reputedly the largest state housing trust fund in the nation. SHIP is funded by a single dedicated revenue source: a portion of the state's documentary tax on the transfer of deeds. The amount of revenue was originally set at $.10 for every $100 of property value, with 50 percent for state housing programs and 50 percent for SHIP. In 1995, an amount equivalent to an additional $.10 per $100 was set aside, with 69 percent for SHIP and 31 percent for state housing programs. SHIP funds are distributed as a block grant to all 67 Florida counties and to cities of 50,000 or more. Funds are distributed on the basis of population.

challenge their actions in court." He did admit that permitting communities to count apartments renting for $1,500 per month as affordable "was an oversight" and the housing elements should therefore not have been approved.

In general, therefore, Whoriskey found that Florida's growth management law has failed to encourage these communities to actively encourage affordable housing. Pembroke Pines and Coral Springs, in particular, reflect an unwillingness to develop flexible land regulations that would permit a variety of housing types and densities, let alone require a minimum number of affordable units as with inclusionary zoning. The state's failure to require communities to zone land in urban areas for minimum densities and for a reasonable number of multifamily units has resulted in jurisdictions such as these having the ability to practice exclusionary zoning—certainly contrary to the Florida Growth Management Act.

It therefore appears that, without a more assertive mandate by the state's growth management legislation to incorporate affordable housing strategies such as inclusionary zoning, local jurisdictions will create and implement plans without making affordable housing available in a fashion that is commensurate with growing population demand. While the Florida Legislature has provided a growth management framework that can incorporate affordable housing and even encourage inclusionary zoning, without a mandate to assure affordable housing, local jurisdictions lack the incentive to adopt inclusionary zoning or other affordable housing strategies.

Affordable Housing under the 1972 Developments of Regional Impact Law

The State has also failed to push hard on another key growth management tool that has an implicit inclusionary zoning provision. Enacted in 1972 as one of Florida's first growth management acts, the Developments of Regional Impact (DRI) law requires regional planning agencies to review the impacts of "any development which, because of its character, magnitude or location, would have a substantial effect upon the health, safety or welfare of citizens of more than one county" (Environmental Land and Water Management Act, Fla. Stat. § 380.06(1), 1972). In its review, the regional planning agency must consider whether "the development will favorably or adversely affect the ability of people to find adequate housing reasonably accessible to their places of employment .... Adequate housing means housing that is available for occupancy and that is not substandard" (Environmental Land and Water Management Act, Fla. Stat. § 380.06(12)(a) 3, 1972). Although a local government is permitted to issue development orders under the DRI statute, both the regional planning agency and the State may appeal its decisions to the Florida Land and Water Adjudicatory Commission, which consists of the Governor and his Cabinet (Environmental Land and Water Management Act, Fla. Stat. § 380.06(25)(f), (h), 1972).

The DRI statute thereby provides a potentially powerful tool that allows both the regional planning agency and the State to assure that new large-scale developments that generate jobs also provide the housing that the employees filling those jobs require. In this sense, the statute is very similar to the concept of housing linkage programs pioneered by such cities as San Francisco and Boston, where laws were adopted in the 1980s that required office developers to provide affordable housing that was 'linked' to the demand for such housing generated by the new jobs (Boston Redevelopment Authority, 2000). But the DRI statute is also consistent with inclusionary zoning in that it requires that affordable housing be located reasonably accessible to job-generating, mixed-use development. Under the DRI statute, a large scale mixed-use development located in the outer fringes of a metropolitan area could be required to provide affordable housing nearby or within the development. Finally, because the DRI statute is enforceable at the regional and state levels, it can be used to make certain that (a) housing solutions commensurate with regional housing need can be generated and (b) developers are able to run from one jurisdiction to another, but not hide from the responsibility to assure affordable housing accompanies the jobs they create.

In an apparent attempt to assure that the DRI law generated affordable housing in large scale developments, the State concentrated its attention in the mid-1990s on developing a rule for determining the actual unmet housing need associated with such projects. Under the Adequate Housing Uniform Standard Rule (Fla. Admin. Code Ann. ch. 9J-2.048, 1994), which implements the affordable housing provisions of the DRI statute, the local government must make a determination as to whether the increase in housing demand generated by the new jobs in a DRI development can be met by existing, affordable dwelling units. To the extent that this increased demand cannot be met, the development must provide mitigation in the form of affordable housing, contributions to an affordable housing trust fund, or rental or homeownership subsidies to the development's lower income employees. Extra credit is given for affordable housing developed on the site of the new development or accessible to the new development via mass transit.

Despite implementation of the DRI Affordable Housing Rule throughout the state, very little affordable housing has resulted. In North Florida, where the St. Joe Company, the state's largest private land owner, is actively developing significant portions of its vast timber tracts and coastal areas, the DRI rule has resulted in relatively few affordable dwelling units in that company's developments. For example, SouthWood, a new urbanist development in Tallahassee, is projected to include 4,700 dwelling units on 3,200 acres, as well as a commercial town center, golf course, and recreational areas ("Arvida," 2000). Numerous, primarily state jobs are located at SouthWood in the Capital Circle Office Center, which houses five State of Florida departments in 15 buildings and employs between 3,600 and 3,900 professional, clerical, and maintenance workers.

Located four miles southeast of downtown Tallahassee's Florida Capital Complex, the Capital Circle Office Center was developed on land formerly owned by St. Joe in a largely undeveloped area. With SouthWood home prices well above what many state workers can afford and no existing housing nearby, employees at the Office Center must commute significant distances to reach their place of employment. On the surface, therefore, SouthWood appeared to be a primary candidate for significant affordable housing mitigation to meet the terms of the state's DRI Affordable Housing Rule. Instead, only 48 new homes were considered necessary, falling below the DRI Housing Rule's 50-unit threshold. St. Joe agreed to pay $550,000 to a City of Tallahassee Housing Trust Fund over a 20-year period of time. Similar scenarios have played out in St. Joe's other large-scale developments in North Florida as well as other DRIs in Florida where reportedly only one DRI has resulted in contributions by the developer for affordable housing (B. Piaweh, personal communication, July 22, 2005; Ziewitz and Wiaz 2004, pp. 177–181).

Working against the incorporation of affordable housing strategies, of course, is the fact that in many neighborhoods and jurisdictions in Florida residents do not see affordable housing contributing to the quality of life they seek in their communities; they display NIMBY (Not in My Backyard) attitudes toward affordable housing (Florida Housing Coalition, 2000; Ross, 2001). In response, neighbors as well as local governments identify environmental or traffic problems alleged to be induced by affordable housing as an excuse for denying permission to develop (B. Piawah, personal communication, July 22, 2005). Florida's requirement that infrastructure, such as roads, to support new development be "concurrent" with that development has been used to deny affordable housing, and thus far the Legislature has not exempted affordable housing from infrastructure concurrency requirements.

Is the Climate Changing? The Tallahassee Experience

Given the fact that the State of Florida has not pushed for effective affordable housing strategies at the local level, it should not be surprising that local governments in Florida have not embraced the clearest, most straightforward approach to assuring that an adequate supply of affordable housing is available to meet growing population needs: inclusionary zoning. This is in contrast to states such as California, New Jersey, Massachusetts, Rhode Island, and Connecticut, where a total of 525 jurisdictions have adopted inclusionary zoning. In one fashion or another, each of these states mandates that local jurisdictions have an obligation to provide affordable housing (Porter, 2004a).

Only one jurisdiction in Florida, Tallahassee, has adopted an inclusionary zoning ordinance. One of the most liberal cities in the state,[5] Florida's capital city adopted inclusionary zoning as an ordinance in April 2005, effective October 1, 2005 ("City: Build It," 2005). Rapidly rising housing prices, however, have propelled a number of Florida jurisdictions to look closely at inclusionary zoning. In turn, the Florida Home Builders Association, alert to government regulation of construction, filed suit in February 2006, challenging the ordinance. It is believed this action was taken, in part, to stem the possible tide of other jurisdictions adopting inclusionary housing ordinances. The question becomes, therefore, whether the tide against inclusionary housing in Florida is changing, and if so, what factors, including shifts in growth management policy, appear to account for this change. This question will be addressed by examining the development of Tallahassee's ordinance, as well as the potential for adoption by other communities in the state.

Tallahassee's inclusionary zoning ordinance was over fifteen years in the making. When the city's planning agency began the comprehensive planning process under the state's 1985 Growth Management Act, planners encouraged citizens to participate

---

5   In the 2004 Presidential election, 61.7 percent of voters in Leon County (where Tallahassee is located) cast a ballot for John Kerry. Statewide, George Bush won 52.1 percent of the popular vote. Retrieved 16 July 2005 from http://www.usatoday.com/news/politicselections/vote2004/countymap.htm.

in drafting suggestions for each of the state-mandated plan elements, including a housing element. A number of local affordable housing activists, as well as representatives of the local building community, participated in a series of citizen-led and staff-supported workshops that provided input into the local planning agency's deliberations. Among the citizen recommendations that found their way into the city's housing element, adopted July 1990, was Policy 1.2.4, which required that seven percent of dwelling units in new developments of 50 units or more be set aside for lower income households. Alternatively, developers were permitted to pay $5,894 *in lieu* of each affordable dwelling unit they were required to build (Tallahassee-Leon County Planning Department, 1990). In return for developing affordable units under this provision, developers would receive a density bonus of up to ten percent.

Adoption of this policy, however, suffered from two key shortcomings. First, the inclusionary zoning policy only applied to the City of Tallahassee. Tallahassee and Leon County share one planning department and one local planning agency and have for the most part developed a joint comprehensive plan.[6] In 1990, however, notable differences were contained in the housing elements of the city and the county. As of 2006 the only part of the comprehensive plan, as amended, that has never been jointly adopted by both the City and the County Commissions is the housing element.

The second shortcoming was that no land development regulations were adopted by the City of Tallahassee to implement the inclusionary zoning policy. This became an issue in 1996 when Tallahassee was preparing its Evaluation and Appraisal Report (EAR). Florida's GMA requires all jurisdictions in the state to prepare an EAR that assesses how well the community is meeting the goals outlined in the comprehensive plan, what new issues, conditions, or state, regional, or local policies have emerged in the community that require attention, and what changes are needed to make certain that the community is meeting both original goals as well as new challenges (Local Planning Act, Fla. Stat. § 163.3191, 1985). The 1996 Tallahassee-Leon County EAR called special attention both to the inadequate supply of affordable housing in the Tallahassee-Leon County area and to the over-concentration of affordable housing in a few areas, chiefly the community's Southside, thereby resulting in "...segregating our population by income, leading to inequities in schools and the isolation of different socio-economic groups from one another which can lead to an 'us versus them' mentality" (Tallahassee-Leon County Planning Department, 1996). The EAR's recommendations were influenced in part by the consultancy of David Rusk, former mayor of Albuquerque, New Mexico, who has published widely on the impact of economic segregation on urban vitality (Rusk, 2001).

The EAR and its recommendations on economic segregation were adopted by both the Tallahassee City Commission as well as the Leon County Commission in September 1997. This was followed up by both commissions adopting an amendment in July 1998 to the 2010 Comprehensive Plan to revise Housing Element Policy 1.2.4 on inclusionary zoning so that it better achieved dispersal of affordable housing

---

6   Tallahassee is the only incorporated jurisdiction in Leon County, making cooperation between municipal and county government more likely than where there are multiple municipal jurisdictions. Nevertheless, city and county often disagree and historically efforts to unify the two governments have failed in voter referenda ("In Athens, Ga.," 2003).

throughout the community. Planning staff analysis criticized existing inclusionary zoning policy because most developers had used the option of paying a fee *in lieu* rather than building the required housing, resulting in relatively little affordable housing in new developments. Moreover, only about $400,000 was collected in fees. The staff proposed either having no *in lieu* fee or adopting a graduated fee that provided more incentive to actually construct affordable housing. Further, planning staff recommended that the inclusionary zoning policy only apply to developments constructed in census tracts with median family income higher than the Leon County median income. Finally, the staff recommended that the revised inclusionary zoning policy be backed by a public ordinance that detailed how the policy would be implemented (Tallahassee-Leon County Planning Department, 1998). The version of Policy 1.2.4 adopted by both City and County Commissions generally followed the planning staff's recommendations, including an *in lieu* fee that varied positively in amount with the difference between the average price of housing in the development and the housing price affordable to a low income household.

Despite the relatively bold and unusual stance against economic segregation made by both City and County Commissions, progress from an adopted housing element to a joint city-county inclusionary zoning ordinance did not come quickly. It took over two years to develop a draft ordinance that could be taken to the local planning agency for review of consistency with the comprehensive plan (Tallahassee-Leon County Planning Department, 2003a). Finally, in late 2000, the draft ordinance was taken to the Leon County Board of Commissioners for consideration, after the city of Tallahassee had decided to let the ordinance be considered by Leon County first (Biblo, 2005). After voting in November 2000 that the ordinance should apply to developments of 25 or more units, of which five percent should be set aside for affordable housing, the County Commission scheduled a public hearing for February 21, 2001. At this point, however, the likelihood that the County Commission would adopt an inclusionary housing ordinance began to diminish. After hearing testimony, the Commission voted to continue the hearing and asked that staff provide additional information in response to a number of issues that had been raised. Subsequent public hearings took place on March 20 and April 10, at which point the Commissioners tabled consideration of the ordinance until staff could confer with "the Chamber of Commerce and other interest groups and develop appropriate recommendations" (Leon County Board of Commissioners, 2001). By May, the County Commissioners had appointed a committee of citizens that was charged with recommending revisions to the draft inclusionary zoning ordinance (Tallahassee-Leon County Planning Department, 2003a).

The committee, which consisted of both affordable housing advocates and representatives of the development community, met six times through October 2001, but did not present its recommendations to the County Commission until May 14, 2002. The committee's two factions could not agree upon a recommendation and instead presented two alternative proposals, one of which, supported by housing advocate Jaimie Ross, would apply to all developments of 25 units or more while the other, favored by Ted Thomas, a local businessman, only applied to developments of 100 dwelling units or more. The Commissioners did not adopt either recommendation, but instead instructed Ms. Ross and Mr. Thomas to work with staff in developing a new ordinance. Once again, the County Commissioners scheduled public hearings, for June and July 2003. At both of these meetings, the opponents, typically from the development community, outnumbered the proponents, and at the second public hearing staff noted Chamber of Commerce opposition to the ordinance. The County Commissioners voted unanimously to send the ordinance back to staff "to come up with something acceptable, realistic, and incentive-based" (Leon County Board of Commissioners, 2003). This never happened, however, as on May 11, 2004, the Leon County Commissioners in a joint meeting with the Tallahassee City Commissioners voted to delete the housing element policy that required Leon County to implement a inclusionary zoning ordinance and substituted for that a policy that requires the county to establish incentives for *voluntary* provision of affordable housing in new developments (Tallahassee-Leon County Planning Department, 2006).

Previously, at a December 2003 joint city-county workshop on the comprehensive plan, the County Commission had announced its decision to proceed with a voluntary ordinance. At the same meeting, the Tallahassee City Commission signaled its intention to go in a different direction, instructing staff to prepare a mandatory inclusionary housing ordinance for consideration (Tallahassee-Leon County Planning Department, 2003b). Once presented a draft ordinance, the City Commissioners moved relatively fast to debate and adopt it. In part, this urgency was attributable to the 1998 amendment to the comprehensive plan's housing element. Although the pre-1998 comprehensive plan housing element language on inclusionary zoning was rather brief, it was considered by the City Attorney as direct and implementable. In contrast, the 1998 amendment's language was considered general and unenforceable without a corresponding amendment to the city's land development regulations (City of Tallahassee, 2004). Consequently, city staff had ceased enforcing the existing inclusionary zoning policy and was no longer collecting *in lieu* fees as had been the case between 1990 and 1998. The City Manager and the senior member of the City Commission were each disturbed to learn of this "staff self directed action" and therefore saw adoption of an ordinance as key to making certain that the comprehensive plan policy on inclusionary zoning was actually enforced. Both of these city officials subsequently played a leadership role in pushing the ordinance to adoption (A. A. Biblo, personal communication, June 9, 2006; A. Favors, personal communication, December 30, 2003).

Despite firm leadership, the City of Tallahassee did not conduct public hearings on a revised inclusionary zoning ordinance until October and December 2004. At those hearings, it became clear that the city's real estate development community was strongly opposed to the city's adoption of inclusionary zoning, just as they had been with the county. The Greater Tallahassee Chamber of Commerce voiced its opposition in part by contracting with a private consultant who presented a report which claimed that inclusionary zoning would place a burden on developers by reducing their profit margins, increasing the cost of non-inclusionary zoning units, and depreciating the property values of dwellings located near the inclusionary units (Ennis, 2004; Fishkind and Associates, 2004). Representatives of the Tallahassee Chamber of Commerce, the Florida Home Builders Association, the Tallahassee Board of Realtors, the Tallahassee Builders Association, and the Northeast (Tallahassee) Business Owners Association expressed opposition that mirrored the conclusions of the consultant's report. Fourteen of 23 speakers at the December 15, 2004, public hearing represented these groups and each spoke against adoption of the ordinance. At the same hearing, only five individuals spoke directly in favor of the ordinance.

Despite widespread opposition from the development community, on April 13, 2005, the Tallahassee City Commission voted unanimously in favor of adopting an inclusionary zoning ordinance ("City: Build It," 2005; City of Tallahassee, 2005). Perhaps because of the opposition, however, the ordinance was modest in its requirements and generous in its benefits in comparison with inclusionary ordinances adopted in other US communities. The population targeted by the ordinance consists of households earning between 70 and 100 percent of area median income. According to the planning's staff own research, inclusionary zoning ordinances in other jurisdictions typically target households earning less than 80 percent of area median income, the US Department of Housing and Urban Development's low income cap, whereas Tallahassee's ordinance clearly applies to middle income households as well.[7] In Tallahassee, only new developments of 50 dwelling units or more are required to set aside units for affordable housing and then only in census tracts where median income is higher than the countywide median family income. Consequently, smaller developments are exempt from the ordinance. In the 13 jurisdictions researched by the Tallahassee Planning Department, only two had thresholds this high, and a study of inclusionary zoning ordinances in 75 California jurisdictions found only three with thresholds that were 50 units or higher (Calavita and Grimes, 1998). Tallahassee's ordinance also applies to all Developments of Regional Impact (DRIs), Critical Planning Areas, and Target Planning Areas, which tend to be larger scale developments.

In Tallahassee, a minimum of 10 percent of dwelling units must be set aside for affordable housing. This percentage is at the low end of the range that runs from 10 to 25 percent for other jurisdictions in the city's comparison sample but there are many jurisdictions in California that also require 10 percent (Calavita and Grimes, 1998). Like many jurisdictions, Tallahassee permits developers to pay a fee *in lieu* of building the required affordable housing. In Tallahassee, the fee rises from $10,000 per dwelling unit for projects with a median price of $176,000 to $25,000 per unit for projects with a median price of $360,000. *In lieu* fees vary significantly in other jurisdictions in the city's comparison sample, ranging from the extreme of permitting no buyouts (Fairfax County, Virginia) or permitting no buyouts in projects larger than a minimum-sized development such as four (Boulder, Colorado) or eight units (Davidson, North Carolina), to requiring developers to pay fees that vary from 100 percent (Longmont, Colorado) to four percent (Annapolis, Maryland) of the cost of an affordable housing unit. With the cost of an affordable unit under the ordinance set at nearly $160,000, *in lieu* fees in Tallahassee run between 6.2 and 15.6 percent

---

7   Jurisdictions included in the Tallahassee-Leon County Planning Department's analysis were: Montgomery County, Maryland; Annapolis, Maryland; Fairfax County, Virginia; Loudon County, Virginia; Davidson, North Carolina; Chapel Hill, North Carolina; Highland Park, Illinois; Boulder, Colorado; Denver, Colorado; Longmont, Colorado; Santa Fe, New Mexico; San Diego, California; Davis, California (Tallahassee-Leon County Planning Department, n.d.). These jurisdictions were selected on the basis of several key characteristics: location in the South, adoption of an ordinance in the recent past, and similarity of key features to what was being proposed in Tallahassee. They were also selected in part to show that Tallahassee's ordinance was relatively mild (A. A. Biblo, personal communication, October 4, 2006).

of the cost of an affordable dwelling unit, thereby placing the city's ordinance at the less stringent end of the spectrum in this regard.

With regard to long-term affordability, in Tallahassee if the original purchaser sells the unit within ten years, it must be made available at the city's designated affordable housing price to an eligible purchaser. Other jurisdictions have varying requirements for preservation of long-term affordability, from none to perpetuity—with ten years located at the lower end of the range. Of the 13 jurisdictions with which Tallahassee compared itself, eight had longer periods of affordability.

Finally, Tallahassee's inclusionary zoning ordinance provides significant incentives to developers: a 25 percent density bonus, alleviation of setback and buffering requirements, expedited development review, and an exemption from transportation concurrency requirements. In addition, developers are permitted to suggest other incentives that they may feel are needed for their development (City of Tallahassee, 2005). Compared to inclusionary zoning jurisdictions in California at least, Tallahassee is above average in providing incentives to developers for inclusionary dwelling units (Porter, 2004b).

Despite its modest requirements and generous incentives, Tallahassee's inclusionary zoning ordinance was the first such ordinance adopted south of North Carolina and one of relatively few adopted outside the big four states of California, Connecticut, New Jersey, and Massachusetts. Its importance, therefore, does not lie in its boldness, but in the fact that it was adopted at all. Tallahassee's inclusionary zoning ordinance faced significant opposition from the local development community. Why, especially in the face of Leon County's unwillingness to pass an ordinance opposed by the same development community, did these Commissioners adopt an inclusionary zoning ordinance?

Certainly, adoption was not influenced by the State of Florida. Despite the legislative language that encourages inclusionary zoning, the DCA never provided technical assistance or encouraged the city of Tallahassee to adopt an inclusionary zoning ordinance. Moreover, DCA did not object when Leon County elected to remove its inclusionary zoning policy from its comprehensive plan (W. Tedder, personal communication, July 8, 2005).[8] This stands in contrast to such states as California, Massachusetts, and New Jersey, which have taken a much more proactive stance in encouraging local jurisdictions to adopt inclusionary zoning (Porter, 2004b).

If state growth management policy did not encourage Tallahassee's adoption of inclusionary zoning, it certainly seems that rapid rises in housing prices did make a difference. Home prices in Tallahassee have been traditionally lower than the national average, but beginning in the late 1990s they began to rise rapidly, with the mean sales price for homes doubling between 1998 and 2005 to a peak of $251,000 ("Real Estate Cools Off," 2006). With home prices and demand for homes peaking in 2005, it is not surprising that interest in providing some form of relief to middle to lower-middle income homebuyers became more salient to the City Commissioners. Several months before he voted to adopt the inclusionary zoning ordinance, Tallahassee City

---

8   DCA does have a link to the Tallahassee ordinance on its web site at http://www.dca. state.fl.us/fdcp/dcp/affordablehousing/index.cfm although the link is incorrectly identified as Leon County's Inclusionary Housing Ordinance.

Commissioner Andrew Gillum stated, "Affordable housing is approaching a crisis stage in our community" ("A Developing Crisis," 2005). The ordinance's preamble cites the rapid rise in housing prices and expresses special concern for the lack of affordable housing in the city's more affluent northeastern neighborhoods (City of Tallahassee, 2005).

Rapidly rising home prices do not fully explain, however, why Tallahassee City's Commissioners elected to adopt inclusionary zoning, especially given that Leon County's Commissioners, facing the same "crisis" in affordable housing, voted not to. What distinguishes Tallahassee is that some leading developers were willing to come forward, at least in private, to negotiate with the city over an ordinance. In particular, the two principals of one of the city's larger residential developers were identified as "the tipping point" for their willingness to meet privately with planning staff to hammer out details that would meet their needs as developers of residential subdivisions (A. A. Biblo, personal communications, November 8, 2004, June 9, 2006). According to one of the city commissioners, "we had people [developers] working with us the whole time."[9] At least two other locally well-known development stakeholders—one a realtor and the other a builder—also lent their quiet support to the inclusionary ordinance. Consequently, even though in public meetings the development community appeared united in opposition, in private an influential minority had decided to negotiate rather than fight the ordinance. Given that the ordinance contains some significant development incentives, it appears that these members of the development community were motivated by what they could get in return for their willingness to negotiate. But they were also motivated by the fact that home prices had risen by so much in Tallahassee that the ability of their children, for example, to live in a desirable neighborhood was threatened and the ordinance therefore seemed like an effective means to address this problem.

The effectiveness of development community opposition to the ordinance was also negated by efforts to expand Tallahassee's Urban Services Area (USA) boundary at the same time that the inclusionary zoning ordinance was being considered. In the summer of 2004, the Residential Land Availability and Affordability Commission, representing some 35 realtors, developers, and environmentalists, urged City and County Commissioners to address the rising cost of housing by making more land available for development. The Commission's report begged the question of whether its members' concern for affordable housing would lead them also to support the inclusionary zoning ordinance, ("City, County Reject Proposal," 2004; "Group Wants More Lots," 2004).

The answer came in October 2004, when the Availability and Affordability Commission took an official stance against the ordinance, contending that inclusionary zoning would be opposed by neighborhood associations and would place an unfair burden on the development community (N. Fleckenstein, personal communication, October 13, 2004).[10] This opposition backfired, however, as it belied

---

9   The Tallahassee-Leon County planner responsible for the inclusionary zoning ordinance is not aware of these two developers approaching Leon County government with similar ideas (A. A. Biblo, personal communication, November 7, 2006).

10   Two months later, however, the Commission's proposals for expanding the Urban Services Area were rejected by the City and County Commissions in a joint meeting. The

the Commission's motive for seeking to increase the supply of developable land and discredited realtor opposition to the ordinance in the eyes of the City Commission (W. Tedder, personal communication, July 8, 2005). Two of the three city commissioners interviewed for this study cited this incident as indicating that the opposition was not sincere about supporting affordable housing.

This sense was reinforced at the December 2004 public hearing, when a City Commissioner asked the Chamber of Commerce president, "Are there any circumstances under which the Chamber of Commerce could support inclusionary housing?" When she could not answer one way or another, it was apparent that she, as well as others, did not wish to negotiate the ordinance. The fact that other members of the development community were less visibly willing to negotiate encouraged City Commissioners to discount what opponents in the development community had to say about inclusionary housing (W. Tedder, personal communication, July 8, 2005). Moreover, the City Commissioners respected information that planning staff collected on how inclusionary zoning succeeded in other communities without destroying the development sector. The availability of solid information from the planning staff was taken more seriously by the City Commissioners than the rhetoric emerging from much of the development community. According to one of the Commissioners interviewed for this study, the opposition could not specifically demonstrate how they would lose money under the inclusionary zoning ordinance.

Finally, important to adoption of Tallahassee's inclusionary zoning ordinance was the work of Tallahassee-Leon County planning staff. One planner, Adam Antony Biblo, devoted considerable time over eight years attempting to craft an ordinance that would be adopted by local government. His supervisor, Wayne Tedder, who became Planning Director in January 2004, was also instrumental in making certain that the planning staff pushed hard for the ordinance. While initially unfamiliar with the inclusionary zoning concept, Tedder was also affected by the fact that he had once lived in a mobile home and therefore resented developers saying they didn't want certain types of people in their community (A. A. Biblo, personal communication, June 9, 2006; W. Tedder, personal communication, July 8, 2005).

Since adoption of the ordinance, the city and its Commissioners have continued to work hard at selling the importance of inclusionary housing. The city has cooperated very closely with a new urbanist developer, who voluntarily agreed to meet the inclusionary zoning requirements prior to the law's enactment in October 2005. This resulted in a July 2006 press conference at the development that was used to sell the idea that cooperative developers could make the new ordinance work to their benefit ("Florida's Capital," 2006).

The City of Tallahassee also prepared a television show for the city's cable channel, in which a Commissioner, the City Attorney, and the Planning Director explained the program to the public. On this show, as well as elsewhere, the city emphasized the importance of giving the occupants of inclusionary zoning units access to new schools and public facilities. In retrospect, this was a very useful rhetorical device. In

---

County Commission, reflecting its greater sympathy to development concerns, split three-three (with one absence) on the proposals, while the City Commissioners voted unanimously (with two absences) against them.

early discussions of the legislation before the Leon County Commission, at least one housing advocate had justified inclusionary zoning on the basis that it would provide improved access to the better schools in the community. One County Commissioner, perhaps disingenuously, responded with incredulity that there was variation in the quality of schools. This had the effect of making support for inclusionary zoning a direct challenge to the quality of all schools in the district—a stance that was bound to invite criticism and erode support. By focusing on access to new schools and public facilities, the City Commissioners effectively found a way to talk about inclusionary zoning's promise for giving residents better access to quality services without implying that services in some areas were inferior to those in other areas.

Since enactment of the inclusionary zoning ordinance in October 2005, two key events have taken place that may have significant impact on the future of inclusionary zoning in Florida. First, in February 2006, the Florida Home Builders, along with the Tallahassee Builders Association, the Tallahassee Board of Realtors, and several related parties, filed suit in state court against the Tallahassee ordinance. The lawsuit claims that the ordinance violates the US Constitution because it does not bear a reasonable relationship to public health, morals, safety, or general welfare, it represents a taking of private property, and it violates the Florida Constitution by serving as an unauthorized tax on real property (Fla. Home Builders Ass'n *et al.* vs. City of Tallahassee, 2006). The suit appears to be a serious attempt by the Florida Home Builders not only to kill the Tallahassee ordinance, but also to prevent proliferation of inclusionary zoning to other jurisdictions.

The other key event is Palm Beach County's March 2006 adoption of an interim inclusionary ordinance that requires seven percent of all new projects larger than 10 units to be affordable ("Palm Beach County," 2006). Palm Beach is the third largest county in the state. Six other jurisdictions in the state, including Sarasota, Miami-Dade County, and Pinellas County, also have inclusionary zoning ordinances under consideration ("Florida Jurisdictions," 2006).

Clearly, there is significant interest in the state in following Tallahassee's lead, with the legal challenge posing an uncertain threat to the expansion of inclusionary zoning in Florida. In late October 2006, a Leon County circuit court judge dismissed the case without prejudice, stating that the developers failed to demonstrate that the inclusionary zoning ordinance had caused any actual harm ("Builders Consider Refiling," 2006; Order Granting Motion to Dismiss, 2006). In late November 2006, however, plaintiffs filed an amended complaint. Consequently, the legal uncertainty surrounding inclusionary zoning in Florida has not ended.

Conclusions and Policy Implications

Although inclusionary zoning is not the only strategy that should be strongly considered by Florida communities for providing affordable housing for all its population, it certainly is a very appropriate strategy in a state where so many places are experiencing rapid growth. Moreover, the GMA specifically names inclusionary zoning as a regulation that local governments should be employing in meeting their affordable housing goals. Finally, because the GMA requires local communities to adopt housing policies that avoid the geographic over-concentration of low income households, it seems that inclusionary housing is a necessary policy. In order to give all population groups full access to these newly developing areas and the services they provide, an inclusionary zoning policy seems imperative.

Given all this, the fact that the state's growth management apparatus, led by the state's governors and the Florida Department of Community Affairs, has been indifferent to the adoption of inclusionary zoning strongly suggests that Florida's implementation of growth management has failed to take affordable housing seriously. This is seen in the state's record for advancing affordable housing under the two key pieces of relevant land use legislation: the 1985 Growth Management Act and the 1972 Developments of Regional Impact legislation. Under these laws, the state of Florida has failed to push local governments very hard to adopt creative or innovative approaches to developing affordable housing and has been satisfied with generalized approaches to affordable housing that lack any real teeth. While the state has been among the nation's leaders with regard to developing an important housing trust fund, the State Housing Initiatives Partnership, these funds have not been spent to motivate local governments to aggressively address affordable housing problems for all present and future residents.

The Tallahassee case demonstrates that under the right circumstances, inclusionary zoning can be adopted in Florida in spite of indifference by the state and opposition from the real estate industry. While it is difficult to generalize from a single case, Tallahassee's experience suggests that in the context of rapidly rising housing prices, which have characterized the entire state as well as Tallahassee, it is possible to portray affordable housing as a problem that affects a relatively broad segment of the population. Tallahassee was not able to adopt an ordinance until after housing prices had greatly accelerated in the first five years of the century. Furthermore, the rapidly rising housing prices in the remainder of the state have created an atmosphere in which inclusionary zoning is now being considered by a number of significant jurisdictions ("Florida Jurisdictions," 2006).

Despite rapidly rising housing prices, it can also be expected that Florida home builders and realtors will object strenuously in every jurisdiction attempting inclusionary zoning. Nevertheless, the Tallahassee experience demonstrates that while the real estate community is influential, it is not always united. Particularly if the jurisdiction designs an ordinance that contains specific benefits to developers, as has Tallahassee, it is less likely that developers will be able to demonstrate that they will lose money. By enabling at least some developers to benefit from inclusionary zoning, jurisdictions will be able to divide the real estate community, thereby weakening opposition and making the opponents appear unreasonable and uncooperative. No doubt a sympathetic elected body is also necessary. Certainly, the Tallahassee City Commission was more eager to embrace inclusionary zoning than the Leon County Commission. But even a sympathetic commission must have support from the community, as well as the means to negate or discount what is likely to be vociferous opposition from the development community. The Tallahassee case demonstrates that this can be done.

Note

This is a substantially revised version of a chapter that appeared in Gerrit Knaap, Hubert Haccoû, Kelly Clifton and John Frece (eds) (2007), *Incentives, Regulations and Plans: The Role of States and Nation-States in Smart Growth Planning*, Northampton, MA: Edward Elgar Publishing.

References

A developing crisis: Commission set to address the soaring cost of housing. (2005, February 13). *Tallahassee Democrat*.

Anthony, J. (2003). The effects of Florida's Growth Management Act on housing affordability. *Journal of the American Planning Association, 69*(3), 282–295.

Arvida, a St. Joe Company, breaks ground on SouthWood. (2000, September 19). Retrieved July 14, 2005, from http://www.arvida.com/southwood/news.asp?id=13

Biblo, Adam Antony, 2005, "Adoption of Tallahassee's Inclusionary Housing Ordinance", Florida State University Department of Urban and Regional Planning Professional Topics Speaker Series, November 1.

Boston Redevelopment Authority. (2000, May). *Survey of linkage programs in other US cities with comparisons to Boston.* Retrieved July 13, 2005, from http://www.cityofboston.gov/bra/PDF%5CPublications%5C/pdr_534.pdf)

Builders consider refiling lawsuit against the city. (2006, November 20). *Tallahassee Democrat*.

Burchell, R., and Galley, C. (2000). Inclusionary zoning: Pros and cons. *New Century Housing, 1*(October), 3–12.

Calavita, N., and Grimes, K. (1998). Inclusionary housing in California. *Journal of the American Planning Association, 64*, 150–169.

Carlson, D., and Mathur, S. (2004). Does growth management aid or thwart the provision of affordable housing? In A. Downs (Ed.), *Growth management and affordable housing: Do they conflict?* (pp. 20–66). Washington, DC: Brookings Institution Press.

City: Build it, or we'll make you. (2005, April 15). *Tallahassee Democrat*.

City, county reject proposal to build on St. Joe Co. land. (2004, December 10). *Tallahassee Democrat*.

City of Tallahassee. (2004, May 12). *City commission agenda item—inclusionary housing strategies, Attachments 1 and 2.*

City of Tallahassee. (2005, April 13). Inclusionary Housing Ordinance, Ordinance no. 04-O-90AA. Retrieved August 8, 2006, from http://www.talgov.com/planning/pdf/af_inch/104o90aa.pdf.

Connerly, C. E., and Muller, N. A. (1993). Evaluating housing elements in growth management comprehensive plans. In J. M. Stein (Ed.), *Growth management: The planning challenge of the 1990s* (pp. 185–199). Newbury Park, CA: Sage Publications.

Connerly, C. E., and Smith, M. (1996). Developing a fair share housing policy for Florida. *Journal of Land Use and Environmental Law, 12*(1), 63–102.

Ennis, Erin (2004) Testimony before Tallahassee City Commission Public Hearing, December 15.

Environmental Land and Water Management Act, Fla. Stat. §§ 380.012–10 (1972).

Fishkind and Associates (2004, December 15). *Economic Analysis of the Proposed Inclusionary Housing Ordinance on the City of Tallahassee.*

Fla. Admin. Code Ann. ch. 9J–2.048 (1994).

Florida Affordable Housing Study Commission. (1999). *The Affordable Housing Study Commission final report 1999.* Tallahassee, FL: Author.

Fla. Home Builders Ass'n *et al.* vs. City of Tallahassee, No. 06–CA–579 (2d Cir. 2006).

Florida Housing Coalition. (2000). *Creating inclusive communities.* Tallahassee, FL: Author.

Florida jurisdictions watch challenge to Tallahassee law. (2006, April 23). *Tampa Tribune.* Retrieved July 24, 2006, from www.knowledgeplex.org.

Florida's capital first to implement workforce housing program. (2006, July 14). *RISMedia.* Retrieved September 20, 2006, from http://www.rismedia.com/index. php/article/articleview/15205/1/1/

Focus: Key West, tropical backwater, heats up. (1986, December 28). *New York Times.* Retrieved October 14, 2006, from http://query.nytimes.com/gst/fullpage. html?res=9A0DE5D61E3AF93BA15751C1A960948260&sec=&pagewanted=p rint.

Group wants more lots for housing. (2004, June 29). *Tallahassee Democrat.*

In Athens, Ga., the drive for 'one community' came in a revolutionary fashion. (2003, October 12). *Gainesville Sun.*

Lawrence, B. (2001, Spring). New Jersey. *NIMBY Report*, 2–4.

Leon County Board of Commissioners. (2001, April 19). *Minutes.* Retrieved from http://www.clerk.leon.fl.us/index.php?section=2&server=cvweb&page=finance/ board_minutes/minutes/2001.html#4

Leon County Board of Commissioners. (2003, July 8). *Minutes.* Retrieved from http://www.clerk.leon.fl.us/index.php?section=2&server=cvweb&page=finance/ board_minutes/minutes/2003.html#7

Local Government Comprehensive Planning and Land Development Regulation Act, Fla. Stat. §§ 163.3161–3242 (1985).

Montgomery County, Maryland. (n.d.) *The moderately priced dwelling unit program Montgomery County, Maryland's inclusionary zoning ordinance: Program summary and background.* Retrieved August, 28, 2004, from http://www. montgomerycountymd.gov/mcgtmpl.asp?url=/Content/DHCA/housing/housing_ P/mpdu/summary.asp.

Nelson, A. C., Pendall, R., Dawkins, C. J., and Knaap, G. J. (2002), The link between growth management and housing affordability: The academic evidence. Retrieved July 24, 2004, from http://www.brookings.edu/es/urban/publications/ growthmang.pdf.

Order Granting Motion to Dismiss Plaintiff's Amended Complaint, Fla. Home Builders Ass'n *et al.* vs. City of Tallahassee, No. 06–CA–579 (2d Cir. Oct. 27, 2006).

Palm Beach County sets workforce housing rules. (2006, March 31). *South Florida Business Journal*. Retrieved September 20, 2006, from http://www.bizjournals. com/southflorida/stories/2006/04/03/story8.html?from_rss=1

Pendall, R. (2000). Local land use regulation and the chain of exclusion. *Journal of the American Planning Association, 66*(2), 125–142.

Porter, D. R. (2004a). The promise and practice of inclusionary zoning. In A. Downs (Ed.), *Growth management and affordable housing: Do they conflict?* (pp. 212–248). Washington, DC: Brookings Institution Press.

Porter, D. R. (2004b). *Inclusionary zoning for affordable housing*. Washington, DC: Urban Land Institute.

Pyatok, M. (2004). Comment. In A. Downs (Ed.), *Growth management and affordable housing: Do they conflict?* (pp. 253–260). Washington, DC: Brookings Institution Press.

Real estate cools off, but less so in Tallahassee. (2006, July 15). *Tallahassee Democrat*.

Ross, J. (2001). Focus on social equity. Retrieved July 26, 2004, from http:// www.1000fof.org/Affordable_Housing/Social_Equity.asp.

Rusk, D. (2001). *Inside game/outside game: Winning strategies for saving urban America*. Washington, DC: Brookings Institution Press.

Tallahassee-Leon County Planning Department. (1990, July 16). *Tallahassee-Leon County 2010 comprehensive plan*. Retrieved from http://www.talgov.com/ planning/compln/comp_plan.cfm.

Tallahassee-Leon County Planning Department, (1996, December 31). *Summary of planning trends and issues for the Evaluation and Appraisal Report (EAR) of the Tallahassee-Leon County 2010 comprehensive plan*. Retrieved from http://www. talgov.com/planning/pdf/compln/earrpt.pdf.

Tallahassee-Leon County Planning Department. (1998, March 11). *City housing policy 1.2.4 and glossary definition of affordable housing*.

Tallahassee-Leon County Planning Department. (2003a). The Leon County Inclusionary Housing Ordinance (website no longer available).

Tallahassee-Leon County Planning Department. (2003b). *Joint city-county workshop on 2004-1 amendments to the 2010 Tallahassee-Leon County comprehensive plan—minutes*.

Tallahassee—Leon County Planning Department. (2006, July). *2006 statistical digest*. Retrieved from http://www.talgov.com/planning/pdf/support/statdgst06. pdf.

Tallahassee—Leon County Planning Department. (n.d.). *City of Tallahassee housing element*. Retrieved July 23, 2006, from http://www.talgov.com/planning/pdf/ compln/hous.pdf.

Tallahassee—Leon County Planning Department. (n.d.). *Inclusionary housing: What the other communities are doing*.

Whoriskey, P. (1999, March 8.) Cities find ways around affordable housing rules. *Miami Herald*.

Ziewitz, K., and Wiaz, J. M. (2004). *Green empire: The St. Joe Company and the remaking of Florida's panhandle*. Gainesville: University Press of Florida.

Chapter 17

Documenting the Rise of Impact Fees in Florida

Gregory S. Burge and Keith R. Ihlanfeldt

Few areas in the United States rival Florida in population growth over recent decades. Since 1975, Florida has accommodated roughly 10 million new residents (more than doubling the population) with no signs of slowing. This has placed intense pressure on local governments to expand public infrastructure to serve their growing communities. Furthermore, there is widespread agreement that the costs of these expansions are not fully recovered by the additional property tax revenue generated by most residential development (Dorfman, 2004; Downing and Gustely, 1977; Smith and Henderson, 2001). Hence, communities facing rapid growth are left with a large fiscal burden that is placed directly on the shoulders of existing residents. Local governments in Florida have reacted by developing a number of new policy tools that are intended to alleviate the burdens of rapid residential growth. One approach has been to place restrictions on economic development through the use of monopoly zoning powers (e.g., large-lot or open space zoning), building permit caps, urban service boundaries, and other devices. A second approach has been to have new development pay part of the cost of the public infrastructure that it requires in the form of development impact fees. This chapter focuses on Florida's experience with impact fees, one of the most innovative and controversial of the new policy tools designed to cope with rapid population growth.

Impact fees are one-time levies, predetermined by a formula adopted by a local government unit, that are assessed on property developers during the construction permit approval process. Although the specific timing of imposition varies somewhat across communities, most local governments in Florida levy the fees against new development projects as a necessary condition for receiving permit approval. Revenues from impact fee assessments are earmarked for the public infrastructure needed to provide urban services (i.e., roads, schools, parks and other recreational areas, library services, police services, and water and sewer services) that are necessary to serve new development adequately. Generally, impact fees apply to all types of development (i.e., residential, commercial, and industrial), but there are exceptions.[1]

Impact fee programs must be created through official public ordinances that are passed by the board of county (city) commissioners. This occurs only after

---

1 For example, educational impact fees generally apply only to residential developments.

the local government has commissioned an independent study to 1) investigate the relationship between the community's infrastructure needs and recent growth, and 2) to recommend a specific amount at which the impact fee should be set.[2] The fact that a study has been commissioned does not always mean the jurisdiction will eventually adopt the recommended fee levels or even that impact fees will be implemented. Local governments are subject to various political pressures as they proceed after a fee-level study has been completed and heated debates continue over whether impact fees should or should not be implemented. A recent example is Baker County, which commissioned a study looking into school impact fees. The recommended amount was $5,068 per new home. However, at the subsequent County Commission meeting, a nearly five hour debate erupted over the amount of the fee and whether or not it should be implemented. "Following lengthy discussion which included polling each commissioner and the representatives in attendance, the new amount of $1,500 was settled on" (Burnsed, 2005, p.1). The adoption of a fee level below the recommendation has been a common occurrence and underscores the role that local politics plays in the adoption process. Changing existing impact fee levels or adding new categories of fees is a complicated political process that frequently yields outcomes that are difficult to predict.[3]

The remainder of this chapter is organized as follows: Section II describes a recently constructed comprehensive panel database on county level impact fees in Florida. These data were used to construct the map and tables presented in this chapter and have also been used by the authors in a series of recent studies investigating the effects of impact fees on housing markets. This section also examines early impact fee use in Florida by documenting the pioneering programs used by Broward and Palm Beach counties and closes by briefly comparing the nature of impact fee programs in Florida to those found in other parts of the country. Section III tracks the rapid spread of impact fees over recent decades and highlights the role that a series of critical court cases and Florida's 1985 Growth Management Act had in encouraging their use. Section IV documents the current state of impact fees in Florida and presents summary information on their levels across counties. Turning next to the effects of impact fees, Section V presents a new theoretical model that links fees to housing supply and price and then discusses empirical evidence relevant to the predictions of this model. Conclusions are stated in Section VI.

---

2   The exception is the case of water/sewer impact fees. Because they are generally not called impact fees by name and because they are usually imposed by the water/sewer system itself rather than by the local government, fees that pay for offsite water/sewer costs are much easier to implement and in many cases do not require commissioned studies or public ordinances. Hence, much of the present discussion of implementation practices applies only to non-water/sewer fees.

3   Impact fee levels can also change over time without the passage of a new ordinance if an existing ordinance specifies those changes. For example, in the case of the Baker County school impact fee mentioned above, the current proposal sets the fee at $1,500 for the first two years, with 20 percent increases each year thereafter. It is also common for impact fees to be phased in. This practice involves charging only a fraction of the eventual fee amount for short periods of time following implementation.

Early Impact Fee Use in Florida

Florida's already fast rate of population growth accelerated during the 1970s. By the end of that decade, more than 3.5 persons lived in Florida for every one resident in 1950. Florida's accelerated growth forced local governments across the state to expand infrastructure systems more rapidly than ever before. Due to both constitutional and statutory restrictions as well as potential repercussions, local governments found it increasingly difficult to continue to rely on property tax rate increases to finance additional infrastructure. Given this historical context, it is not surprising that impact fees first surfaced in Florida during the late 1970s.

In order to study the evolution and effects of impact fees in Florida, we constructed a unique panel database of impact fees in Florida counties. For each county, the database contains detailed information on fee levels dating back to the inception of impact fees within the county. The methodology for obtaining these rates varied across counties but generally involved contacting county planning and growth management officials to obtain current and historical impact fee schedules, usually through copies of either administrative records or actual impact fee ordinances. Impact fees were grouped into two categories: water/sewer fees and non-water/sewer fees. Non-water/sewer impact fees can be used for a variety of services including roads, schools, parks, libraries, police, fire, and jails. Road, school, and park fees are the most common and also tend to be the largest in magnitude. Because impact fee charges can depend on either the number of bedrooms or interior square footage of a home, all impact fee amounts were standardized to correspond to the amount charged for a three bedroom, 1,800 square foot single-family home.[4]

The first Florida county to levy impact fees was Broward County in 1977. Residential impact fees were small—$125 for our standard sized new home—and were used to pay for park services. Two years later, Broward extended its program to include an educational impact fee that was just over $300. Palm Beach was another pioneering county, adopting a small ($300 per new home) road impact fee in 1979. These amounts, even after adjusting for inflation, are much smaller than what these same counties charge today. Starting with one or just a couple of fees set at low levels is not unique to Broward and Palm Beach counties; in fact, most early adopters in Florida followed this pattern. This is not surprising since the legality of impact fees was still very much in question at the time. However, even years after a series of court cases clearly established the legality of impact fees in Florida, the majority of counties adopting impact fees for the first time still choose to test the waters with one or two small fees, usually for either roads, parks, or schools, or for water/sewer services. Only after a period of time with initially low impact fees do counties increase fees beyond a nominal level or expand their coverage to additional services. While explaining this tendency is not a goal of this chapter, there is always

---

4   For the case of water/sewer fees, some counties have services provided to the entire county (including unincorporated areas) by a large city located within the county. Water/sewer fees charged by these systems are used since the city's system is equivalent to a county level of government offering the same services.

the concern that impact fees will completely stifle development. When this proves not to be the case, counties respond by increasing their fees.

A caveat regarding our panel database is that adoption dates for early non-water/ sewer impact fee programs are more easily verified than initial adoption dates for water/sewer impact fees. As opposed to road, school, park, and other types of impact fee programs that must be created through county ordinances, water/sewer impact fees can be created through more informal measures, and record keeping concerning their early presence and levels is often suspect. For example, contrary to the official dates given in Table 17.1, there is some evidence that water/sewer impact fees existed in both Orange and Pinellas counties prior to 1980.

Since many rapidly growing areas across the United States use impact fees, impact fee programs in Florida can be compared to those found in other parts of the country. One characteristic of Florida's programs that distinguishes them from those in other states is that counties, rather than cities, dominate the impact fee arena.[5] Impact fees in Florida are imposed by county governments and are countywide in their application. This largely reflects the fact that over half of Florida's population resides in unincorporated areas according to the 2000 US Census. While cities may charge impact fees for services not covered by the county (or charge different rates for a particular service), this practice is rare and city fees are in all cases small relative to those at county levels.

Measured by within-state coverage, impact fee programs are far more common in Florida than in most other parts of the United States. However, most other states do have at least some high growth areas where impact fees are imposed. Arizona, Colorado, Maryland, Georgia, Illinois, North Carolina, New Mexico, Oregon, and Washington all have what could be considered moderate levels of impact fee use, while most other states have at least some sporadic imposition of impact fees. Average residential impact fees are high in Florida compared to most areas in the US, save California. As seen in Figure 17.1, only California, Maryland, and Oregon have higher average residential non-water/sewer impact fees than Florida. Detailed descriptive statistics on impact fee levels are provided below in Section IV; however, it is worth noting here that among Florida counties charging impact fees in 2003, the average total impact fee for our standard sized single-family home was near $6,000. The highest fees are found in Collier County and equal $14,220. Another distinguishing characteristic of impact fees in Florida is that counties often change their fee schedules. This occurs as additional services receive their own impact fees. As noted above, counties typically start with a single impact fee (e.g., for roads) and then add fees for other services as time passes. Fee schedules also frequently change as the result of fees being raised for services already covered. In fact, evidence of recent increases in impact fees can be found by comparing the average non-water/ sewer fees in 2003 of our database ($3,082) to the corresponding 2005 average ($4,947) coming from Figure 17.1. To summarize, impact fees in Florida are: 1) widespread in use, 2) countywide in their application, 3) well established—Florida

---

5   Florida and Maryland are the only states where impact fee collection is largely handled at the county level (Duncan and Associates, 2005). Colorado is the one state where both levels of government play a significant role.

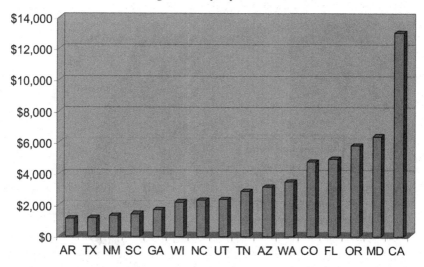

Figure 17.1   Statewide average non-water/sewer impact fees, 2005
Source: Duncan, 2005

was one of the pioneering states in terms of impact fee use dating back to the 1970s, 4) non-trivial in magnitude, and 5) frequently changing in levels and coverage within counties. All of these factors combine to make Florida an ideal case study for learning more about the factors underlying the adoption and growth of impact fees and the effects fees have on local communities.

The Rapid Spread of Impact Fees in Florida

Following the initial introduction of impact fee programs in Broward and Palm Beach counties, the first half of the 1980s was a relatively quiet period during which the vast majority of Florida counties were reluctant to implement impact fee programs. Counties were waiting for questions surrounding the legality of impact fees to be resolved in the courts. While water/sewer impact fee programs can be traced back to the early 1980s for a handful of counties (including Brevard, Duval, Hernando, and Manatee), the only adoption of non-water/sewer impact fees during the first half of the 1980s was in Orange County in 1983. This inactivity may have led one to wonder at the time if impact fees would ever take root in Florida. However, beginning in 1985, a literal explosion of impact fee programs swept across the state. Whereas only a handful of counties used impact fees prior to 1985, with most charging only water/sewer fees, nearly 30 counties adopted impact fees for the first time over the five year time span of 1986–1990. Simply put, over this short time period, impact fee programs went from being an anomaly to being the norm. By 1990 over half of Florida's counties, and the majority of its urbanized areas, had impact fee programs in place.

Rapid growth in adoptions began in 1985 when five counties (Collier, Hillsborough, Indian River, Lake, Lee and Pasco) first adopted impact fees. Subsequent years saw even more activity. Figure 17.2 visually documents the pattern of impact fee use present in Florida by January 1, 2000, after much of this rapid expansion had taken place.

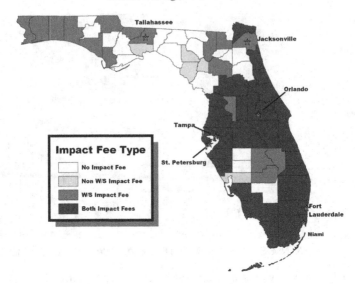

Figure 17.2   Florida county impact fee usage, 2000

A clear geographic pattern of impact fee adoption over the past three decades is also identifiable in looking at our data. South Florida counties clearly led the way, with all but a pocket of inland rural counties having programs prior to 1990, with most counties using both types of impact fees. Impact fees came to North Florida later. Only during the 1990s have many counties in North Florida adopted impact fees for the first time (and North Florida still lags behind South Florida in terms of overall impact fee usage). In examining these adoption patterns, an important question arises: why the explosion during this particular period?

Unlike many other states, Florida has never passed a specific enabling act or law that authorizes local governments to impose impact fees on new development. Rather, because Florida was a pioneering state in the impact fee movement, Florida law governing impact fees evolved through a series of critical court cases dating back to the late 1970s and early 1980s.[6] Collectively, these court cases have ensured that properly designed impact fee programs are protected under the policing powers already assigned to local governments in Florida. The decisions in these cases held that impact fees are *not* legally considered to be a tax upon new development as long as there is a sufficient connection between the level of the impact fee and the capital costs necessitated by growth. Prior to these court cases, counties viewed the legality

---

6   These cases include: *Contractors and Builders Association of Pinellas County* v. *City of Dunedin*, 329 So.2d 314 (Fla. 1976); *Hollywood, Inc.* v. *Broward County*, 431 So.2d 606 (Fla. 4th DCA 1983); and *Home Builders and Contractors Association of Palm Beach County, Inc.* v. *Board of County Commissioners of Palm Beach County,* 446 So.2d 140 (Fla. 4th DCA 1983). While these cases do enable local governments to impose impact fees upon new development, they also ensure that all impact fee ordinances in Florida must satisfy a "rational nexus" test which requires (1) a clear connection between new growth and the need for new capital facilities, (2) fees that are proportional to the costs of providing the new public infrastructures, and (3) accrual of the benefit from the new public facilities to the payer of the fee.

of impact fees with a great deal of uncertainty. Once local governments in Florida were assured that properly crafted impact fee programs would stand up in court, these programs became widespread.

Another contributing factor to the growth in impact fee use over the last half of the 1980s was Florida's 1985 Growth Management Act (GMA). Although the GMA is not impact fee enabling legislation per se, the language of the Act recognizes the general validity of impact fees and encourages their use by local governments: "This section shall be constructed to encourage the use of innovative land development regulations which shall include impact fees [Fla. Stat. Section 163.3203(3)]." Given the timing of the GMA endorsement and the enabling court decisions, the considerable latent interest that existed in impact fees during the early 1980s bore actual programs in the latter half of the decade after the uncertainty surrounding impact fee programs was resolved.[7]

The Current State of Impact Fees in Florida

Our panel database reveals that 49 of Florida's 67 counties (nearly three fourths) have charged impact fees at some time in their history. Water/sewer fees have been used by 44 counties, while 35 counties have implemented non-water/sewer impact fee programs. No county was found to have moved away from using water/sewer impact fees once a program of that type had been adopted. Of the 35 counties that have used non-water/sewer impact fees, 32 still have their original programs in place. Alachua, Leon, and Putnam counties abandoned their impact fee programs after relatively short periods of time. We now briefly outline the experience of Alachua County, a case where impact fees were adopted in 1992, revoked in 1994, and adopted again a decade later.

Alachua County was a latecomer in terms of impact fee use in Florida. A 1986 editorial article said that "while other Florida cities and counties adopted impact fees as a means of insuring that new growth would pay for itself, Alachua County commissioners waited. Now, with the county facing a budget crunch, impact fees are still barely in the talking stage" (Schell, 2004, p. 1). After years of debate over the merits of impact fees and largely motivated by the increasing costs of widening existing arterial roads, Alachua County did adopt a road impact fee that was first imposed on March 26, 1992. However, the amount was very low, at only $611 per single family home. The movement towards adoption was laden with strife. One commissioner drew the ire of homebuilders because she insisted the discussions of the proposed impact fee program take place in meetings open to the public— something the homebuilders vehemently opposed.

Once in place, the program was criticized from all sides. Some commissioners felt the fees were too small and immediately pushed for large increases. Developers complained they had a negative effect on economic growth and were keeping businesses away. Largely due to political pressures, the program was soon abandoned

---

7 Jeong (2006) empirically investigates the determinants of impact fee adoptions in Florida counties over the years 1977 to 2001. His results indicate that the GMA had a positive effect on impact fee adoption rates.

and fees were officially removed by the end of 1994. Still, the debate raged on. After a full decade without non-water/sewer impact fees in place, Alachua has recently revisited the issue and passed a set of impact fees that were adopted on September 28, 2004, and were first charged on March 28, 2005. While still relatively small (just over $2,250 for an average sized single-family home), they are considerably larger than the previous levels and include charges for park and fire services in addition to the road fee. Most recently, the county has commissioned a study of the effects the recently passed fees have had on the community. One commissioner has said, "Once we start looking back on the analysis, I don't think all of the bad things people said would occur have happened, I don't see where development has stopped" (Swirko, 2006, p. 3). Given its tumultuous experience with impact fees, Alachua County is a good example of how the relative strength of different interest groups over time can determine the eventual policy outcome.

Table 17.1 provides summary information on impact fee levels in Florida counties for two years—1993 and 2003—as well as each county's population in 2000 and its growth rate between 1990 and 2000. The second column indicates the year each county first used impact fees of either type.[8] As mentioned in Section III, a majority of counties implemented their programs in the second half of the 1980s, with several latecomers adopting during the 1990s. One obvious trend is apparent when comparing these dates with the population figures included in the third column: Larger counties tended to adopt impact fees earlier than did smaller counties. This is the result of two factors. First, because they have larger and more specialized staffs, larger counties already have many of the human capital resources needed to implement and maintain impact fee programs. Smaller counties, on the other hand, generally need to hire specialized workers to develop their impact fee programs. Second, impact fee programs have large fixed cost components, so that small and large counties must devote similar amounts of real resources to their development and implementation. However, even when facing similar growth rates, large counties receive greater revenue streams from impact fee programs than do smaller counties since the number of new residential units constructed is much higher. A simple benefit/cost approach to the question of when a county is likely to implement an impact fee program suggests that larger counties reach the implementation threshold sooner than smaller counties, *ceteris paribus*.

Table 17.1 provides fee levels for both of our defined categories of impact fees for 1993 and 2003. Looking first at water/sewer fees, the final row of Table 17.1 indicates these charges, averaged across counties, have remained relatively stable between 1993 and 2003, reaching an average of nearly $3,000 by 2003.[9] However, variation in fee levels across counties is considerable. The highest levels of charges are in

---

8   It should be noted that for a small number of counties, early water/sewer impact fee rates were not obtainable and therefore may in fact have predated the implementation year listed in Table 17.1. We specifically note this possibility for five counties where data collection efforts implicate the presence of early water/sewer impact fees, but we cannot confirm the exact starting point.

9   After indexing for inflation, mean water/sewer impact fee levels are nearly identical in both time periods.

Table 17.1 Single-family home impact fee levels in Florida counties: 1993 and 2003

County	Implementation Year[b]	Population[c] 2000	Growth Rate[c] 1990–2000	Water/Sewer Impact Fees[d] 1993	Water/Sewer Impact Fees[d] 2003	Non-Water/Sewer Impact Fees[d] 1993	Non-Water/Sewer Impact Fees[d] 2003
Alachua	1992[e]	217,955	20.0%	1,250	1,476	611	0
Baker	1991	22,259	20.4%	1,000	1,000	0	0
Bay	1987	148,217	16.7%	1,718	1,718	0	0
Brevard	1981	476,230	19.4%	4,160	4,160	947	1,512
Broward	1977	1,623,018	29.3%	2,536	3,016	918	2,465
Charlotte	1986	141,627	27.6%	3,232	4,913	1,884	2,510
Citrus	1987	118,085	26.3%	2,133	3,565	1,433	3,131
Collier	1985	251,377	65.3%	2,240	5,530	4,200	8,690
Columbia	1993	56,513	32.6%	0	1,725	0	0
Dade	1989	2,253,362	16.3%	1,162	2,447	3,240	5,038
Dixie	1986	13,827	30.6%	0	0	350	525
Duval	1981	778,879	15.7%	312	1,283	0	0
Escambia	1990	294,410	12.0%	1,674	2,108	0	0
Flagler	1990	49,832	73.6%	NA	2,842	646	1,133
Gadsden	1996	45,087	9.7%	0	450	0	0
Gilchrist	1999	14,437	49.3%	0	0	0	1,441
Gulf	1999	14,560	15.9%	0	650	0	0
Hernando	1980	130,802	29.4%	1,792	2,125	1,470	4,878
Highlands	1994	87,366	27.7%	0	4,618	0	0
Hillsborough	1985	998,948	19.8%	3,828	3,829	2,138	2,063
Indian River	1985	112,947	25.2%	4,100	4,096	1,133	1,523
Jackson	1997	46,755	13.0%	0	2,375	0	0
Lafayette	1986	7,022	25.9%	0	0	150	300
Lake	1985	210,527	38.4%	2,538	3,280	1,186	3,551
Lee	1985	440,888	31.6%	2,225	2,875	2,702	6,247
Leon	1989	239,452	24.4%	3,550	4,137	271	0
Manatee	1981	264,002	24.7%	2,285	2,423	1,742	4,925
Marion	1990	258,916	32.9%	2,600	2,600	640	1,685
Martin	1987	126,731	25.6%	2,380	3,600	1,846	5,212
Monroe	1986	79,589	2.0%	2,500	2,625	2,095	1,534
Nassau	1987	57,663	31.2%	284	1,166	933	1,008
Okaloosa	1988	170,498	18.6%	2,000	2,000	0	0
Okeechobee	1995	35,910	21.2%	0	959	0	0
Orange	1983[e]	896,344	32.3%	3,260	3,582	2,272	5,122
Osceola	1989	172,493	60.1%	3,713	3,870	2,280	4,087
Palm Beach	1979	1,131,191	31.0%	2,400	3,000	3,127	5,976
Pasco	1985	344,765	22.6%	2,145	2,056	1,969	4,896
Pinellas	1986[e]	921,495	8.2%	1,690	2,572	1,632	1,777
Polk	1990[e]	483,924	19.4%	2,560	4,239	847	1,168
Putnam[a]	1995	70,423	8.2%	0	0	0	0
Saint Johns	1988	123,135	46.9%	2,636	3,373	1,346	2,630
Saint Lucie	1986	192,695	28.3%	NA	3,026	1,240	3,622
Santa Rosa	1995	117,743	44.3%	0	2,283	0	0
Sarasota	1988	325,957	17.3%	4,093	4,362	3,373	2,967
Seminole	1987[e]	365,199	27.0%	3,090	3,091	2,671	2,671
Sumter	1987	53,345	68.9%	2,188	2,700	0	0
Volusia	1986	443,343	19.6%	2,369	3,343	2,334	3,078
Wakulla	1989	22,863	61.0%	0	0	564	1,247
Walton	1995	40,601	46.3%	0	6,427	0	0
Average[f]			23.5%	2,402	2,898	1,642	3,082

[a] Putnam County is included since they passed a set of impact fees that were first charged in May of 1995 but later revoked as was described in the text.

[b] Implementation year is defined as the calendar year when impact fees were first collected rather than the date they were first passed by County Ordinance.

[c] County population and population growth rates are taken from US Census figures.

[d] Fees reflect January 1st values and are standardized across counties for a 3 bedroom, 1800 square foot home.

[e] Water/sewer impact fees were in place for an indeterminate length of time prior to this date.

[f] Simple average using only counties with positive levels for average impact fee calculations. Average population growth rate reflects Florida's statewide average as reported by the Census and is not the mean of the counties included in this table.

Collier ($5,530) and Walton ($6,427) counties in 2003. Comparatively low levels of water/sewer fees are found in Duval, Gadsden, Gulf, Nassau, and Okeechobee counties. In all of these counties fees are less than $1,300.

In contrast to the trend in water/sewer fees, Table 17.1 shows that the average amount of non-water/sewer fees has increased dramatically, nearly doubling in nominal terms over the 1993–2003 period. Many counties added new types of impact fees to already existing programs and/or increased existing fees over this period. Variation in impact fee levels across counties is even greater than for water/sewer fees. While the mean level of impact fees among counties charging positive rates in 2003 is similar for non-water/sewer and water/sewer fees ($3,082 and $2,898, respectively), the standard deviation of non-water/sewer fees ($1,959) is considerably larger than for water/sewer fees ($1,291). Collier, Dade, Lee, Martin, Orange, and Palm Beach counties all charged non-water/sewer fees that were over $5,000 in 2003 and several other counties were near this level. At the same time, Alachua, Dixie, Lafayette, and Leon Counties all charge (or previously charged, in the cases of Alachua and Leon) fees that are less than $611.

In addition to the significant variation in non-water/sewer fee levels across counties, these fees vary a great deal over time within counties. They have doubled (or more than doubled) from 1993 to 2003 in 14 counties, with particularly large increases in Broward, Collier, Hernando, Lake, Lee, Marion, Martin, and Saint Lucie counties. While declines in fees are rare, Monroe and Sarasota counties did register slight declines in their non-water/sewer impact fees over this period.[10]

While the magnitudes of residential impact fees provide one indication of how important impact fees are within a community, they do not tell the whole story. Because commercial and industrial impact fees also play a large role in determining overall impact fee collections within a community, another approach toward measuring the importance of impact fees is to consider the contributions they make to local government budgets. Table 17.2 presents several measures of how important impact fee revenues are to local government budgets: per capita impact fee revenues (column 2), the ratio of total impact fee collections to total ad valorem property tax collections (column 3), and the ratio of total impact fee collections to total local government revenue (column 4). For simplicity, nine representative counties were selected for inclusion in Table 17.2, with three counties being characterized as falling into each of three different relative intensity use groups.[11] Charlotte, Marion, and Sarasota counties represent "high" intensity impact fee use counties. Per capita impact fee revenues average over $200 in this group, with total impact fee revenues generally approaching one half the level of ad valorem property tax revenue.

---

10 Largely in response to the rapidly increasing trend in impact fee levels, the Florida Impact Fee Review Task Force was recently created. Its purposes are twofold: 1) to document current trends in impact fee usage, and 2) to make recommendations to the Governor and Legislature concerning future directions for impact fee use in the state.

11 For clarity of comparison, only counties that charged both categories of impact fees were considered for inclusion in Table 17.2. The nine selected counties represent the three counties with the highest reliance on impact fees, the three counties with the lowest reliance on impact fees, and three counties that were near the mean levels for our constructed reliance measures.

Table 17.2   Relative intensity of impact fee use for selected Florida counties

County	2002–2003 Impact Fee Revenue Per Resident	Impact Fee to Ad Valorem Property Tax Ratio	Impact Fee to Total Revenue Ratio
High Impact Fee Use			
Charlotte	$202.45	0.441	0.079
Marion	$146.30	0.474	0.069
Sarasota	$276.86	0.567	0.092
Medium Impact Fee Use			
Indian River	$105.28	0.176	0.032
Lee	$140.14	0.216	0.039
Saint Johns	$97.52	0.179	0.050
Low Impact Fee Use			
Hillsborough	$36.89	0.066	0.011
Manatee	$40.28	0.073	0.017
Nassau	$66.21	0.113	0.027

Note: All revenue figures are taken from the fiscal year ending 2003 (Oct 1, 2002 to Sept 30, 2003) Florida local government detailed revenue reports available through the Florida Department of Financial Services at http://www.fldfs.com/localgov/downloads.html. All figures reflect totals after summing both county and city level local government revenues. County population levels used to construct per resident impact fee collections are US Census estimates for July 1, 2003.

Although impact fee revenues are a significant portion of local government revenues for this group, no county in Florida registered impact fee collections that are more than 10 per cent of local government revenues. This suggests that the potential exists to use impact fees more intensively, even among counties that are already using impact fees at comparatively high levels.

"Medium" intensity counties included in Table 17.2 are Indian River, Lee, and Saint Johns counties while "low" intensity counties are Hillsborough, Manatee, and Nassau. Within the medium intensity use group, impact fees seem to generate slightly less than twenty percent of the revenue that property taxes bring in and to constitute anywhere from three to five percent of local government revenues. Among low intensity counties, these ratios drop even lower, with impact fee revenues generally less than ten percent of property tax collections and only one or two percent of total revenues. Within these counties, there is clearly room to use impact fees much more intensively.

The final question addressed in this section is whether impact fees in Florida have been set at levels that fully cover the marginal cost to the community of accommodating residential growth. In theory, impact fees can be used to force development to "pay its own way" such that all development is revenue neutral. Under this approach, two values are of interest: 1) the net present value of current and future expenditures necessitated by the development and 2) the net present value of future tax revenues to be garnered from the development. Impact fees set at the difference between the former and the latter insure development that is revenue neutral.[12] Additionally, it is reasonable to assume Florida's communities are capable of estimating the fiscal impacts of residential development. Work by Downing and Gustely (1977) shows that fiscal impact analysis models were available in Florida by the late 1970s. More recently, the Department of Community Affairs has formalized and adopted the Fiscal Impact Analysis Model (FIAM), a tool used to measure the financial implications of specific developments or of alternative land use scenarios.

In practice, however, impact fees in Florida are set well below the full marginal cost level. The Florida Legislative Committee on Intergovernmental Relations addressed this issue in their 1986 report, *Impact Fees in Florida*.[13] The report cited two estimates of the average marginal cost of infrastructure necessitated by a new single-family home in Florida: $10,865 from a 1973 study (Downing and Gustely, 1977) and $22,000 from a 1985 study (Frank, 1985). The report concluded that impact fees did not even approach this level in any location in Florida at that time. Bringing both figures forward to current dollars and comparing them to the impact fee levels reported in Table 17.1, we see the same is still true today. Therefore, impact fees in Florida lessen, but do not eliminate, the fiscal burden that residential development imposes on the community.

The Effects of Impact Fees on Urban Housing Markets

Having documented Florida's experience with impact fees and outlined the current trends in impact fee use across the state, we turn to a discussion of the effects that impact fees have on urban housing markets. Two competing arguments have surfaced in the literature. The development community has argued that impact fees are a tax on residential development that will reduce the construction of new homes (especially within the starter home market) and exacerbate problems of housing affordability. A number of early investigations of the effects of impact fees (Downing and McCaleb, 1987; Huffman, Nelson, Smith, and Stegman, 1988; Singell and Lillydahl, 1990; Snyder, Stegman, and Moreau, 1986) voiced similar concerns. The second argument found in the literature comes to the exact opposite conclusion. Based on the assumption that exclusionary zoning and other forms of exclusionary regulation are motivated (at least in part) by the fiscal deficit that lower income residential housing places on a community, various scholars (Altshuler and Gomez-Ibáñez, 1993; Gyourko, 1991; Ladd, 1998) have suggested that impact fees reduce the level of exclusion, allowing more low income housing to be built within suburban areas.

The literature is consistent in asserting that impact fees will cause the price of housing to rise within communities, but there is little agreement as to whether higher prices are driven by reductions in supply or increases in demand. In an influential recent theoretical investigation, Yinger (1998) presents a convincing argument that house price increases are due to an increase in the demand for housing within the community rather than a reduction in the supply of new homes, but the issue is far

---

12 Note that the difference need not be positive. Communities have long competed for desirable commercial and retail development by offering property tax abatements.

13 At the time the report was prepared, the committee was known as the Florida Advisory Council on Intergovernmental Relations.

from being definitively resolved. The literature has also been divided on the issue of whether impact fees slow or speed up residential development, especially the construction of smaller homes in the suburbs. In the remainder of this section we present a theoretical model that connects impact fees to the level of exclusionary land use regulation within a community and discuss the results of some recent empirical investigations that we have done to test the predictions of our theory.

Theory[14]

Consider a single local taxing jurisdiction and assume that the market for single-family homes is in long-run competitive equilibrium, where demand and supply prices are equal and characterized by:

$$V = C + P_L * Q_L + F_T + F_U + A$$

where V is the price per unit of housing stock. The right hand side includes all of the costs incurred in building one unit of housing stock: C is the opportunity cost of construction, allowing for a normal rate of return; P_L and Q_L are land price per acre and lot size in acres, respectively; F_T and F_U are impact fees earmarked for public services financed by property taxes (non-water/sewer) and user fees (water/sewer), respectively; and A is project approval costs imposed on the developer by local government (i.e., explicit fees, compliance costs, and time delays), other than impact fees.

For a number of reasons, impact fees are expected to have a direct effect on housing prices and on project approval costs. The dependence of housing prices on impact fees follows from housing demand theory. This theory posits that market values for single-family homes will be affected by any future liabilities associated with holding the property, including future property tax payments. Impact fees shift a portion of the costs of new public infrastructure projects away from homeowners to developers. This shift causes future property tax rates to fall, assuming the local government must balance its budget and does not change the per capita level of public services. In turn, this reduction in future property taxes lowers a specific liability of holding residential property and increases housing values (V).[15]

A functional relationship will also exist between impact fees and other project approval costs if a community excludes single-family housing based upon a fiscal motive. If the perception exists that these developments do not generate enough additional property tax revenue to cover the costs of providing them with public services, the community may adopt exclusionary zoning and other exclusionary

---

14 For a more detailed version of the theoretical model presented here, see Burge and Ihlanfeldt (2006a). Also, this discussion focuses on impact fees and single-family housing. For a similar discussion of impact fees and their effect on multi-family housing, see Burge and Ihlanfeldt (2006b).

15 A practical way to think about this effect is that impact fees provide an insurance policy against future property tax rate increases caused by the need to finance additional public infrastructure for new residents.

regulations.[16] These regulations increase the developer's costs of obtaining project approval. For example, if less land is zoned for single-family homes, it becomes increasingly likely that the developer must obtain a rezoning in order to build. Rezonings lengthen the project review process and increase the developer's compliance costs.[17] Since impact fees mitigate the fiscal deficit imposed on the community, their presence may cause the community to zone more land for residential improvements. Impact fees may also make variances/rezonings in favor of residential developments easier to obtain, or reduce the developer's project approval costs in other ways. Hence, increases in impact fees are expected to reduce the level of other project approval costs.

A simple framework for thinking about the possible effects of impact fees on housing construction is easily constructed. On the one hand, impact fees add a direct cost (the fee itself) to the development process but, on the other hand, will add two indirect benefits: the higher demand driven equilibrium price found in the market and the reduction in project approval costs. Construction will increase (decrease) if the increase in the price of homes caused by the reduction in property taxes is greater (smaller) than the impact fee itself, less other savings in project approval costs. Because there is no reason, *a priori*, to conclude that either the costs or benefits to developers will always dominate the other, the original question of whether impact fees are a growth limiting or a growth enabling mechanism becomes an empirical question. However, before reviewing recent empirical studies, there are several specific predictions generated by our theoretical model that are worth noting.

First, because the property tax is an *ad valorem* tax, house price increases will be greater for more expensive homes. In fact, increases in impact fees are expected to cause larger absolute price increases for large (expensive) homes but are also expected to cause similar *percentage* increases in price for homes of all sizes. Second, an increase in impact fees will lead to larger tax savings in rapidly growing communities as compared to slowly growing communities. This is important to keep in mind since population growth rates vary widely across individual communities in Florida. Third, because less expensive homes generate a greater fiscal deficit than more expensive homes, the savings in project approval costs is expected to be greater for small homes than for larger, more expensive homes.[18] However, this may only be true in income homogeneous communities. Within mixed income communities, voter constituencies better reflect the interests of low income households, making it politically difficult to selectively exclude lower income housing. This means that, in comparison to suburban areas, the level of preexisting exclusionary barriers is expected to be low within both central cities and rural areas. It is therefore within suburban areas that impact fees have the greatest likelihood of expanding single-family home construction.

---

16 After reviewing the evidence, Burchell et al. (1998) conclude that most residential development has a negative fiscal impact on the home community. See Ihlanfeldt (2004) for a review of the evidence on various forms of exclusionary land use regulation.

17 Compliance costs are payments to engineers, surveyors, and attorneys in order to satisfy specific rules and regulations that govern changes in the local jurisdiction's land use map.

18 Consider, for example, the exclusionary practices of large-lot zoning and low density requirements. These regulations influence the approval costs for expensive homes very little, while creating significant barriers for smaller, more affordable homes.

A final prediction of our theory is that water/sewer impact fees are expected to have different effects than non-water/sewer fees. While the direct costs (the fee payment itself) are the same for both types of fees, a given increase in water/sewer impact fees is expected to generate fewer benefits to developers than a similarly sized increase in non-water/sewer fees. This expectation is based on local politics. In the absence of water/sewer impact fees, the costs of offsite infrastructure expansions will be recovered through higher base rates that are uniformly distributed across users. In contrast, the costs for other types of infrastructure expansion projects are borne by property owners in proportion to the values of their homes, given that the property tax is an *ad valorem* tax. Property owners, and especially more affluent owners, have far more political power within local communities than does the average water/sewer customer.[19] Hence, the exclusion of single-family housing (notably affordable homes) from suburban communities is likely to be driven more by the desire of local government officials to keep property tax rates low than by a desire to lower water/sewer user charges.[20] So while both types of impact fees should lessen fiscal exclusion within communities, the savings in project approval costs is expected to be greater for non-water/sewer fees. Impact fees used to fund services otherwise covered by property taxes are therefore more likely to generate savings in project approval costs that, when combined with the new higher market prices, are large enough to exceed the direct costs of the fees themselves, resulting in increased single-family home construction.

Empirical Evidence

As mentioned above, resolving the debate over whether impact fees limit or facilitate residential growth is largely an empirical matter. Here we summarize the findings of three recent investigations by one or both of us that have used impact fee data coming from Florida counties.

Ihlanfeldt and Shaughnessy (2004) used monthly time-series data from Dade County to investigate the effects of impact fees on the price of both new and existing housing. Their data covered the period from January 1985 to December 2000 and captured a great deal of variation in impact fee levels following Dade's initial adoption of an $879 road impact fee in 1989.[21] Their empirical approach consisted

---

19 See Peterson (1981) for support of the idea that affluent property owners play a major role in the determination of local public policies. As he notes, "it is the contribution to the fiscal base of local government that is crucial, not the number of votes the entity casts in local elections. A city concerned about its economic interests does not consider each taxpayer's benefit/tax ratio equally but in proportion to his contribution to the local coffer" (p. 36).

20 Furthermore, the connection between water/sewer rates and new development may be less apparent to the average homeowner than that between property taxes and new development, because the free-rider issue has been highlighted by local media as it applies to services financed by property taxes and not water/sewer services financed by user charges. One way in which the free rider issue gets highlighted is that it comes to the forefront in public debates over whether a school bond referendum should be supported. No similar event highlights increases in user fees resulting from expansions of water/sewer systems.

21 By the end of the investigated time period, nominal impact fees had increased to $5,239 for an average sized new single family home and covered schools, parks, police, and fire services.

of two stages. In the first stage, monthly constant-quality price indices for both new and existing homes were constructed using both hedonic and repeat-sales regression techniques. The second stage regressed each constructed index on impact fees and an extensive set of control variables. They found that a $1 increase in impact fees raised the constant-quality price of new and existing homes by $1.64 and $1.68, respectively. Their estimated impact fee regression coefficients were not statistically different from each other and were also not significantly different from $1, meaning a one-for-one resulting price increase for new and existing homes was not ruled out. In separate regressions they estimated the present value of the future property tax savings associated with impact fee increases to be about $1.20 for each $1 of additional impact fees. Hence, impact fee increases were found to increase housing prices by roughly the same amount that they generate in property tax savings. Their results provide evidence to support the idea that impact fees add value for housing consumers that is capitalized into home prices rather than operating merely as an excise tax that may be passed on to consumers (given the right market conditions).

The other two recent studies using impact fee data from Florida have been done by the two of us (Burge and Ihlanfeldt, 2006a, 2006b). Both of these studies directly modeled the supply effects of impact fees and investigated whether impact fees reduce or increase the number of affordable housing opportunities within Florida's urban areas. Each study separately examined the effects of water/sewer and non-water/sewer impact fees. While those interested in the effects of residential impact fees on single family and multi-family housing construction are encouraged to see Burge and Ihlanfeldt (2006a, 2006b) for a more detailed discussion, some important findings are summarized below.

The first study used panel data from 41 Florida counties spanning the years 1993–2003 to investigate the effect that impact fees have on single-family home prices and construction rates (Burge and Ihlanfeldt, 2006a). All empirical models were segmented into three size categories (small, medium, and large homes) so the effects of impact fees on more affordable housing opportunities (small homes) may be directly estimated. Using a two stage estimation procedure that employs repeat-sales regressions to construct constant-quality house price indices in the first stage, we found evidence to support the idea that non-water/sewer impact fees raise prices through an increase in the demand for housing rather than through a reduction in the supply curve coming from the monetary cost of the fees. Results from estimating both fixed effect and random trend price equations showed that a $1 increase in non-water/sewer impact fees increased the price of small, medium, and large homes by roughly the same percentage, a result that was consistent with the idea that impact fees increase demand by reducing homeowners future property tax burdens.[22]

More importantly, the study investigated the effect that impact fees have on the number of newly constructed homes falling into each size category. Using panel data estimation techniques, including random trend models that allow us to control for unobservable heterogeneity across Florida counties, we found that non-water/sewer

---

22 The estimated magnitudes of the price increase associated with a $1 increase in non-water/sewer impact fees coming from the small, medium, and large homes equation were $0.39, $0.82, and $1.27, respectively. Only the first estimate was significantly different from one.

impact fees increased the construction of small homes within inner suburban areas in Florida (where a majority of Florida's population resides) and medium and large homes within all types of suburban areas. We also computed the change that would occur in the number of new home completions for each size category of homes if the average inner or outer suburban area adopted the average level of non-water/ sewer impact fees observed in our data and found these marginal effects to be non-trivial in magnitude. The largest percentage increase in construction came from the small homes equation for the inner suburbs, an area where exclusionary barriers to affordable housing are perceived to be the most stringent. Collectively, the findings from both the price and construction models are consistent with the argument that non-water/sewer impact fees reduce exclusionary barriers and can, under certain circumstances, actually increase the amount of affordable housing opportunities.

In the other study (Burge and Ihlanfeldt, 2006b) we investigated the effects of impact fees on multi-family housing construction and found a similar set of results. Using a number of panel data estimation techniques, where the random trend model was again preferred, we found that non-water/sewer impact fees significantly increased the stock of multi-family housing within inner suburban areas. No significant effect on multi-family construction was found for central city and outer suburban areas where exclusionary barriers to affordable housing are believed to be less stringent. Water/sewer impact fees, on the other hand, were found to reduce the construction of multi-family housing throughout the entire metropolitan area. This is consistent with the prediction from the theoretical model that non-water/sewer fees have the greater potential to reduce other project approval costs and suggests that only non-water/sewer impact fees reduce the developer's cost of obtaining project approval by enough to overcome the cost of the fees themselves.

Conclusion

This chapter has outlined Florida's experience with impact fees over the past three decades. We first discussed how, when faced with unprecedented levels of population growth and intense pressure on local infrastructure systems, a small group of counties in Florida initially tested the waters with impact fee programs in the late 1970s and early 1980s. After the outcomes of three critical court cases clearly established the legality of impact fee programs and the 1985 Florida GMA encouraged their use, other communities across the state (but particularly in South Florida) quickly began to follow suit. The result is that well over two-thirds of the counties in Florida, including almost all metropolitan counties, presently levy impact fees upon residential development. In addition to documenting the rapid increase in the use of impact fees, recent trends in their magnitudes were also discussed. Section IV outlined how water/sewer impact fees have remained relatively constant over the years while non-water/sewer impact fees, due to increases in individual fees and the expansion of impact fee programs, have increased rapidly. However, in spite of their growth, impact fees remain well below the marginal cost of providing public infrastructure to new development. There is, therefore, ample opportunity for Florida counties to use them more intensively. However, before moving towards more intensive use of impact fees to help alleviate the burden of rapid residential growth, it is critical to understand the effects that impact fees have on urban housing markets. In particular, important questions concerning the effect that impact fees have on the affordability and availability of single-family starter homes and apartments remain contentious issues, even as impact fee programs continue to increase rapidly in size and scope.

In Section V we presented a simple theoretical model linking impact fees to the price and supply of housing that generated a number of interesting predictions, including the idea that non-water/sewer impact fees may actually increase the construction of smaller single-family homes and other types of affordable housing in suburban areas if they are able to reduce other regulatory barriers. The results of some recent empirical studies that used impact fee data from Florida suggest that non-water/sewer impact fees break down exclusionary barriers within suburban areas and are therefore a growth enabling mechanism in these areas rather than a growth limiting mechanism.

From a policy perspective, the implication of our work is clear: If the goal is to increase the stock of affordable housing within Florida's inner suburban areas, the Legislature should encourage communities to adopt non-water/sewer impact fees as an alternative to other regulatory barriers meant to slow residential construction, but discourage the use of water/sewer impact fees. One caveat is that because our research used impact fee data from Florida, where it is generally agreed that impact fee levels are set far below the full marginal cost of residential construction in most communities, it is important to note that our results may or may not accurately describe the effects of impact fees on construction rates or home values as they begin to approach levels that meet (or exceed) the full fiscal burden of affordable housing. More generally, our work suggests that *small* steps towards reducing the fiscal burden of affordable housing may go a long way towards reducing the anti-growth or anti-affordable-housing sentiment present within Florida communities. Hence, alternatives to impact fees that also reduce the reliance on the property tax as a method of financing expansions in local public infrastructure may also stimulate the construction of affordable housing within suburban areas. Additionally, because the empirical evidence suggests that impact fees increase the construction of affordable housing within inner suburban areas, where jobs for lower-skilled workers have been growing the fastest, impact fees may actually help alleviate spatial mismatch problems that exist within many urban labor markets.[23]

Acknowledgements

The financial assistance of the Federal Home Loan Bank of Atlanta, the US Department of Housing and Urban Development, and the Lincoln Land Institute is gratefully acknowledged.

---

23 See Ihlanfeldt (2006) for a review of the empirical evidence concerning the spatial mismatch hypothesis.

References

Altshuler, A, A., and Gomez-Ibáñez, J. A. (1993). *Regulation for revenue: The political economy of land use exactions.* Washington, DC: Brookings Institution and Cambridge, MA: Lincoln Institute of Land Policy.

Burchell, R. W. (with Shad, N. A., Listokin, D., Phillips, H., Downs, A., Seskin, S., et al.). (1998). *The costs of sprawl—revisited.* Washington, DC: National Academy Press.

Burge, G. S., and Ihlanfeldt, K. (2006a). Impact fees and single-family home construction. *Journal of Urban Economics* 60, 284–306.

Burge, G. S., and Ihlanfeldt, K. (2006b). The effects of impact fees on multifamily housing construction. *Journal of Regional Science.* 46, 5–23.

Burnsed, M. (2005, November 24). School impact fee set. *Baker County Standard,* p.1.

Dorfman, J. H. (2004). *The local government fiscal impacts of land uses in Hall County: Revenue and expenditure streams by land use category.* Athens, GA: Dorfman Consulting.

Downing, P. B., and Gustely, R. D. (1977). The public service costs of alternative development patterns: A review of the evidence. In P. B. Downing (Ed.), *Local service pricing policies and their effect on urban spatial structure.* Vancouver: University of British Columbia Press.

Downing, P. B., and McCaleb, T. S. (1987). The economics of development exactions. In J. E. Frank and R. M. Rhodes (Eds.), *Development exactions* (pp. 42–69). Washington, DC: Planners Press.

Duncan and Associates. (2005). 2005 national impact fee survey, annual survey report. Retrieved 1/20/1006, from http://www.impactfees.com.

Florida Advisory Council on Intergovernmental Relations. (1986). *Impact fees in Florida.* Tallahassee, FL: The Council.

Florida Statues, Section 163.3161–3203, (1985). *Local Government Comprehensive Planning and Land Development Act of 1985.*

Gyourko, J. (1991). Impact fees, exclusionary zoning, and the density of new development. *Journal of Urban Economics 30,* 242–256.

Huffman, F. E., Nelson, A. C., Smith, M. T., and Stegman, M. A. (1988). Who bears the burden of development impact fees? *Journal of the American Planning Association 54,* 49–55.

Ihlanfeldt, K. R. (2004). Exclusionary land-use regulations within suburban communities: A review of the evidence and policy prescriptions. *Urban Studies 41*(2), 261–283.

Ihlanfeldt, K. R. (2006). A primer on spatial mismatch within urban labor markets. In R. Arnott and D. McMillen (Eds.), *A companion to urban economics* (pp.404–417). Malden, MA: Blackwell Publishing, Inc.

Ihlanfeldt, K. R., and Shaughnessy, T. M. (2004). An empirical investigation of the effects of impact fees on housing and land markets. *Regional Science and Urban Economics 34*(6), 639–661.

Jeong, M.-G. (2006). Local choices for development impact fees. *Urban Affairs Review 41*(3), 338–357.

Ladd, H. F. (1998). *Local government tax and land use policies in the United States: Understanding the links*. Cambridge, MA: Lincoln Institute of Land Policy.

Peterson, P. E. (1981). *City limits*. Chicago: The University of Chicago Press.

Schell, S. (2004, November/December). Penny Wheat steps down after 16 years on county commission. *Gainesville Iguana*, 19, #2.

Singell, L. D., and Lillydahl, J. H. (1990). An empirical examination of the effect of impact fees on the housing market. *Land Economics 66* (1), 82–92.

Smith, L. J., and Henderson, P. (2001). Cost of community services study for Brewster, Massachusetts: A report on the fiscal implications of different land uses. Orleans, MA: Association for the Preservation of Cape Cod.

Snyder, T. P., Stegman, M. A., and Moreau, D. H. (1986). *Paying for growth: Using development fees to finance infrastructure*. Washington, DC: Urban Land Institute.

Swirko, C. (2006, January 29). County revisits raising impact fees. *Gainesville Sun*, pp.3.

Yinger, J. (1998). The incidence of development fees and special assessments. *National Tax Journal 51*, 23–41.

PART IV
Conclusion

Chapter 18

The 1985 Florida GMA: Satan or Savior?

Timothy S. Chapin, Charles E. Connerly, and Harrison T. Higgins

A 1992 special issue on growth management in the *Journal of the American Planning Association* was provocatively titled "Growth Management: Satan or Savior?" The collection of articles in that volume described numerous state programs and discussed the promise of these programs, while simultaneously noting that these processes might yield outcomes not foreseen or intended by program designers. Collectively these articles suggested that state growth management was likely to create new problems even as other problems were addressed. While based largely on early returns from these state initiatives, this collection of articles suggested that state-mandated growth management was likely a useful and appropriate planning approach but one with years of growing pains and policy refinements in front of it.

Perhaps not surprisingly, then, the chapters in this book yield a very similar conclusion. These chapters paint Florida's growth management approach as neither "Satan" nor "savior". Instead, these chapters provide a decidedly mixed view of the influence and impacts of the state's growth management approach on local development patterns, infrastructure provision, and local government finance. Where Florida's approach has yielded major impacts has been in the area of policy and program innovation. While the literature often portrays Florida's growth management system as a state-controlled, top-down system, in reality local governments in Florida have been encouraged and empowered to innovate for land use planning, environmental planning, infrastructure planning, and fiscal planning purposes. Below we discuss these and other major findings and contributions from the chapters in this book.

Growth Management Has Not Been a "Savior"

One of the most robust findings from this collection of chapters is that most of Florida's growth-related problems have not been solved by the choices made in implementing the state's growth management and comprehensive planning model as established by the 1985 GMA. The work of Sanchez and Mandle (Chapter 6), Carruthers et al. (Chapter 7), and Knaap and Song (Chapter 9) illustrate that low density sprawl still predominates in the state. As detailed by Chapin and Connerly (Chapter 5), citizens still view traffic and water supply issues as major impediments to a better quality of life in the state. Similarly, Nelson et al. (Chapter 12) find that quality of life improvements have not followed from the implementation of growth management in the state. Deyle et al. (Chapter 11) find that the state's population continues to surge in coastal areas, those areas most prone to hurricane-related impacts, such as flooding and wind damage. In short, no chapter provides conclusive evidence of policy success on any single front. Despite its comprehensiveness, or perhaps because of it, the evidence suggests that Florida's growth management approach has so far been unsuccessful in yielding more compact and more dense development patterns, fewer environmental impacts, more efficient provision of and use of infrastructure, and reduced local government fiscal stress.

Growth Management Has Not Been "Satan" Either

While the evidence suggests that state's system has not solved Florida's growth-related problems, there is strong evidence that growth management has not halted development in the state and irretrievably harmed the state economy. While some had predicted growth management would be a death knell for the development industry, growth management did not put the brakes on the Florida economy. Nicholas and Chapin (Chapter 4) reveal that the state has continued growing at rates similar to those experienced pre-GMA. Similarly, Chapin (Chapter 8) finds that the economies of the state's largest cities have experienced no deleterious impacts related to growth management, with some evidence suggesting that these urban economies fared somewhat better than those of their peers in the 1990s. While the national recession of the early 1990s had led some to conclude that the 1985 GMA was harming the overall state economy, the longer-term evidence indicates that the Florida economy has continued to thrive, even given the "shackles" of growth management.

However, while not "Satan," the news is not all good for proponents of growth management. Holcombe (Chapter 14) suggests that certain anti-sprawl policies function as exclusionary practices in many Florida counties. Connerly (Chapter 16) finds that Florida's approach has done little to help with the state's affordable housing problem, although he concludes that this outcome is largely a function of a lack of commitment to this issue by the state. Higgins and Paradise (Chapter 18) rather than containing sprawl sometimes contributes to it. In their reviews of the history of growth management in the state, Pelham (Chapter 2) and Ben-Zadok (Chapter 3) indicate that Florida's growth management system has routinely generated perverse outcomes related to the "wicked problems" associated with growth.

Florida as a Hotbed of Planning Innovation

One area where there is strong evidence of success lies in the planning innovations that can be attached to the Florida growth management process. The art of comprehensive planning has been advanced in Florida, and the state is seen as a model for comprehensive plan development in the United States. Florida is also a national leader in planning for natural hazards, especially relating to hurricanes and coastal development, advancements largely attributable to the 1985 act and the review of local plans by DCA (see Deyle, Chapin and Baker, Chapter 11). While concurrency was not invented in Florida, the broad implementation of concurrency and the numerous refinements to the policy, especially in the area of transportation, make the state a national leader on this front (see Steiner, Chapter 13). Florida is also a national leader in local government infrastructure financing, with impact fees, special districts, and local option taxes being implemented in Florida in ways that are now being copied in other states (see Nicholas and Chapin, Chapter 4, and Burge and Ihlanfeldt, Chapter 17).

These are just a few of the many planning innovations that can be attributed, at least in part, to the state's comprehensive, mandated growth management approach. It is important to note that Florida's growth management program has spawned innovation because it was designed from the very beginning to do just that. The state provided broad policy mandates but left the implementation of these mandates to local governments. As a result, local governments tackled similar planning problems in very different ways, resulting in policy variation and sometimes policy innovation. While DCA has retained a role in reviewing these locally generated policy innovations, occasionally rejecting them as not adequately addressing the state's interests, the state has allowed planners, public officials, and land developers great leeway in determining local, context specific solutions to problems identified in the 1985 GMA.

Growth Management in Florida Didn't Begin in 1985

Some analysts have attempted to assess the Florida growth management model as if it were a straightforward, one-time intervention, one in which comprehensive planning went from inactive to active with the passage of the 1985 GMA. However, as detailed by Pelham (Chapter 2) and Ben-Zadok (Chapter 3), and reinforced by other chapters, this book helps to illustrate that the 1985 GMA wasn't one great leap forward but rather one (albeit one rather large) step forward in a long progression towards a state approach to managing growth. The 1985 GMA was built upon almost two decades of state legislation that brought planning to the nation's fastest growing state. This book remains largely silent on the impact of the state's Development of Regional Impact process (better known as the DRI process), primarily because the statutory language on DRIs has been in place since 1972. Similarly, many analysts have overlooked the fact that the Florida legislature originally mandated local comprehensive planning in 1975, also providing for limited state review of these plans.

Policy Variation is the Rule Not the Exception

Another finding that appears in a number of chapters is one of variation in the implementation of the 1985 GMA by local governments. Holcombe (Chapter 14) finds that some counties have adopted some form of urban growth boundary, while others have not pursued this policy for managing growth. Similarly, Steiner (Chapter 13) hints at very different methods for addressing transportation concurrency issues, a variety of methods enabled and promulgated by the state government. On the financial side, Burge and Ihlanfeldt (Chapter 17) find widely differential use of impact fees and Nicholas and Chapin (Chapter 4) illustrate that local governments have attempted to pay for growth using a variety of approaches. Song (Chapter 10) even goes so far as to account for this policy variation in her multivariate analyses, finding that local policy variations do impact housing spillovers. Ben-Zadok (Chapter 3) labels the Florida growth management as characterized by "discretionary implementation," where local governments are constrained by broad policy guidelines but with the ability to tailor growth management to local conditions.

Given the state's commitment to comprehensive planning and the promulgation of minimum criteria for comprehensive plans, the Florida growth management system is often viewed by analysts as a "one size fits all" approach. However, as detailed in this book, the state's system is better understood as a state-mandated but locally implemented system. Under this model, broad state policy mandates (such as concurrency and consistency) and minimum standards for comprehensive plans are responded to and translated into policy by local governments in very different ways. Having spent years reviewing (and occasionally generating) comprehensive plan content, we have found no two comprehensive plans to be alike in their organization, presentation, level of detail, and choice of policy interventions. Unlike its representation in much of the literature as a command and control, top-down model, the Florida approach actually enables and embraces diversity in local comprehensive plans and the choice of local policies for managing growth.

Growth Management Implementation Remains an Ongoing Process

The chapters in this book illustrate that implementation of the 1985 act has been an iterative, ongoing process. In the area of transportation concurrency, in particular, the state has tinkered with and refined this policy, seemingly annually. Steiner (Chapter 13) documents the many exceptions and exemptions that have been created so that transportation concurrency can be achieved while simultaneously promoting the larger state goal of compact urban development. Pelham (Chapter 2) details many of the refinements to the state's system over the years, while Ben-Zadok (Chapter 3) provides a framework for understanding how these refinements have proceeded during the post-GMA implementation period. When viewed together, these chapters illustrate that the practice of planning in Florida and the state's growth management process has been in a state of almost constant evolution since 1985.

A System Plagued by Broken Promises

A key finding from this set of chapters is the large number of "broken promises" associated with the Florida growth management approach. The biggest broken promise, documented by Nicholas and Chapin (Chapter 4), was the legislature's failure to adequately fund desperately needed infrastructure improvements in the state. As originally designed, Florida's growth management approach rested upon a combination of adequate funding to support growth and local, regional, and state comprehensive planning to manage that growth. While the comprehensive plan portion of the model has generally been implemented at the local level, the state has largely ignored its funding obligations, instead passing along revenue generation responsibilities to local government, while simultaneously limiting local government's ability to raise these funds. Only in the last few years has the state made preliminary steps towards reconnecting with the state-mandated, state-funded model articulated in the original 1985 legislation, primarily through Senate Bill 360. While DCA staff and planners throughout the state continue to work through the details of implementing SB360, there is general agreement that this legislation represents a major change to the state's growth management approach.

Pelham (Chapter 2) also documents the failure of the state to commit to and take direction from the State Comprehensive Plan (Chapter 187 Florida Statutes), an integral part of the State-Regional-Local framework originally established by the 1985 legislation. As documented by Pelham, the State's Comprehensive Plan has had very little impact upon local comprehensive plans and subsequent major plan updates. The 1985 legislation required consistency between state agency plans and policies and the State Comprehensive Plan, but the State Plan has generally been ignored by state agencies. While the state has insisted upon comprehensive planning and the adoption of regulations to promote the plan at the local level, Florida's legislative and executive branches have resisted any attempts to see their hands tied by a comprehensive plan. The most compelling evidence of the failure of the State Comprehensive Plan is that it has not been updated since 1985, despite a requirement in the original legislation that it be revisited annually by the legislature.

Beyond these two fundamental failures, Florida's growth management approach has been plagued by other broken promises. Regional planning was to be a major element of the state's growth management approach, but the state's Regional Planning Councils have seen their powers eroded over the years, to the point that they are now largely advisory bodies. The state has also experienced wildly inconsistent commitment by governors and legislators to state planning and plan review through the Department of Community Affairs.

In sum, a number of chapters in this book illustrate that the Florida approach that was lauded nationally in the mid-1980s was not the system that ultimately was implemented by the state. When viewed from afar, the Florida approach seems to have most, if not all of the elements required of a successful growth management model. However, the reality has been that the system's state and regional layers have never been fully implemented and the funding commitment from the governor and legislature lasted all of six months.

Evaluation Remains Ongoing

One last key finding from these chapters is that twenty years may be too soon for an assessment of the impacts of the 1985 GMA. Several of the chapters hint at positive changes in areas that we would expect some impact. For example, Nelson et al. (Chapter 12) suggest that quality of life may be turning a corner in Florida. Deyle et al. (Chapter 11) conclude that growth has continued to occur in coastal areas of the state but at rates lower than the pre-GMA period. The central city indicators employed by Chapin (Chapter 8) are suggestive of greater progress in Florida's central cities when compared to their peer cities. Higgins and Paradise (Chapter

15) document the substantial state investment in conservation lands. However, these findings are generally suggestive and far from conclusive. In most cases the authors indicate that another ten years of data (at least) are required before a firmer set of conclusions can be derived.

Why is so little evidence available twenty years after passage of the 1985 GMA? The answer lies in part on the slow implementation of the state mandates. While the act was passed in 1985, most local governments didn't have a comprehensive plan in place until around 1990. Land development regulations then had to be brought in line with these plans, a process that still remains ongoing in many of the state's communities.

Even given the adoption of local comprehensive plans and the reworking of the local land development code, many areas had already established allowable densities for already built and soon-to-be built portions of their jurisdictions. By law, newly developed comprehensive plans had little to say about already permitted projects, those projects that had navigated through the local land use planning process. Similarly, the 1985 legislation allowed "vested" development, projects with an executed development agreement in place at the time new comprehensive plans were passed by local governments, to bypass any new land development regulations. As a consequence, even given full commitment by local leaders to the vision articulated in a comprehensive plan, the opportunity to influence development outcomes was largely limited to those portions of the community not already developed or with permitted/vested development rights.

Final Thoughts

When we began the process of organizing a symposium to assess the 1985 Growth Management Act in the year of its twentieth anniversary, we hoped to support the work of some of the leading growth management scholars to generate clear findings from the Florida growth management experiment. Our hope was that this work could offer some evidence as to the effectiveness of the 1985 GMA, with a secondary goal of providing policymakers with clear policy direction in the area of growth management. For those readers that have digested every chapter in the book, you are probably aware that a set of clear findings and distinct, defensible policy directions are generally not to be found. Instead, while some chapters are suggestive of certain policy directions, firm, clear conclusions concerning the effectiveness of the state's growth management approach are few and far between.

Where there is little debate is that many of Florida's growth-related problems continue to plague the state, even after over twenty years under the growth management regime established by the landmark 1985 Growth Management Act. Traffic congestion, potable water supply shortages, the development of environmentally sensitive lands, and local government fiscal problems can be found in all corners of the state. For those proponents of the 1985 act that suggested that the comprehensive planning model would yield very different development outcomes, evidence for this view is not to be found in this book. Instead, the evidence presented in this volume indicates that development continues at the urban fringe, urban centers continue to lag behind suburban areas, and coastal development remains the norm. Growth management has not solved the problems associated with the 850 net new state residents that arrive each day in the state.

However, these chapters do offer some limited evidence that growth management and comprehensive planning have made their mark on the state in positive, if limited ways. Planning has a foothold in the state like never before and residents are still supportive of growth management despite continued problems with everyday concerns like traffic congestion. Environmental problems remain, but the state is a national leader in land conservation and natural hazards planning, while water quality and stormwater issues are not the major problems they were pre-GMA. The state has also experienced success in linking new development and the costs of growth through impact fees and special districts. Growth management appears to have been effective in narrow ways, some foreseen and others unforeseen by the drafters of the 1985 GMA.

When viewed holistically, the Florida growth management experiment is probably best understood as being the epitome of the concept of incrementalism. The chapters in this book reinforce the long-held view that planning is indeed an incremental process, one that generates changes in small ways that over time aggregate into major changes. This book illustrates that these increments occur on *both* the policy development side and the development outcome side. Growth management policy has been evolving in the state for over twenty years and there is every reason to believe these incremental changes will continue in the next twenty years. Similarly, the suggestive findings concerning both the positive and negative impacts of the GMA may become more concrete findings in another twenty years.

In closing, we hope this book serves as a springboard for a greater discussion of the merits and limitations of state-mandated growth management and comprehensive planning. Given the relative paucity of research on growth management effectiveness, we also hope this book spurs further empirical research on the influence of growth management policies on development outcomes, environmental impacts, infrastructure provision, and local government fiscal practices. While it is clear that a final verdict on the effectiveness of Florida's growth management approach is decades in making, we believe the chapters in this book provide an initial evaluation of the nation's most far-reaching and most ambitious attempt to manage growth.

Index

Printed in the United States
by Baker & Taylor Publisher Services